D1591991

Experimental Neutron Scattering

Experimental Neutron Scattering

B. T. M. Willis
University of Oxford, U.K.

C. J. Carlile
Lund University, Sweden

OXFORD
UNIVERSITY PRESS

Great Clarendon Street, Oxford OX2 6DP

Oxford University Press is a department of the University of Oxford.
It furthers the University's objective of excellence in research, scholarship,
and education by publishing worldwide in

Oxford New York

Auckland Cape Town Dar es Salaam Hong Kong Karachi
Kuala Lumpur Madrid Melbourne Mexico City Nairobi
New Delhi Shanghai Taipei Toronto

With offices in

Argentina Austria Brazil Chile Czech Republic France Greece
Guatemala Hungary Italy Japan Poland Portugal Singapore
South Korea Switzerland Thailand Turkey Ukraine Vietnam

Oxford is a registered trade mark of Oxford University Press
in the UK and in certain other countries

Published in the United States
by Oxford University Press Inc., New York

First Published 2009

British Library Cataloguing in Publication Data
Data available

Library of Congress Cataloging in Publication Data
Data available

Typeset by Newgen Imaging Systems (P) Ltd., Chennai, India
Printed in Great Britain
on acid-free paper by
Antony Rowe, Chippenham

ISBN 978–0–19–851970–6

1 3 5 7 9 10 8 6 4 2

Preface

The first systematic experiments in neutron scattering were carried out in the 1940s on fission reactors built for the nuclear power and nuclear weapons programmes. Since then the community of neutron users has expanded enormously. Crystallographers were amongst the first to exploit the new technique, and they were soon followed by condensed-matter physicists interested in the magnetic and vibrational properties of crystals. In the 1960s chemists were attracted into the field, and later biologists, using the wide range of wavelengths provided by cold neutron sources. Engineers and earth scientists are the most recent recruits to the club.

Because of the requirement to produce an intense source of neutrons, neutron scattering can only be performed at a few central facilities. According to a report of the European Neutron Scattering Association there are now more than 4,000 part-time and full-time users of neutron scattering in Europe alone. Over half of them are part-time users: they visit central installations, such as the Institut Laue Langevin (Grenoble), ISIS (Oxford), Saclay (Paris), Hahn-Meitner Institut (Berlin), FRM-II (Munich) or Paul Scherrer Institut (Villigen), for short periods to carry out experiments on samples that they also examine with other techniques in their home laboratories.

Our aim in this book is to present a broad survey of the experimental basis of neutron scattering in a form suitable to newcomers to the field, and to provide a foundation upon which they can build before referring to more specialized publications or conference proceedings. Most of the individual chapters are on topics which are covered in one or more monographs devoted to that specific topic. Both authors are physicists and we append a glossary of special terms for the benefit of researchers from other disciplines.

In the early years of neutron scattering nuclear reactors were the principal neutron sources, but today accelerator-based pulsed sources have attained equal prominence. Reactor fluxes are limited by heat transfer problems, whereas pulsed sources are not so severely constrained. The most intense source of reactor neutrons, worldwide, is at the Institut Laue Langevin (ILL) in Grenoble, France, while the world's most productive pulsed neutron source is the ISIS Facility of the Rutherford Appleton Laboratory in the United Kingdom. As authors of this book we have extensive experience in working at the ILL and at ISIS, and in our text we shall refer to many of the instruments and techniques used there. Since the book is considered to be a students' manual rather than a research monograph, we beg the indulgence of those who are more familiar with working at other Institutes.

The book is divided into three parts. Part I deals with the fundamental physical principles of neutron scattering and with the types of apparatus required for experimental work. Part II covers experimental methods, such as powder diffraction and small-angle scattering, in which the scattered neutrons are detected without their energy being determined: these are methods in *neutron diffraction*. Part III is concerned with experiments in *neutron spectroscopy*, or inelastic scattering, where the change of energy of the scattered neutrons is measured. We give examples at two levels of experimental measurements: classical experiments illustrating a particular technique, and state-of-the-art experiments describing the achievement of the technique at the time of writing. The Appendices at the end of the book provide tables and other information, which may be required by the experimentalist during the course of his or her investigation.

With the availability of high-brilliance X-ray synchrotron sources, such as the European Synchrotron Radiation Facility (ESRF) in France, DIAMOND in the United Kingdom, SOLEIL in France, the Advanced Photon Source in the United States and the SPRING 8 machine in Japan, there will undoubtedly be increasing interest in complementary studies using X-rays and neutrons. The determination of hydrogen-atom positions in molecular crystals, the measurement of phonon dispersion relations and the study of magnetic structures can now be contemplated with X-ray sources and are no longer restricted to neutron scattering. Throughout the text we shall refer to various aspects of this complementarity between X-rays and neutrons.

The book has developed from the *Oxford Summer Schools on Neutron Scattering*, which have been jointly run by the two authors at two-yearly intervals since 1988. We are particularly grateful to Alberto Albinati, Andrew Boothroyd, Bruce Forsyth, Gerry Lander, Sean Langridge, Ross Stewart, and Heinrich Stuhrmann, who gave their time freely as lecturers at these Schools. Andrew Boothroyd and Gerry Lander have read the entire text and made many improvements and corrections. Comments on individual chapters have been made by Alberto Albinati, Julia Higgins, Mike O'Riordan, Michael Pendlebury, Jeff Penfold, Helmut Rauch, Alan Soper, Gordon Squires, Ross Stewart, Andrew Wildes and Chick Wilson. Patrick Feuillye, Sam Willis and Russell Brown gave help in preparing the diagrams. Soenke Adlung, Senior Editor of the Oxford University Press, has exercised both tact and patience in ensuring final publication.

The book is dedicated to our ever-supportive wives, Margaret and Gill, and to the hundreds of students at these Summer Schools. These students represent the future of neutron scattering.

Terry Willis and Colin Carlile

Oxford & Lund
March 2009

Contents

Part I

General considerations

Over 70 years of neutron scattering

1

1.1 The discovery of the neutron

In the 1920s James Chadwick undertook a series of experiments in search of the neutron, whose existence had been predicted by Rutherford in his Bakerian lecture to the Royal Society of London in 1920. It was not until 1932 that Chadwick finally achieved his goal, announcing the possible discovery of the neutron (Chadwick, 1932a) and confirming its existence later in the same year (Chadwick, 1932b).

The breakthrough came after Bothe and Becker (1930) discovered an unusually penetrating radiation produced by the alpha particle bombardment of beryllium. Irène and Frédéric Joliot-Curie (Curie and Joliot, 1932) then found that this radiation was capable of ejecting high-speed protons from hydrogenous materials such as paraffin or water. The Joliot-Curies believed that the radiation consisted of very high frequency gamma rays, and that the protons were ejected by a process analogous to the Compton effect. Chadwick recognized, however, that there were two serious difficulties with this gamma-ray hypothesis: (i) the number of protons observed by the Joliot-Curies were two orders of magnitude higher than that calculated from the theoretical Klein–Nishina formula for the scattering of gamma rays by protons (as derived from quantum electrodynamics) and (ii) the energy calculated from the same formula was much less than the observed energy. Chadwick realized that both these difficulties could be overcome by assuming that the unknown radiation consists of neutrons rather than gamma rays.

Chadwick confirmed the neutron hypothesis after 10 days of frenetic activity in which he extended the measurements of the Joliot-Curies to scattering by light elements. His apparatus is illustrated in Fig. 1.1(a), and a photograph of the original chamber is shown in Fig. 1.1(b). Chadwick carried out experiments with paraffin wax, and then replaced the paraffin wax successively with a number of light elements (helium, lithium, beryllium, boron, carbon, nitrogen, oxygen and argon) in solid form or as a gas introduced into the ionization chamber. He found that the 'beryllium radiation' ejected particles from all these light elements and he also determined the maximum energy of these recoil nuclei from their range in air. By combining the measurements for hydrogen and nitrogen and assuming that the unknown radiation consisted of neutral

(a)

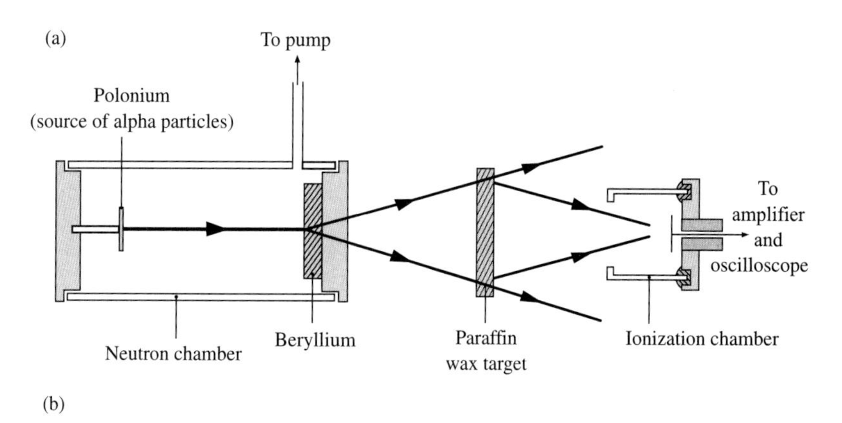

(b)

Fig. 1.1 (a) The apparatus used by James Chadwick in his discovery of the neutron. Alpha particles from a polonium source bombarded a block of beryllium to produce neutrons by the (α,n) reaction. These neutrons then knocked out protons from a sheet of paraffin wax and were counted in the ionization chamber. (b) Chadwick's neutron chamber in which alpha particles from a polonium source at one end bombarded a beryllium source at the opposite end. The vertical pipe was attached to a vacuum pump. (From the *Photographic Archives of the Cavendish Laboratory, Cambridge.* Courtesy of G. L. Squires.)

particles of mass m_{n}, he obtained the result

$$m_{\mathrm{n}} = 1.15 \text{ a.m.u}$$

with an estimated error of about 10%. (The atomic mass unit, a.m.u., is defined as one twelfth of the mass of $^{12}_{6}\mathrm{C}$.) Independent but consistent values of m_{n} were obtained by using the energy measurements for other pairs of light elements. Chadwick concluded that the radiation consisted of neutrons $^{1}_{0}\mathrm{n}$ of unit mass and zero charge, which were ejected from the beryllium nucleus by alpha particles in a reaction represented by

$$^{4}_{2}\mathrm{He} + {}^{9}_{4}\mathrm{Be} \longrightarrow {}^{12}_{6}\mathrm{C} + {}^{1}_{0}\mathrm{n} \tag{1.1}$$

Later, he derived a better estimate of the mass of the neutron by replacing the beryllium target in Fig. 1.1 with a target of powdered boron to yield neutrons from the reaction

$$^{4}_{2}\mathrm{He} + {}^{11}_{5}\mathrm{B} \longrightarrow {}^{14}_{7}\mathrm{N} + {}^{1}_{0}\mathrm{n} \tag{1.2}$$

An improved estimate was possible, because the mass of $^4_2$He, $^{11}_5$B and $^{14}_7$N in eqn. (1.2) were known from mass-spectrograph measurements but not the mass of $^9_4$Be in eqn. (1.1). Using the Einstein equation $E = mc^2$, the energy balance requirements applied to eqn. (1.2) give

$$(m_{\text{He}} + m_{\text{B}})c^2 + T_{\text{He}} \longrightarrow (m_{\text{N}} + m_{\text{n}})c^2 + T_{\text{N}} + T_{\text{n}} \qquad (1.3)$$

where T denotes the kinetic energy. T_{He} was obtained from the range of the alpha particles, T_{N} was calculated from the momentum conservation law and T_{n} from the maximum recoil energy of the proton. The remaining unknown in eqn. (1.1) is m_{n}, which was evaluated as

$$m_n = 1.0067 \pm 0.0012 \text{ a.m.u} \qquad (1.4)$$

This result compares remarkably well with the much later value given by Eidelman (2004) of

$$m_{\text{n}} = 1.008664915 \pm 0.000000001 \text{ a.m.u} \qquad (1.5)$$

It is interesting to note the speed of events that took place in early 1932. In January 1932 Chadwick received the paper of the Joliot-Curies, which had been published earlier that month in *Comptes Rendus*. From 2 February to 12 February Chadwick carried out his experimental measurements, and on 17 February he submitted his letter to *Nature* (entitled *Possible existence of a neutron*), which was published on 27 February. The full paper (*The existence of a neutron*) was published in the June 1932 issue of *The Proceedings of the Royal Society.*

1.2 Early history of neutron scattering

Four years after the discovery of the neutron, von Halban and Preiswerk (1936) and Mitchell and Powers (1936) showed that *thermal neutrons*, with a de Broglie wavelength comparable to interatomic distances, can be diffracted by crystalline matter. The experiment of von Halban and Preiswerk was on polycrystalline iron, and that of Mitchell and Powers was on single crystals of magnesium oxide. The apparatus of Mitchell and Powers is shown in Fig. 1.2. They used a cylindrical array of large MgO crystals whose $(1, 0, 0)$ planes were at a glancing angle of $22°$ to the neutron beam. This is the Bragg angle for reflecting the neutrons at the peak of the velocity distribution of thermal neutrons. By comparing the observed counts with those recorded by offsetting the crystals from the Bragg condition, they were able to demonstrate the enhanced intensity caused by diffraction from MgO.

In general, the scattering of neutrons by atoms is a nuclear process, but in the case of magnetic atoms with unpaired electrons there is additional scattering arising from the interaction between the magnetic moment of the neutron and that of the atom. From the early days, there was intense interest in this additional scattering. The first theoretical study by

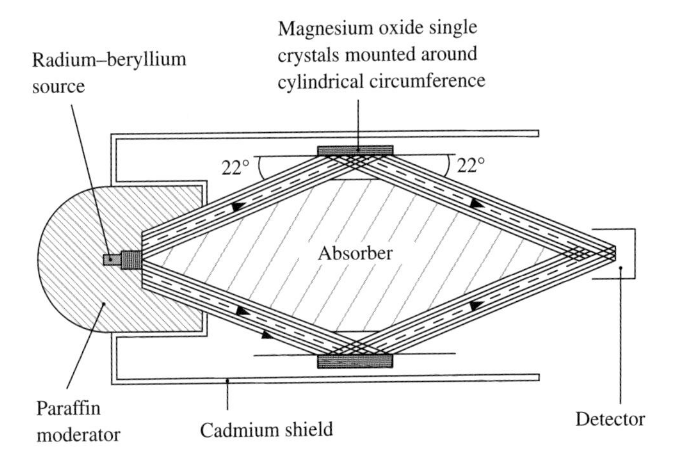

Fig. 1.2 The apparatus used by Mitchell and Powers (1936) for demonstrating the diffraction of neutrons.

Bloch (1936) assumed a classical dipole–dipole interaction between the neutron and the magnetic atom, but Schwinger (1937a) quickly pointed out Bloch's error in failing to use the correct Dirac value of the electronic current density of the scattering atom. From the observed ratio of the scattered intensity of slow neutrons by *ortho-* and *para-*hydrogen, Schwinger (1937b) also showed that the neutron spin is $1/2$. Frisch *et al.* (1938) succeeded in partially polarizing a neutron beam by transmitting it through iron; they then measured the precession of the neutron polarization in an applied field, and concluded that the magnitude of the magnetic moment of the neutron is about two nuclear magnetons and is negative. This negative sign implies that the direction of the magnetic moment of the neutron is opposite to that of its angular momentum. Alvarez and Bloch (1940) provided the first accurate measurement of the neutron magnetic moment as 1.93 ± 0.02 nuclear magnetons. Finally, a comprehensive theory of magnetic scattering was published by Halpern and Johnson (1939) and this work has served as the basis for all subsequent theoretical and experimental works on magnetic scattering.

The early scattering experiments were carried out with a radium–beryllium neutron source, which produces a 'white' beam of neutrons with a Maxwellian distribution of velocities. For quantitative studies much more intense neutron sources were required, capable of providing collimated and monochromatic neutron beams. The construction of the first nuclear fission reactor (Chicago Pile Number One, CP-1) in 1942 opened the way for the development of such sources.

CP-1 was built in strict secrecy under the direction of Enrico Fermi on the campus of the University of Chicago. Channels were machined into 4,500 graphite bricks, which were then assembled into a stack, and 5.5 tons of natural uranium metal and 36 tons of uranium oxide were inserted into the channels. The purpose of the graphite was to slow down, or moderate, neutrons emitted by the uranium, whose fission cross-section is considerably enhanced at slow neutron velocities.

Figure 1.3 shows the neutron intensity in CP-1 on that historic afternoon of 2 December 1942. The neutron intensity rises and then levels off

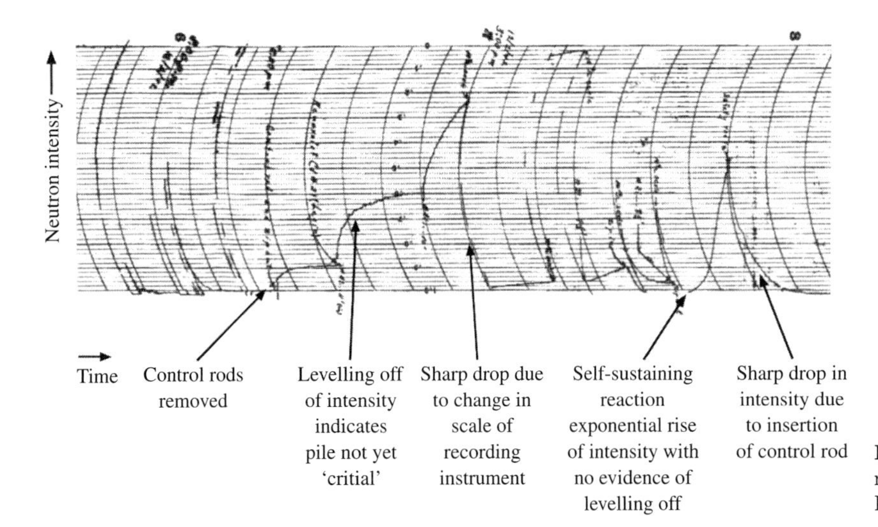

Fig. 1.3 Neutron intensity in CP-1 as registered by a chart recorder. (*After* Rhodes, 1988.)

after each successive neutron-absorbing control rod is withdrawn, but eventually the pile achieves criticality with the neutron population doubling every 2 minutes. The reaction came to a stop at 3.53 p.m. when the safety rods were inserted. Figure 1.4 is a copy of an artist's illustration of the event; the original drawing was made with ink derived from the graphite used in CP-1.

Fermi's experiment laid the foundation of one of the most important scientific developments of the 20th century, and when CP-1 was assembled at the University of Chicago, there was every reason to believe that it was the first such nuclear fission reactor to exist on earth. It was subsequently discovered that Fermi's experiment had been anticipated in nature two billion years earlier! The extraordinary sequence of events leading to the discovery of this natural nuclear reactor in a uranium deposit at Oklo in Gabon, Africa, is vividly described by Cowan (1976).

In 1943 additional reactors were constructed: a heavy-water moderated reactor CP-2 in the Argonne Forest close to Chicago (now the site of the Argonne National Laboratory), and a graphite-moderated reactor at the Oak Ridge National Laboratory (ORNL) in Tennessee, USA. The principal purpose of these reactors was to produce plutonium for the Manhattan project, but they were also used in the first neutron scattering experiments. The neutron flux in the cores of these reactors was about 10^{12} n/cm^2/sec, and it was now possible to do experiments with collimated neutron beams. A neutron diffractometer was built by W. H. Zinn (1947) at the Argonne National Laboratory, and a similar instrument at the same laboratory was employed by Fermi and Marshall (1947) to measure the neutron scattering amplitudes of 22 elements ranging from hydrogen to lead.

At the Oak Ridge National Laboratory, Wollan set up a two-axis diffractometer and recorded the first neutron diffraction pattern of NaCl

Fig. 1.4 On 2 December 1942, mankind first achieved a self-sustaining nuclear chain reaction. This is a copy of the lithograph of the event made by Leo Vartanien. The large portraits are, from left to right, Leo Szilard, Arthur H. Compton, Enrico Fermi and Eugene P. Wigner. (Courtesy of Argonne National Laboratary.)

in 1946. The ORNL soon became the main research centre in the world for neutron scattering. The powder diffraction technique was used to study magnetic materials, and Shull and Smart (1949) confirmed the prediction of Néel that the spins of the manganese ions in MnO are arrayed antiferromagnetically. Shull and Wollan (1951) extended the list of scattering amplitudes to 60 elements and isotopes, and this compilation formed the basis for future systematic studies in neutron scattering. Unlike X-ray scattering factors, neutron scattering amplitudes vary erratically with atomic number and, in the absence of a proper theory of nuclear forces, they must be determined experimentally. For example, hydrogen and uranium (which have vastly different X-ray amplitudes) were found to have comparable neutron amplitudes, and this feature enabled Rundle (1951) to determine all the atomic positions in uranium hydride using neutron diffraction.

Another landmark in the history of neutron scattering was the development of the technique of slow neutron spectroscopy by Brockhouse, which was used to study the excitations of atoms in condensed matter. The energy of a thermal neutron is comparable to the quantum of energy, or phonon, in a normal mode of vibration of a crystal. When a neutron exchanges energy with a single phonon, there is an appreciable change of neutron energy and this change can be readily measured together with the wave vector of the phonon. This provides a method for determining the phonon dispersion relationships of crystals, and the first such measurements were carried out on aluminium using a triple-axis spectrometer built at the Canadian Chalk River reactor (Brockhouse and

Stewart, 1955). The basic design of Brockhouse's spectrometer is still in use today, but with focusing monochromators and analysers and with automatic computer control.

There are other important landmarks in the history of neutron scattering. They include the use of polarized neutrons with a triple-axis spectrometer in carrying out 'polarization analysis' (Moon *et al.*, 1969), the invention of the neutron spin-echo spectrometer (Mezei, 1972) and the development of spherical neutron polarimetry (Tasset *et al.*, 1988). A summary of the early history of work with polarized neutrons is described by Moon (1999).

There are now nearly 300 research reactors in operation throughout the world, including 85 in developing countries. With only a few exceptions, these research reactors were designed primarily for testing materials under irradiation and for the production of isotopes rather than for neutron scattering. In 1965 the first reactor designed solely for scattering experiments was commissioned at the Brookhaven National Laboratory on Long Island, USA. The flux in the beam tubes of this HFBR (high-flux beam reactor) was 100 times greater than in any earlier reactor: this was made possible by using an under-moderated core surrounded by a heavy-water reflector where the maximum flux occurred. A few years later an even more powerful dedicated reactor was built at a new laboratory, the Institut Laue-Langevin (ILL), Grenoble, France. This HFBR began its operation in 1971 and the ILL soon became the principal international centre for neutron scattering. A third dedicated reactor (Orphée) was commissioned at Saclay, France, in 1980 and a fourth such reactor FRM-II at Munich, Germany, in 2003.

Instrument components (see Chapter 5) have also undergone significant and progressive improvements. A particularly important development was due to Maier-Leibnitz and Springer (1963), who showed that neutrons could be guided over long distances by total external reflection inside metal-coated glass tubes and without significant loss of beam flux. Ten of these neutron guides were installed at the ILL, bringing neutrons out of the containment shell of the reactor and into a large experimental hall where there was more space for experiments and a considerably reduced background of radiation. A second neutron guide hall was installed at the ILL in 1991.

There is an upper limit to the neutron flux that can be generated with a nuclear reactor, largely because the maximum power density in the core is determined by the rate at which the heat can be removed. With present-day technology, it is unlikely that the neutron flux could be pushed much further than 10^{15} n/cm^2/sec. With an accelerator-based pulsed source, the peak neutron flux in the pulse can be higher than that from a nuclear reactor, particularly at the upper energy limits of the thermal range, whilst the mean power density is kept at a level compatible with the rate of heat extraction from the target. Assuming that the entire flux of the white beam can be used by the employment of time-of-flight instrumentation, a pulsed source with a power output of

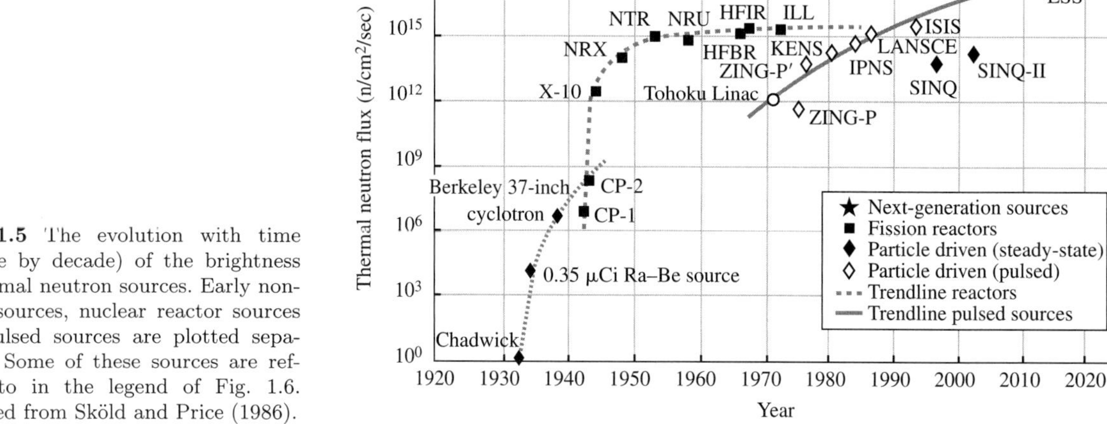

Fig. 1.5 The evolution with time (decade by decade) of the brightness of thermal neutron sources. Early non-fissile sources, nuclear reactor sources and pulsed sources are plotted separately. Some of these sources are referred to in the legend of Fig. 1.6. Updated from Sköld and Price (1986).

1 MW can yield an effective flux which is equivalent to that given by a 50 MW reactor. Therefore, appropriately designed instruments operating on pulsed sources offer an alternative procedure for carrying out scattering experiments.

Figure 1.5 illustrates the evolution, decade by decade, of the effective neutron flux produced by charged-particle sources, reactors, and pulsed sources. It seems that reactor sources have reached a limiting flux indicated by the horizontal line, whereas pulsed sources continue to show an upward trend, thereby holding out the possibility of worthwhile improvements of source strength in the future.

Pulsed neutron sources go back to the 1950s, when electron linear accelerators were adapted for generating neutrons by photofission in a heavy metal target. These sources were used for measuring the neutron cross-sections of materials required for nuclear power plants. The development of pulsed nuclear reactors rather than pulsed accelerators, with the aim of exploiting the high peak flux but low time-averaged power, has been pioneered in Russia. The IBR-30 pulsed reactor at the Joint Institute for Nuclear Research in Dubna began its operation in 1960 and was succeeded 20 years later by the IBR-2 reactor. IBR-2 has a time-averaged power of 4 MW and gives an impressive instantaneous neutron flux of 10^{16} n/cm^2/sec. A major disadvantage of the pulsed reactor source is the long pulse, typically 200 μsec, which is much longer than the microsecond pulses achieved with particle accelerators. Such long pulses do not lend themselves naturally to the building of high-resolution instruments.

Second-generation pulsed sources based upon the spallation reaction have been built at the Argonne National Laboratory, at the High Energy Physics Laboratory KEK in Tsukuba (Japan) and at the Paul Scherrer Institute (Switzerland). Neutrons are produced by the proton spallation process, in which protons from a synchrotron accelerator bombard a heavy metal target. At Los Alamos in New Mexico, a proton linear accelerator has been coupled with a storage ring to produce narrow bursts of protons, with each proton in the burst generating about 20 neutrons. The spallation neutron source ISIS at the Rutherford Appleton Laboratory, UK, has been in operation since 1986 and, at present, is the world's most productive pulsed source with a power of 160 kW.

1.3 Future prospects

In the early years of the 21st century we find that neutron scattering research is at a crossroads. Apart from FRM-II at Munich and the new source at Lucas Heights, Australia, no new reactor sources have been brought on line since the early 1980s when Orphée at Saclay, France, and Dhruva at Trombay, India, began their operation. A reasonable estimate of the maximum lifetime of a reactor is 40–50 years, and there may be less than 10 research reactors for neutron scattering still in operation by 2020. There is a long period, of perhaps 10 or more years, from the initial approval for a new source to its final commissioning. A further 10 years could easily be needed in making the scientific and political case beforehand. The Advanced Neutron Source in Oak Ridge, USA, employing a 350 MW reactor, was the technological successor to the ILL source and its cancellation in 1995 was a symbolic event. The supply of highly enriched uranium fuel, and its subsequent reprocessing or storage, is becoming increasingly difficult to secure for political reasons, and the environmental climate still remains hostile to the building of more nuclear power reactors, despite the dependence on them by France, Japan and developing nations for the supply of electricity. For environmental reasons associated with the production of the greenhouse gas CO_2 in fossil-fuelled power stations, this sentiment is beginning to change.

Pulsed sources offer a way forward out of these difficulties. It is now recognized that pulsed source instruments in many areas of neutron scattering can equal the best reactor instruments. Figure 1.6 shows the location of the principal pulsed and reactor sources throughout the world. The 1.4 MW Spallation Neutron Source at Oak Ridge, USA, and the 0.6 MW spallation source at the Japanese Hadron facility J-PARC, Tokai-mura, are now operational. The proposal for a 5 MW European Spallation Source, ESS, is currently waiting for the funding and site issues to be resolved.

The present state of neutron scattering is apparently healthy. The neutron user community is growing, thanks in part to organizations such as the European Neutron Scattering Association, which is drawing in new users and encouraging new fields. It is clear, however, that at least one new next-generation source needs to be fully operational and instrumented in each of the major geographical regions of the world, coupled with significant investment in instrumentation at existing sources, in order to secure the future. Decisions in the years up to 2010 will be critical in determining the future health of neutron scattering.

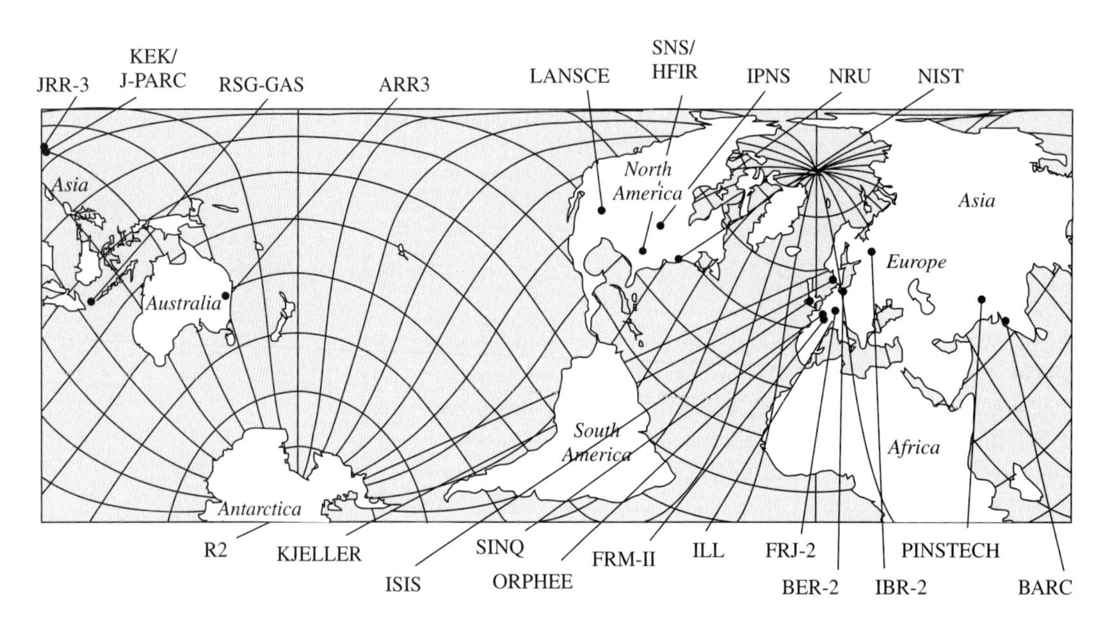

Fig. 1.6 The geographical distribution of the principal centres of research past and present using neutron scattering. BARC: Dhruva at Mumbai, India; BER-2: 10 MW reactor at Berlin, Germany; FRJ-2: 23 MW reactor at Julich, Germany; FRM-II: 20 MW reactor at Munich, Germany; ARR3: 10 MW reactor at Lucas Heights, Australia; HFIR: 100 MW reactor at Oak Ridge, TN, USA; IBR-2: pulsed reactor at Dubna, Russia; ILL: 57 MW reactor at Grenoble, France; IPNS: pulsed source at ANL, IL, USA; ISIS: pulsed source at Oxford, UK; JRR-3: 20 MW reactor at JAERI, Japan; J-PARC: 0.6 MW spallation source at JAERI; KEK: pulsed source at Tsukuba, Japan; KJELLER: 2 MW reactor at Kjeller, Norway; LANSCE: pulsed source at Los Alamos, NM, USA; NIST: 20 MW reactor at Washington, DC, USA; NRU: 130 MW reactor at Chalk River, Canada; ORPHEE: 14 MW reactor at Saclay, France; PINSTECH: reactor at Islamabad, Pakistan; RSG-GAS: 30 MW reactor at Serpong, Indonesia; R2: 50 MW reactor at Studsvik, Sweden; SINQ: 1 MW continuous spallation source at Villigen, Switzerland; SNS: 1.4 MW Spallation Neutron Source at Oak Ridge, TN, USA.

References

Alvarez, L. W. and Bloch, F. (1940) *Phys. Rev.* **57** 111–122. "*A quantitative determination of the neutron moment in absolute nuclear magnetons.*"

Bloch, F. (1936) *Phys. Rev.* **50** 259–260. "*On the magnetic scattering of neutrons.*"

Bothe, W. and Becker, H. (1930) *Z. Physik.* **66** 289–306. "*Künstliche Erregung von Kern-γ-Strahlen.*"

Brockhouse, B. N. and Stewart, A. T. (1955) *Phys. Rev.* **100** 756–757. "*Scattering of neutrons by phonons in an aluminium single crystal.*"

Chadwick, J. (1932a) *Nature* **129** 312. "*Possible existence of a neutron.*"

Chadwick, J. (1932b) *Proc. R. Soc. Lond. A* **136** 692–708. "*The existence of a neutron.*"

Cowan, G. A. (1976) *Sci. Am.* **235** 36–47. "*A natural fission reactor.*"

Curie, I. and Joliot, F. (1932) *C. R. Acad. Sci.* **194** 273–275. "*Émission de protons de grande vitesse par les substances hydrogénées sous l'influence des rayons-γ très pénétrants.*"

Eidelman, S. (Editor) (2004) *Phys. Lett. B* **592** 861–862. "*Biennial review of particle physics: Baryon particle listings.*"

Fermi, E. and Marshall, L. (1947) *Phys. Rev.* **71** 666–677. "*Interference phenomena of slow neutrons.*"

Frisch, O. R., von Halban, H. and Koch, J. (1938) *Phys. Rev.* **53** 719–726. "*Some experiments on the magnetic properties of free neutrons.*"

von Halban, H. and Preiswerk, P. (1936) *C.R. Acad. Sci. Paris* **203** 73–75. "*Preuve experimentale de la diffraction des neutrons.*"

Halpern, O. and Johnson, M. H. (1939) *Phys. Rev.* **55** 898–923. "*On the magnetic scattering of neutrons.*"

Maier-Leibnitz, H. and Springer, T. (1963) *J. Nucl. Energy A/B* **17** 217–225. "*The use of neutron optical devices on beam-hole experiments.*"

Mezei, F. (1972) *Z. Physik.* **255** 146–160. "*Neutron spin echo: a new concept in polarised thermal neutron techniques.*"

Mitchell, D. P. and Powers, N. (1936) *Phys. Rev.* **50** 486–487. "*Bragg reflection of slow neutrons.*"

Moon, R. M. (1999) *Physica B* **267–268** 1–8. "*Early polarised neutron work.*"

Moon, R. M., Riste, T. and Koehler, W. C. (1969) *Phys. Rev.* **181** 920–931. "*Polarisation analysis of thermal neutron scattering.*"

Rhodes, R. (1988) "*The making of the atomic bomb.*" Penguin Books, London.

Rundle, R. E. (1951) *J. Am. Chem. Soc.* **73** 4172–4174. "*The hydrogen positions in uranium hydride by neutron diffraction.*"

Schwinger, J. S. (1937a) *Phys. Rev.* **51** 544–552. "*On the magnetic scattering of neutrons.*"

Schwinger, J. S. (1937b) *Phys. Rev.* **52** 1250. "*On the spin of the neutron.*"

Shull, C. G. and Smart, J. S. (1949) *Phys. Rev.* **76** 1256–1257. "*Detection of antiferromagnetism by neutron diffraction.*"

Shull, C. G. and Wollan, E. O. (1951) *Phys. Rev.* **81** 527–535. "*Coherent scattering amplitudes as determined by neutron diffraction.*"

Skold, K. and Price, D. L. (Editors) (1986) "*Neutron scattering.*" Academic Press: Orlando, FL.

Tasset, F., Brown, P. J. and Forsyth, J. B. (1988) *J. Appl. Phys.* **63** 3606–3608. "*Determination of the absolute magnetic moment direction in Cr_2O_3 using generalized polarisation analysis.*"

Zinn, W. H. (1947) *Phys. Rev.* **71** 752–757. "*Diffraction of neutrons by a single crystal.*"

Neutron properties

2

Neutrons have a number of properties which distinguish them from X-rays. They have a weak interaction with matter and so have high penetrating capacity: walls of furnaces, cryostats, pressure vessels and others present little obstacle to the experimentalist. The atomic scattering amplitude is not linked to the number of shell electrons and so the magnitude of the interaction is different for the different isotopes of an element. The spin of the neutron interacts with the magnetic moments of unpaired electrons, allowing the study of magnetic materials. The energy and wavelength of thermal neutrons are comparable to the thermal energy and spacing of the atoms in solids; therefore, neutron scattering is the natural method for studying excitations, such as phonons and magnons, in crystals. In fields such as small-angle scattering from polymers, reflectometry from surfaces and vibrational spectroscopy of molecules, neutron scattering provides information which is different but complementary to that derived from X-ray or light scattering techniques.

This chapter addresses some of the underlying concepts of neutron scattering and will provide the experimentalist with sufficient information to get started. The quantities measured in an experiment are the scattering cross-sections as defined in Section 2.3. Later we shall quote expressions for some of these cross-sections without giving their full derivations, which can be found in the books of Squires (1996) and Lovesey (1984) on the theory of thermal neutron scattering.

2.1 Wave–particle duality

The neutron can be represented as either a wave or a particle, and this dual description will be used frequently throughout this book. Therefore, studies of neutron diffraction or of the collective excitations in crystals are readily interpreted by assuming that the neutron is a wave; on the other hand, in dealing with the production and detection of neutrons, it is necessary to consider the neutron to be a particle.

A striking example of the wave nature of neutrons is provided by the experiment of Zeilinger *et al.* (1988), who studied the Fraunhofer diffraction of neutrons by using a double slit. We recall that according to Feynman *et al.* (1965), 'The double-slit experiment has in it the heart of quantum mechanics'. Thermal neutrons have a wavelength which is 1,000 times smaller than that of visible light. Even so, by employing very cold neutrons to maximize the *de Broglie wavelength* and by using an optical bench over 10 m long, it was possible to reproduce with neutrons

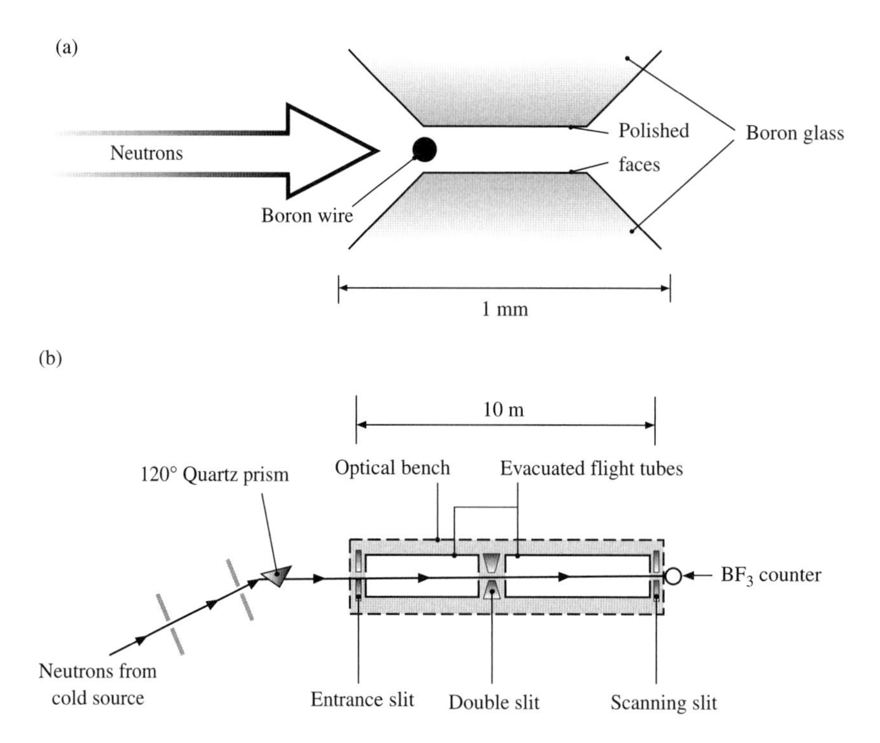

Fig. 2.1 (a) A section through the double slit used by Zeilinger *et al.* (1988) demonstrate the interference of neutron waves: the width of the slit is 150 μm. (b) The layout of the instrument showing the optical bench incorporating the double slit in (a).

the conditions of the Young double-slit experiment with light. In the neutron experiment, as illustrated in Fig. 2.1(a), the double slit was created by placing a boron wire (which is opaque to slow neutrons) between the faces of two plates of polished boron glass. The 120° quartz prism in Fig. 2.1(b), which is used to separate the different wavelengths in the beam by rainbow prism refraction, is inverted with respect to its orientation in the corresponding light experiment because the refractive index of quartz for neutrons is less than unity (see Chapter 11).

The full line in Fig. 2.2 gives the intensity of diffracted neutrons, as calculated from classical diffraction theory. The oscillatory interference pattern, as recorded at different positions of the scanning slit, is in close agreement with that predicted on the assumption that the incident neutron beam has a de Broglie wavelength of 18.45 Å and a bandwidth of 1.40 Å.

2.1.1 Interconversion of units

Experimentalists in neutron scattering often need to convert their measurements between energy and wavelength or frequency or velocity .... Table 2.1 lists the conversion factors between these various units. It is assumed that an energy E corresponds to a temperature T given by $E = k_{\mathrm{B}}T$ where k_{B} is Boltzmann's constant. The table has been

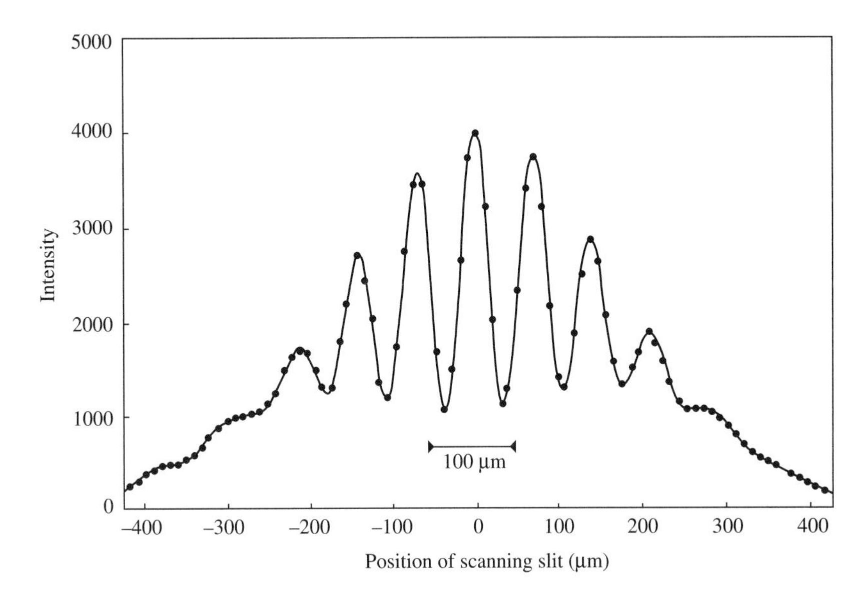

Fig. 2.2 A double-slit diffraction pattern recorded by Zeilinger *et al.* (1988) with the apparatus shown in Fig. 2.1. The full line represents the theoretical prediction, and the points are the experimental measurements.

Table 2.1 Wave–particle relationships

	E (meV)	v (km/sec)	λ (Å)	k (Å^{-1})	f (cm^{-1})	T (K)	ω (THz)
Energy E of 1 meV	1	0.437	9.045	0.695	8.066	11.60	1.519
Velocity v of 1 km/sec	5.227	1	3.956	1.588	42.16	60.66	7.948
Wavelength λ of 1 Å	81.81	3.956	1	6.283	659.8	949.3	124.3
Wave vector k of 1 Å^{-1}	2.072	0.629	6.283	1	16.71	24.04	3.148
Optical frequency f of 1 cm^{-1}	0.124	0.154	25.69	0.245	1	1.439	0.188
Temperature T of 1 K	0.086	0.128	30.81	0.204	0.695	1	0.131
Angular frequency ω of 1 THz	0.658	0.354	11.15	0.564	5.31	7.63	1

derived using the following relations between the units:

$$E = 5.227v^2 = 81.81/\lambda^2 = 2.0723k^2 = 0.1239f = 0.0861T = 0.658\omega \quad (2.1)$$

where E is in meV, v in km/sec, λ in Å, k(=wave vector) in Å^{-1}, $f(=1/\lambda)$ in cm^{-1}, T in K and ω in THz. The table should be read horizontally; for example, in the first line an energy of 1 meV corresponds to a wavelength of 9 Å and in the last line an angular frequency of 1 THz corresponds to an energy of 0.66 meV. The table can be used for deriving the conversion factors between units by recalling that E is directly proportional to f, T, ω, k^2 and v^2 and is inversely proportional to λ^2. Therefore, a velocity of 2,200 m/sec corresponds to an energy of 25.30 meV, a temperature of 293.6 K and a wavelength of 1.80 Å.

The energy E, wavelength λ and wavenumber k are related by

$$E(\text{meV}) = 2.072[k(\text{Å}^{-1})]^2 = \frac{81.81}{\lambda(\text{Å})^2} \approx \left[\frac{9}{\lambda(\text{Å})}\right]^2 \quad (2.2)$$

and the neutron velocity v and wavelength λ by

$$v(\text{m/sec}) = \frac{3{,}956}{\lambda(\text{Å})} \tag{2.3}$$

Some useful rules of thumb worth memorizing are

$$1 \text{ Å} \approx 82 \text{ meV}; 1 \text{ meV} \approx 8 \text{ cm}^{-1} \approx 1.5 \text{ THz} \approx 11.6 \text{ K} \tag{2.4}$$

2.2 Properties of thermal neutrons

Neutrons and protons each consist of up-quarks and down-quarks, bound together by the strong nuclear force. The up-quark has a charge of $+2e/3$ and the down-quark has a charge of $-e/3$. The neutron consists of two down-quarks and one up-quark and is electrically neutral, whereas the proton consists of two up-quarks and one down-quark and has a charge of $+e$. We shall not be concerned with the quark model in this book, because at the thermal energies in neutron scattering experiments the neutron behaves like an elementary particle, and in this spirit its elementary properties are summarized in Table 2.2.

2.2.1 Lifetime

A free neutron is unstable and undergoes radioactive decay. It is a beta-emitter, decaying spontaneously into a proton, an electron and an electron anti-neutrino. Its lifetime can be measured by storing ultracold neutrons inside a neutron bottle (see Chapter 11) and then counting the number of neutrons surviving as a function of time. If $N(0)$ is the number present at time zero and $N(t)$ is the number present after time t, then we can write

$$N(t) = N(0) \exp\left(\frac{-t}{\tau}\right) \tag{2.5}$$

where τ is the 'lifetime' of the neutron, *i.e.* the time after which $1/e$, or about 0.37, of the original population remains. The latest estimate of τ obtained by Arzumanov *et al.* (2000) is

$$\tau = 886 \pm 1 \text{ sec} \tag{2.6}$$

Table 2.2 Principal properties of the neutron.

Lifetime τ	886 ± 1 s
Mass m_n	1.67495×10^{-27} kg
Wavelength λ	1.798 Å for velocity of 2,200 m/s
Energy E	25.30 meV for velocity of 2,200 m/s
Spin S	1/2
Magnetic moment μ_n	$-1.913043(1)$ nuclear magnetons

The 'half-life' $\tau_{1/2}$ or the time taken for half of the neutrons to decay is $\tau \ln 2 = 614$ sec.

In a typical neutron scattering experiment, the time taken by neutrons to traverse the distance from source to detector is much less than their half-life. Thus a neutron of velocity 2,200 m/sec takes 45 ms to travel 100 m, and so the fraction of neutrons decaying in this period of time is vanishingly small. It is safe to assume that the finite lifetime of the neutron is of no practical significance in a scattering experiment.

2.2.2 Mass

Chadwick's experiments leading to the discovery of the neutron involved a measurement of the mass of the neutron (see Section 1.1). There have been many further improvements in the measurement of the mass of the neutron, all based on the technique devised by Chadwick, and m_n is now known to an uncertainty of 1 part in a 10^8. This limiting error is mainly associated with the uncertainty in the value of Avogadro's number. The latest value of the neutron mass (see http://www-pdg.lbl.gov/) is

$$\begin{aligned} m_n &= 1.00866492(1) \text{ a.m.u.} \\ &= 1.67495 \times 10^{-27} \text{ kg} \end{aligned} \tag{2.7}$$

2.2.3 Wavelength

The wavelength λ of a neutron is related to its velocity v through the de Broglie relationship:

$$\lambda = \frac{h}{m_n v} \tag{2.8}$$

where h is the Planck's constant.

The neutrons used for scattering experiments are thermal neutrons produced by a reactor or an accelerator source. In a nuclear reactor the fission process leads to the emission of neutrons with energies up to 10 MeV. These fast neutrons then enter a moderator, a material such as water or liquid hydrogen, which reduces the neutron energy to the thermal region. The moderator should possess a low atomic or molecular mass (to ensure that slowing down to the thermal neutron region occurs after relatively few collisions), a large scattering cross-section for neutrons and a low absorption cross-section. The neutrons emerge from the reactor in thermal equilibrium with the moderator.

The velocity distribution of the neutrons follows the Maxwell–Boltzmann distribution law for the atoms or molecules of a gas:

$$n(v) = \frac{4n_0 v^2}{\sqrt{\pi} v_T^3} \exp\left(-\frac{v^2}{v_T^2}\right) \tag{2.9}$$

where $n(v)\mathrm{d}v$ is the number of thermal neutrons per unit volume with velocities between v and $v + \mathrm{d}v$, n_0 is the total number of neutrons in unit volume, and

$$v_{\mathrm{T}} = \left(\frac{2k_{\mathrm{B}}T_{\mathrm{eff}}}{m_n}\right)^{1/2} \tag{2.10}$$

where T_{eff} is the effective temperature of the moderator. The neutron flux $\phi(v)$ or the number of neutrons passing through unit area per unit time is proportional to v times the neutron density, $n(v)$:

$$\varphi(v) \propto \frac{4n_0}{\sqrt{\pi}}\frac{v^3}{v_{\mathrm{T}}^3}\exp\left(-\frac{v^2}{v_{\mathrm{T}}^2}\right) \tag{2.11}$$

Putting $\phi(\lambda)\mathrm{d}\lambda = \phi(v)\mathrm{d}v$, where $\phi(\lambda)\mathrm{d}\lambda$ is the flux in the wavelength range λ to $\lambda + \mathrm{d}\lambda$, the flux spectrum in terms of wavelength is proportional to

$$\varphi(\lambda) \propto \frac{4n_0}{\sqrt{\pi}}\frac{1}{v_{\mathrm{T}}^3}\left(\frac{h}{m_{\mathrm{n}}}\right)^4\frac{1}{\lambda^5}\exp\left(-\frac{h^2}{2m_{\mathrm{n}}\lambda^2 k_{\mathrm{B}}T}\right) \tag{2.12}$$

The maximum value of the flux (see Fig. 2.3) occurs at a wavelength λ_{m} given by

$$\lambda_{\mathrm{m}} = \frac{h}{(5m_{\mathrm{n}}k_{\mathrm{B}}T)^{1/2}} \tag{2.13}$$

From this equation λ_{m} is 1.14 Å for a moderator temperature of 20°C. This wavelength is of the same order of magnitude as the interatomic distances in solids and so thermal neutrons, as for X-rays, offer an ideal probe for determining the atomic arrangement in condensed matter. In many experiments, it may be desirable to use either longer or shorter wavelengths rather than those obtainable from a moderator at ambient temperature. 'Cold neutrons' with wavelengths from 4 Å to 30 Å can be obtained using a cold moderator of liquid hydrogen at 20 K; shorter wavelengths, down to 0.3 Å, are produced by using a hot graphite moderator at a temperature of 2,000 K. With accelerator-based pulsed sources, the neutrons are under-moderated because the moderators are relatively small and so wavelengths as short as 0.1 Å are available (see Chapter 3).

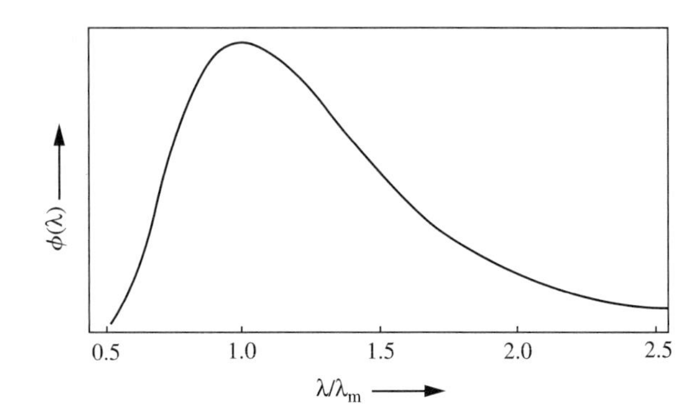

Fig. 2.3 The Maxwell–Boltzmann wavelength spectrum of thermal neutrons as a function of $\lambda/\lambda_{\mathrm{m}}$, showing an effective flux cut-off at about half the peak wavelength λ_{m}.

2.2.4 Energy

de Broglie's relationship between the momentum p ($=m_n v$) of a particle and its wavelength λ is $p = h/\lambda$. Hence the corresponding kinetic energy E, which equals $1/2 m_n v^2$ or $p^2/(2m_n)$, is given by

$$E = \frac{h^2}{2m_n\lambda^2} \tag{2.14}$$

Assuming that $E = k_B T$, we have

$$E = 0.08617T \tag{2.15}$$

where E is in meV and T is in degrees Kelvin. Thus the energy of the neutrons emerging from a water moderator at 20°C is centred around 25 meV, whereas with liquid hydrogen moderator at 20 K this energy is about 2 meV. From Table 2.1, the energy of a 1 meV neutron corresponds to a frequency of 1.519 THz, a wavenumber of 8.066 cm^{-1} or a temperature of 11.60 K.

The atoms in a solid are held together by strong binding forces: when an atom is displaced from its equilibrium position by thermal agitation, waves of atomic displacement are propagated throughout the entire crystal. The energy associated with these waves, or collective excitations, is quantized, and the corresponding quantum of energy is known as a *phonon*. When neutrons are inelastically scattered by exchanging energy with these excitations, a phonon is created or destroyed in the sample and the neutron loses or gains a corresponding amount of energy. Neutrons can also exchange energy with magnetic excitations, with crystalline electric fields, with the vibrations and rotations of molecules, and with diffusional processes of atoms or molecules in solids and liquids. Thermal neutrons can be used for studying all these excitations, because the energy changes involved are of the same order of magnitude as the initial neutron energy.

2.2.5 Magnetic properties

The magnetic moment associated with the spin of an electron is approximately 1,000 times larger than the magnetic moment of a neutron. Nevertheless, the magnetic moment of the neutron (see Table 2.2) is sufficiently large to give rise to an interaction with unpaired electrons in magnetic atoms, which is of comparable strength to the interaction of the neutron with the nucleus. In most atoms the electrons are paired off with other electrons of opposite spin, so that the magnetic moment of one electron is cancelled by the magnetic moment of its partner and there is no magnetic interaction between the neutron and the atom. On the other hand, in compounds containing elements of the first transition series in the Periodic Table, which includes iron, cobalt and nickel, the $3d$ shell contains unpaired electrons: the magnetic field created by these unpaired electrons in the sample interact with the neutron magnetic moment to give magnetic scattering. Magnetic scattering also takes place from compounds containing elements in the rare-earth or actinide groups of the Periodic Table.

Elastic magnetic scattering, or magnetic diffraction, leads to the determination of the magnetic structures of crystals, *i.e.* the arrangement of the magnetic moments of atoms on the crystalline lattice. Inelastic magnetic scattering yields information on the magnetic excitations in the scattering system. A spin wave is such an excitation, in which there are oscillations in the orientation of successive spins on the crystal lattice. Spin waves are quantized in energy units called *magnons*, just as the normal modes of vibration are quantized in energy units called phonons. Spin waves can be studied by inelastic magnetic scattering in a manner similar to the study of lattice vibrations by inelastic nuclear scattering.

2.2.6 Polarization properties

If a magnetic field, or *guide field*, is applied along the path of a neutron beam, then the neutron spins or moments are aligned parallel or antiparallel to the field. The beam is *polarized* when the population of the parallel moments, the *up spins*, differs from the population of the antiparallel moments, the *down spins*. A fully polarized beam is one with all the spins aligned either up or down.

One method of polarizing a neutron beam is to use a magnetized Co–Fe crystal as a monochromator. This crystal has an appreciable cross-section for the *up spin* state of the neutron and an almost zero cross-section for the *down spin* state: by rejecting half the neutrons, a polarized beam is produced. Other polarizing methods, employing reflection from magnetized mirrors or *supermirrors*, and transmission through polarized absorbers such as ^{3}He, are described in Chapter 8.

2.3 Definition of cross-sections

Figure 2.4 illustrates the geometry of a scattering experiment. An incident neutron, specified by its wave vector $\mathbf{k}_\mathrm{i}$, where $|\mathbf{k}| = k = 2\pi\lambda$, is scattered into a new state having a wave vector $\mathbf{k}_\mathrm{f}$. The origin of coordinates is at the position of the nucleus and the neutron is scattered to a point $\mathbf{r}$. The direction of scattering is defined by the azimuthal angle ϕ and by the angle 2θ[1] between incident and scattered beams. Scattering occurs in an elementary cone of solid angle $\mathrm{d}\Omega$. If the scattering is elastic, then the magnitude of the wave vector is unchanged on scattering, *i.e.* $k_\mathrm{i} = k_\mathrm{f}$. If the scattering is inelastic, then there is an exchange of energy between the neutron and the sample: when the neutron gains energy $k_\mathrm{i} < k_\mathrm{f}$, and when it loses energy $k_\mathrm{i} > k_\mathrm{f}$.

[1]We choose the symbol 2θ for the angle between k_f and k_i because, in the case of elastic Bragg scattering, θ is then equal to the Bragg angle θ_B. In Part III, which deals with inelastic scattering, 2θ is replaced by ϕ.

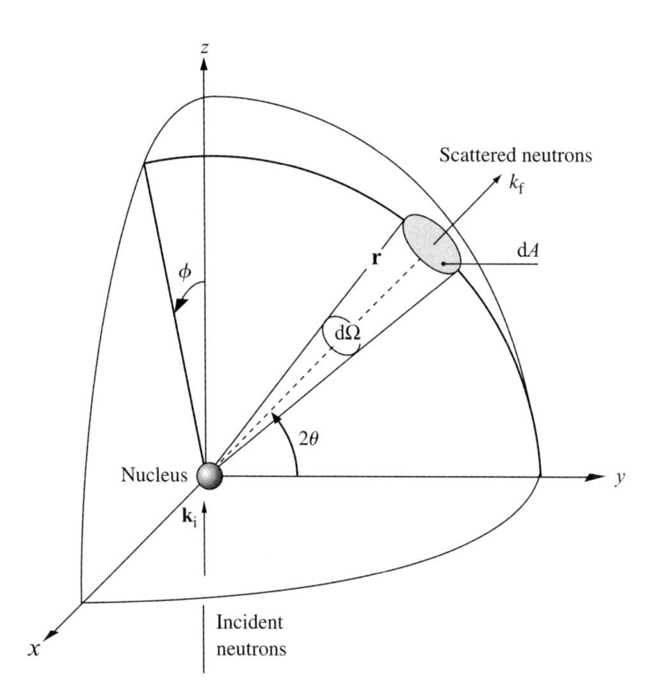

Fig. 2.4 The geometry of a neutron scattering experiment.

After the neutrons have interacted with the sample, the measurement of the scattered neutrons leads to the determination of the neutron scattering cross-section. There is a variety of cross-sections that can be measured with a wide variety of instruments.

2.3.1 Total cross-section

The total cross-section σ_{tot} is defined by

$$\sigma_{\text{tot}} = \frac{\text{number of neutrons scattered in all directions per second}}{\text{incident flux}(I_0)} \tag{2.16}$$

The incident flux I_0 is the number of neutrons striking unit area of the sample in unit time, where the area is taken to be perpendicular to the incident neutron beam.

Let us consider scattering by a single nucleus. An incident plane wave of neutrons travelling in the z-direction is represented by the wave function

$$\psi_{\text{i}} = \mathrm{e}^{\mathrm{i}kz} \tag{2.17}$$

where $k = 2\pi/\lambda$ is the wavenumber. The probability of finding a neutron in a volume $\mathrm{d}V$ is $|\psi_{\text{i}}|^2\mathrm{d}V$. But $|\psi_{\text{i}}|^2 = 1$, so that eqn. (2.17) refers to a density of one neutron per unit volume throughout all space. The flux of neutrons incident normally on unit area per second is

$$I_0 = \text{neutron density} \times \text{velocity} \ = v \tag{2.18}$$

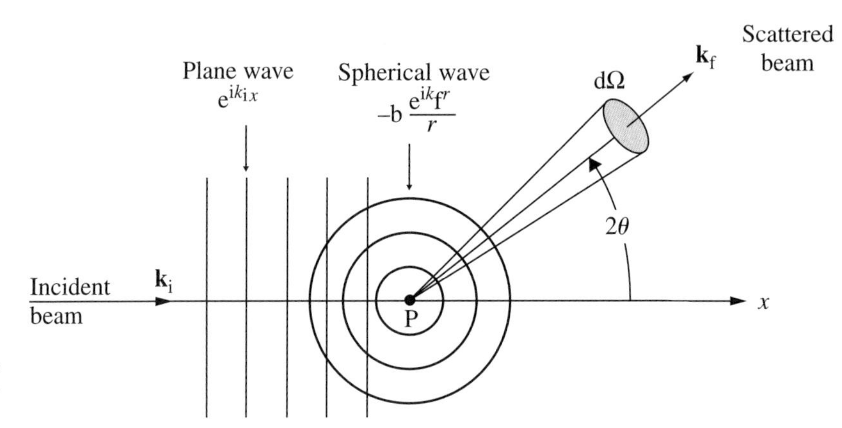

Fig. 2.5 The scattering of a plane wave of neutrons by a single point scatterer at P.

The wavelength of slow neutrons vastly exceeds the nuclear radius, and so there is *s*-wave scattering which is isotropic with no dependence on direction. The wave scattered by an isolated nucleus, as indicated in Fig. 2.5, is of the form

$$\psi_f = -b\frac{e^{ikr}}{r} \tag{2.19}$$

which is the distance from the scattering nucleus and b is known as the 'scattering length' of the nucleus. We assume that the nucleus is fixed so that the scattering is elastic. The minus sign in eqn. (2.19) is adopted to ensure that most values of b for the elements are positive. In the absence of an appropriate theory of nuclear forces, the scattering length is treated as a parameter to be determined experimentally for each kind of nucleus.

The scattering lengths for the elements are listed in Appendix B. In the thermal neutron region b is independent of wavelength, apart from some rare exceptions such as ^{113}Cd which has an (n,γ) resonance in this region.

The scattered flux I_f is $|\psi_f|^2 \times$ velocity $= (b^2/r^2)v$, and the number of neutrons scattered per second is flux × area:

$$\left(\frac{b^2}{r^2}\right) v \cdot 4\pi r^2 = \left(4\pi b^2\right) v \tag{2.20}$$

Hence from (2.18) and (2.20), we get

$$\sigma_{tot} = \frac{I_f}{I_0} = 4\pi b^2 \tag{2.21}$$

which is the effective area of the nucleus viewed by the neutron. The units used for cross-sections are known as barns, where 1 barn $= 10^{-28}$ m^2, and the units used for scattering lengths are fermis, where 1 fermi $= 10^{-15}$ m.

2.3.2 Differential cross-section

The differential cross-section $\mathrm{d}\sigma/\mathrm{d}\Omega$ is defined as

$$\frac{\mathrm{d}\sigma}{\mathrm{d}\Omega} = \frac{\text{number of neutrons scattered per sec into a solid angle } \mathrm{d}\Omega}{I_0 \mathrm{d}\Omega} \tag{2.22}$$

This cross-section is measured by using a *neutron diffractometer*, which is an instrument for recording the scattered intensity as a function of the wave-vector transfer, $\mathbf{k}_\mathrm{i} - \mathbf{k}_\mathrm{f}$, without reference to any changes in energy.

In Fig. 2.5, the solid angle $\mathrm{d}\Omega$ subtended by the detector at the sample is $\mathrm{d}A/r^2$ where $\mathrm{d}A$ is the shaded area. The number of neutrons scattered per second into this solid angle is flux $\times$ area or $|\psi_\mathrm{f}|^2 \times \mathrm{d}A$, and from (2.22) this is also equal to $I_0(\mathrm{d}A/r^2)(\mathrm{d}\sigma/\mathrm{d}\Omega)$. Hence, we have

$$\frac{\mathrm{d}\sigma}{\mathrm{d}\Omega} = |\psi_\mathrm{f}|^2 r^2 \tag{2.23}$$

For the special case of scattering by a single nucleus $|\psi_\mathrm{f}|^2 = b^2/r^2$, and so

$$\frac{\mathrm{d}\sigma}{\mathrm{d}\Omega} = b^2 = \frac{\sigma_\mathrm{tot}}{4\pi} \tag{2.24}$$

The units of $\mathrm{d}\sigma/\mathrm{d}\Omega$ are barns/steradian.

2.3.3 Double differential cross-section

The double differential cross-section relates to those experiments where a change of neutron energy on scattering is measured. It is defined by

$$\frac{\mathrm{d}^2\sigma}{\mathrm{d}\Omega \mathrm{d}E_\mathrm{f}} = \frac{\begin{array}{c}\text{number of neutrons scattered per sec into a}\\ \text{solid angle } \mathrm{d}\Omega \text{ with final energy between } E_\mathrm{f} \text{ and } E_\mathrm{f} + \mathrm{d}E_\mathrm{f}\end{array}}{I_0 \ \mathrm{d}\Omega \mathrm{d}E_\mathrm{f}} \tag{2.25}$$

and has the dimensions of area per unit energy. This is an inelastic scattering cross-section involving a change of energy in the scattering process and the instrument used for measuring it is known as a *neutron spectrometer*.

The numerator in (2.25) is proportional to the bandwidths $\mathrm{d}\Omega$ and $\mathrm{d}E_\mathrm{f}$ and to the number of incident neutrons $|\psi_\mathrm{i}|^2$. The incident flux I_0 is $|\psi_\mathrm{i}|^2 v_\mathrm{i} = |\psi_\mathrm{i}|^2 \hbar k_\mathrm{i}/m_\mathrm{n}$. Inserting these quantities into eqn. (2.25), the cross-section for a system of N atoms (see Van Hove, 1954) is

$$\frac{\mathrm{d}^2\sigma}{\mathrm{d}\Omega \mathrm{d}E_\mathrm{f}} = b^2 \frac{k_\mathrm{f}}{k_\mathrm{i}} N S(\mathbf{Q}, \omega) \tag{2.26}$$

where the quantity $S(\mathbf{Q}, \omega)$ is the *scattering function* of the system. It gives the probability that the scattering changes the energy of the system by an amount $\hbar\omega(= E_\mathrm{i} - E_\mathrm{f})$ and its momentum by $\hbar\mathbf{Q}(=\hbar\mathbf{k}_\mathrm{i} - \hbar\mathbf{k}_\mathrm{f})$. In

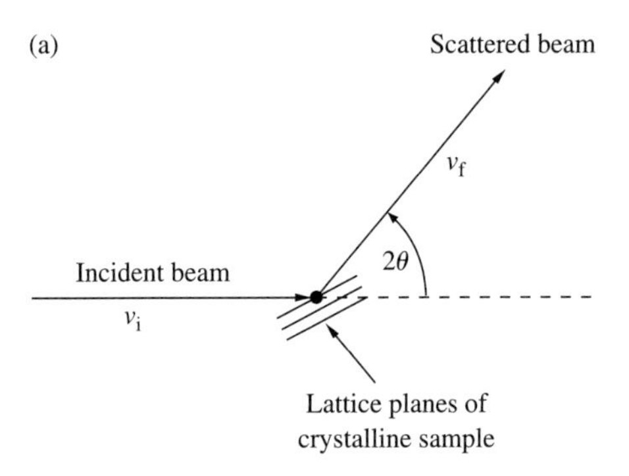

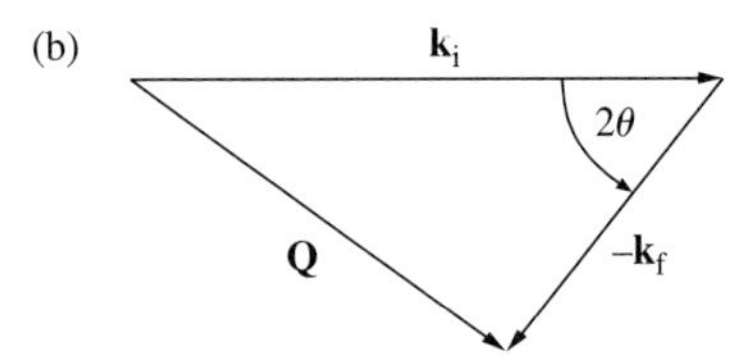

Fig. 2.6 (a) Scattering of a neutron in real space. v_i and v_f are the initial and final velocities of the neutron. (b) Scattering in reciprocal space. $\mathbf{k}_i$ and $\mathbf{k}_f$ are the initial and final wave vectors, and $\mathbf{Q}$ is $\mathbf{k}_i - \mathbf{k}_f$.

speech, $S(\mathbf{Q}, \omega)$ is referred to as 'S of $\mathbf{Q}$ and ω'. The energy transferred in the scattering process is

$$\hbar\omega = E_i - E_f = \frac{\hbar^2}{2m_n}\left(k_i^2 - k_f^2\right) \tag{2.27}$$

From Fig. 2.6, we obtain

$$Q^2 = k_i^2 + k_f^2 - 2k_i k_f \cos 2\theta \tag{2.28}$$

which reduces to

$$Q = 2k_i \sin 2\theta \tag{2.29}$$

for elastic scattering ($k_i = k_f$).

None of the definitions of cross-sections in this section takes into account the possibility that the nucleus of the same element may have different scattering amplitudes. This gives rise to further subdivision of the cross-section into *coherent* and *incoherent* components.

2.4 Coherent and incoherent nuclear scattering

The different isotopes of a chemical element have their own characteristic scattering lengths. Moreover, if the nucleus of a specific isotope has a non-zero spin, each isotope itself has two different scattering lengths, b_+ and b_-; these scattering lengths are associated with the two states of the *compound nucleus*, in which the combined spins of the nucleus and the neutron are either parallel or anti-parallel. The existence of isotope and spin effects gives rise to cross-sections with components of both coherent and incoherent scattering. The coherent scattering gives interference effects and yields space and time relationships between different atoms as, for example, in the collective motion of atoms undergoing lattice vibrations. The incoherent scattering does not give interference effects: it arises from the deviations of the scattering lengths from their mean values and leads to information about the properties of individual atoms.

Consider the scattering from a sample containing an element with several isotopes of non-zero spin. For the jth atom at the position r_j there is an assortment of different scattering lengths b_j because of the effects of nuclear spin and the presence of different isotopes. The differential cross-section for the *elastic scattering* from an assembly of fixed nuclei in the sample is

$$\frac{\mathrm{d}\sigma}{\mathrm{d}\Omega} = \left| \sum_j b_j \exp(\mathrm{i}\mathbf{Q} \cdot \mathbf{r}_j) \right|^2 \tag{2.30}$$

where the summation is over all atoms. The term $\exp(\mathrm{i}\mathbf{Q} \cdot \mathbf{r}_j)$ takes into account the phase difference between a wave scattered in the direction $\mathbf{k}_\mathrm{f}$ by a nucleus at $\mathbf{r}_j$ and a wave scattered by a nucleus at the origin ($\mathbf{r}_j = 0$). Using the relation $|A|^2 = AA^*$, where the asterisk represents complex conjugate, eqn. (2.30) can be rewritten as a double summation:

$$\frac{\mathrm{d}\sigma}{\mathrm{d}\Omega} = \sum_j \sum_{j'} b_j b_{j'} \exp\left[i\mathbf{Q} \cdot (\mathbf{r}_j - \mathbf{r}_{j'})\right] \tag{2.31}$$

In general, b_j will depend on which isotope is at the site $\mathbf{r}_j$ and on the spin states associated with that isotope. Assuming that the isotopes and the spin states of the nuclei are distributed at random amongst all available sites and are uncorrelated between sites, the quantity $b_j b_{j'}$ in eqn. (2.31) can be replaced by the ensemble average $\langle b_j b_{j'} \rangle$, where $\langle\,\rangle$ indicates the mean value. If j and j' refer to different sites ($j \neq j'$) and there is no correlation between the values of b_j and $b_{j'}$, we can write

$$\langle b_j b_{j'} \rangle = \langle b_j \rangle \langle b_{j'} \rangle = \langle b \rangle^2 \tag{2.32}$$

On the other hand, for the same site ($j = j'$), we have

$$\langle b_j b_{j'} \rangle = \langle b_j^2 \rangle = \langle b^2 \rangle \tag{2.33}$$

Thus eqn. (2.31) can be expressed as the sum of two terms, A and B. Term A represents the scattering from different pairs of sites and term B the scattering from the same site:

$$A = \sum_{j,j'}^{j \neq j'} \langle b_j b_{j'} \rangle \exp\left[i\mathbf{Q} \cdot (\mathbf{r}_j - \mathbf{r}_{j'})\right] \tag{2.34}$$

and

$$B = \sum_{j,j'}^{j = j'} \langle b_j b_{j'} \rangle \exp\left[i\mathbf{Q} \cdot (\mathbf{r}_j - \mathbf{r}_{j'})\right] \tag{2.35}$$

The differential cross-section becomes

$$\frac{d\sigma}{d\Omega} = \sum_{j,j'}^{j\neq j'} \langle b\rangle^2 \exp\left[i\mathbf{Q}\cdot(\mathbf{r}_j - \mathbf{r}_{j'})\right] + \sum_j \langle b^2\rangle \tag{2.36}$$

Inserting an additional term for $j = j'$ in the first summation of eqn. (2.36) and subtracting the same term from the second summation gives the final expression

$$\frac{d\sigma}{d\Omega} = \sum_{j,j'} \langle b\rangle^2 \exp[i\mathbf{Q}\cdot(\mathbf{r}_j - \mathbf{r}_{j'})] + \sum_j \left(\langle b^2\rangle - \langle b\rangle^2\right) \tag{2.37}$$

The first summation $\sum_{j,j'}$ in eqn. (2.37) represents the coherent scattering component, in which there is interference between the waves scattered from each nucleus. The coherent cross-section takes the same form as that given by a system where each nucleus possesses the average scattering length $\langle b\rangle$. The second summation $\sum_j$ is the incoherent scattering component, for which there is no interference: its magnitude is determined by the mean-square deviation of the scattering length from its average value. These features are illustrated in Fig. 2.7, where the coherent diffraction peaks from Bragg scattering are determined by the magnitude of $\langle b\rangle^2$ and the incoherent scattering gives a flat background determined by the magnitude of $\sum_j (\langle b^2\rangle - \langle b\rangle^2)$.

The mean value $\langle b\rangle$ of the scattering length of an atom is known as its coherent scattering length $b_{\rm coh}$. The total coherent scattering cross-section $\sigma_{\rm coh}$ of an individual atom is

$$\sigma_{\rm coh} = 4\pi\langle b\rangle^2 = 4\pi b_{\rm coh}^2 \tag{2.38}$$

and the total incoherent scattering cross-section is

$$\sigma_{\rm incoh} = 4\pi\left(\langle b^2\rangle - \langle b\rangle^2\right) \tag{2.39}$$

We have seen that there is both an isotopic and spin dependence of the scattering length. For a nucleus of spin I, the spins of the neutron and the nucleus are either parallel or antiparallel, and there are two compound states with total spins $I + 1/2$ and $I - 1/2$ and with scattering lengths b_+ and b_-. The corresponding weighting factors of the states are w_+ and w_-, given by

$$w_+ = \frac{I+1}{2I+1} \text{ and } w_- = \frac{I}{2I+1} \tag{2.40}$$

The expressions for $\sigma_{\rm coh}$ and $\sigma_{\rm incoh}$ are then

$$\sigma_{\rm coh} = 4\pi(w_+b_+ + w_-b_-)^2 \text{ and } \sigma_{\rm incoh} = 4\pi(w_+b_+^2 + w_-b_-^2) \tag{2.41}$$

Experimentally determined values of $b_{\rm coh}$, $\sigma_{\rm coh}$ and $\sigma_{\rm incoh}$ for the elements and isotopes are presented in Appendix B.

The most striking example of spin incoherence is given by hydrogen ^{1}H for which $\sigma_{\rm incoh}$ is 80.3 barns and $\sigma_{\rm coh}$ is only 1.8 barns. Deuterium ^{2}H,

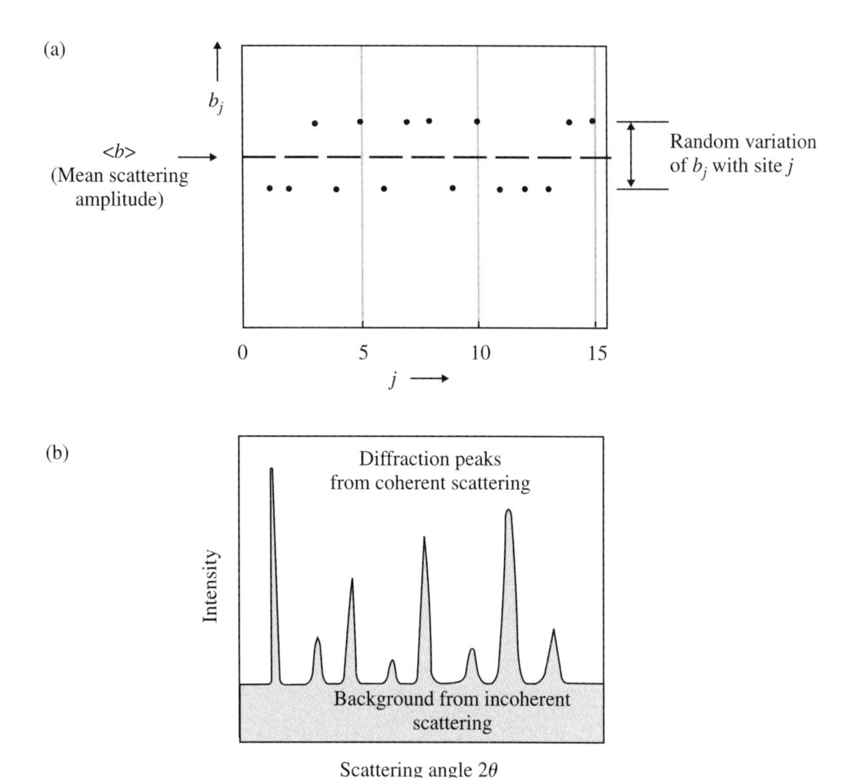

Fig. 2.7 (a) The scattering amplitude at successive atomic sites $j = 1, 2, 3, \ldots$ of a sample containing a single isotope with non-zero spin. (b) A schematic diffraction pattern showing the incoherent background arising from random variations of b_j from atom to atom.

on the other hand, has a value of 2.1 barns for σ_{incoh} and 5.6 barns for σ_{coh}. For vanadium, the scattering is almost entirely incoherent with σ_{coh} less than 1% of σ_{incoh}; therefore, the scattering is uniform and featureless and vanadium can be used as a standard scatterer for the calibration of neutron instruments. For carbon and oxygen, σ_{incoh} is almost zero as both nuclei have zero spin and the natural element consists almost entirely of a single isotope. Curiously, thorium is the only naturally occurring element which gives no incoherent scattering at all; this is because it is the only element which consists of a single isotope with zero spin.

The energy of 1 Å neutrons is ~80 meV which is much less than the chemical binding of the scattering atoms. If the atomic nucleus were free to recoil under the impact of the neutron, then the free atom cross-section σ_f would be diminished in accordance with the expression

$$\sigma_{\text{f}} = \left(\frac{A}{A+1}\right)^2 \cdot \sigma_{\text{b}} \tag{2.42}$$

where A is the atomic mass and σ_{b} is the bound-atom cross-section. The effect of chemical binding on the value of scattering cross-sections is present in all elements. It is most marked for hydrogen, where $\sigma_{\text{f}} = 1/4\sigma_{\text{b}}$, on account of its low atomic mass; for heavier nuclei, the difference between σ_{f} and σ_{b} rapidly diminishes with increasing A. The

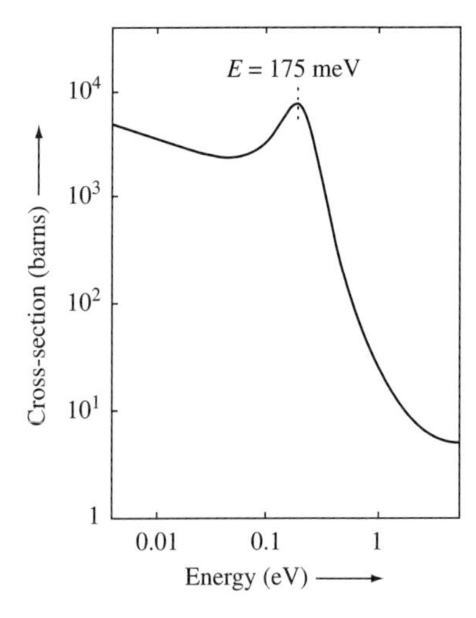

Fig. 2.8 The total cross-section of cadmium versus energy in the thermal neutron range. At neutron energies greater than 250 meV cadmium starts to become transparent to neutrons.

transition from the free-atom cross-section to the bound-atom cross-section is a gradual process and the actual value of the cross-section varies according to the specific material under investigation.

For completeness we consider the absorption cross-section σ_a of an atom, which is defined as the probability per unit time that an incident neutron is removed entirely by nuclear capture processes. Whilst bound-atom scattering cross-sections for thermal neutrons are independent of neutron energy, absorption cross-sections tend to increase inversely with the neutron velocity. For most elements σ_a is proportional to $1/v$ and σ_a is referred to as a 'one-over-vee' cross-section. ^{3}He is a $1/v$ absorber and finds application in neutron gas detectors and neutron polarizers. Deviations from this monotonic behaviour occur when the neutron energy matches the excitation energy of the compound nucleus, leading to an absorption cross-section that rises sharply over a narrow energy range. Cadmium and gadolinium are particularly strong resonant absorbers in the energy range of thermal neutrons. The total cross-section *versus* energy curve for cadmium is shown in Fig. 2.8, where the maximum absorption occurs at an energy of 175 meV or a wavelength of 0.68 Å. In general, resonances become sharper as the neutron energy rises and are most abundant in the keV energy range.

The cross-sections discussed above are cross-sections per atom and are often called microscopic cross-sections. They are measured in units of area, such as the barn which is 10^{-28} m^2. The reader will sometimes find references to macroscopic cross-sections Σ, either scattering or absorption. The macroscopic cross-section is defined as

$$\Sigma = N_0 \sigma \tag{2.43}$$

where N_0 is the number of nuclei per unit volume of the sample. Macroscopic cross-sections are expressed, therefore, in units of inverse length (m^{-1}).

The total effective removal cross-section σ_{eff} of an atom is the sum of the scattering and absorption cross-sections, *i.e.* $\sigma_{\text{eff}} = \sigma_{\text{coh}} + \sigma_{\text{incoh}} + \sigma_a$. The intensity of a neutron beam transmitted through a sample of uniform thickness t is

$$I = I_0 \exp(-N\sigma_{\text{eff}} t) \tag{2.44}$$

N_0 is given by $N_0 = \rho/N_A$ where N_A is Avogadro's number and ρ is the sample density. The attenuation ε of the beam by the sample is

$$\varepsilon = 1 - \frac{I}{I_0} = 1 - \exp\left[-\frac{N_A \rho}{A}\sigma_{\text{eff}} t\right] \tag{2.45}$$

2.5 Inelastic scattering: Van Hove formalism

The differential scattering cross-section $d\sigma/d\Omega$ gives information about the spatial correlations between the atoms in the sample. By measuring the double differential scattering cross-section $d^2\sigma/(d\Omega dE_f)$, we also obtain information about both space and time correlations. Again this cross-section can be expressed as the sum of coherent and incoherent components:

$$\frac{\mathrm{d}^2\sigma}{\mathrm{d}\Omega\mathrm{d}E_\mathrm{f}} = \left(\frac{\mathrm{d}^2\sigma}{\mathrm{d}\Omega\mathrm{d}E_\mathrm{f}}\right)_\mathrm{coh} + \left(\frac{\mathrm{d}^2\sigma}{\mathrm{d}\Omega\mathrm{d}E_\mathrm{f}}\right)_\mathrm{incoh} \tag{2.46}$$

The coherent part is dependent on the correlation at different times between the positions of different nuclei. The incoherent component depends only on the correlation between the positions of the same nucleus at different times and is absent if there is no isotope or spin incoherence in the sample. Coherent inelastic scattering is measured in the determination of phonon dispersion relations (see Chapter 14) and incoherent inelastic scattering in studies of molecular spectroscopy (Chapter 15) and of diffusion (Chapter 16).

The cross-section $\mathrm{d}^2\sigma/(\mathrm{d}\Omega\mathrm{d}E_\mathrm{f})$ can be related, in a general sense and for any scattering system, to certain correlation functions which describe the dynamics of the scattering system. This formulation of the scattering theory is due to Van Hove (1954), who showed that for a monatomic system of N nuclei, the coherent inelastic cross-section is given by

$$\left(\frac{\mathrm{d}^2\sigma}{\mathrm{d}\Omega\mathrm{d}E_\mathrm{f}}\right)_\mathrm{coh} = \langle b\rangle^2 \frac{k_\mathrm{f}}{k_\mathrm{i}} NS\left(\mathbf{Q},\omega\right) \tag{2.47}$$

where $\langle b\rangle$ is the mean scattering length. The quantity $S(\mathbf{Q},\omega)$ [introduced earlier in eqn. (2.20)] is the four-dimensional Fourier transform of a function $G(\mathbf{r},t)$ of the space and time coordinates, $\mathbf{r}$ and t. $G(\mathbf{r},t)$ is known as the time-dependent pair-correlation function of the scattering system:

$$S(\mathbf{Q},\omega) = \frac{1}{2\pi\hbar}\iint G(\mathbf{r},t)\exp\left\{i\left(\mathbf{Q}\cdot\mathbf{r}-\omega t\right)\right\}\mathrm{d}\mathbf{r}\mathrm{d}t \tag{2.48}$$

For a classical system, $G(\mathbf{r},t)$ gives the probability that, with an atom at the origin $\mathbf{r}=0$ at the initial time $t=0$, the same or any other atom will be found at a position $\mathbf{r}$ at a later time t. For a quantum system $G(\mathbf{r},t)$ has to be defined in terms of operators with imaginary components, and it is not possible to give it a simple physical interpretation.

The incoherent part of the double differential cross-section is given by

$$\left(\frac{\mathrm{d}^2\sigma}{\mathrm{d}\Omega\mathrm{d}E_\mathrm{f}}\right)_\mathrm{incoh} = \left\{\langle b^2\rangle - \langle b\rangle^2\right\}\frac{k_\mathrm{f}}{k_\mathrm{i}} NS_\mathrm{inc}\left(\mathbf{Q},\omega\right) \tag{2.49}$$

where

$$S_\mathrm{inc}(\mathbf{Q},\omega) = \frac{1}{2\pi\hbar}\iint G_\mathrm{s}(\mathbf{r},t)\exp\{i(\mathbf{Q}\cdot\mathbf{r}-\omega t)\}\mathrm{d}\mathbf{r}\mathrm{d}t \tag{2.50}$$

$S_\mathrm{inc}(\mathbf{Q},\omega)$ is the incoherent scattering function and $G_s(\mathbf{r},t)$ is the time-dependent self pair-correlation function. Given that there was a particle at the origin at $t=0$, $G_\mathrm{s}(\mathbf{r},t)$ is defined classically as the probability that the same particle will be at the position $\mathbf{r}$ at a later time t.

We note that the double differential scattering cross-section in eqns (2.47) and (2.50) is the product of two factors. The first factor is $\langle b\rangle^2$ $(k_\mathrm{f}/k_\mathrm{i})N$ in (2.47) and is $\left\{\langle b^2\rangle - \langle b\rangle^2\right\}(k_\mathrm{f}/k_\mathrm{i})N$ in (2.49); the factor is of practical importance, as it determines the intensity of the scattered radiation, but is of little intrinsic interest in the experiment. The second factor, the scattering function $S(\mathbf{Q},\omega)$ or $S_\mathrm{inc}(\mathbf{Q},\omega)$, is of primary interest because it contains information about the structure and dynamics of the sample. The aim of all neutron scattering studies is to determine this function in the required ranges of $\mathbf{Q}$ and ω.

2.6 Kinematics of inelastic scattering

Inelastic scattering occurs when the neutrons exchange energy with excitations in the sample. The initial and final energies of the incident and scattered neutron beams are $E_\mathrm{i} = (\hbar^2 k_\mathrm{i}^2/2m_\mathrm{n})$ and $E_\mathrm{f} = (\hbar^2 k_\mathrm{f}^2/2m_\mathrm{n})$ where $\mathbf{k}_\mathrm{i}$ and $\mathbf{k}_\mathrm{f}$ are the initial and final wave vectors. The energy transfer $\hbar\omega\omega$ from the neutron to the sample is

$$\hbar\omega = \frac{\hbar^2}{2m_\mathrm{n}}\left(k_\mathrm{i}^2 - k_\mathrm{f}^2\right) \tag{2.51}$$

The scattering events are characterized by the two quantities $\mathbf{Q}$ and ω. Referring to the scattering triangle in Fig. 2.6(b) we obtain from the cosine rule the expression:

$$Q^2 = k_\mathrm{i}^2 + k_\mathrm{f}^2 - 2k_\mathrm{i}k_\mathrm{f}\cos 2\theta \tag{2.52}$$

or in terms of energy:

$$\begin{aligned}\frac{\hbar^2 Q^2}{2m_\mathrm{n}} &= E_\mathrm{i} + E_\mathrm{f} - 2\sqrt{(E_\mathrm{i}E_\mathrm{f})}\cos 2\theta \\ &= 2E_\mathrm{i} - \hbar\omega - 2\sqrt{E_\mathrm{i}\left(E_\mathrm{i} - \hbar\omega\right)}\cos 2\theta\end{aligned} \tag{2.53}$$

Equations (2.53) determine the range of $(\mathbf{Q},\omega)$ space that can be covered in a scattering experiment. In a *direct-geometry spectrometer* (see Chapter 14), E_i is fixed and the energy transfer between the neutron and the sample is obtained by measuring the scattered intensity as a function of E_f. Figure 2.9 shows the accessible region of $(\mathbf{Q},\omega)$ space for an incident energy $E_\mathrm{i} = 100$ meV and for different scattering angles 2θ.

For an *indirect-geometry spectrometer*, where the final energy E_f is fixed and the initial energy E_i varies, eqn. (2.53) takes the form

$$\frac{\hbar^2 Q^2}{2m_\mathrm{n}} = 2E_\mathrm{f} + \hbar\omega - 2\sqrt{E_\mathrm{i}\left(E_\mathrm{f} + \hbar\omega\right)}\cos 2\theta \tag{2.54}$$

The (Q,ω) loci are obtained by referring to Fig. 2.9 and then reflecting all the curves across the energy axis $\hbar\omega = 0$. We note that there is a larger range of (Q,ω) space available for energy-loss measurements using an indirect-geometry spectrometer in place of a direct-geometry instrument.

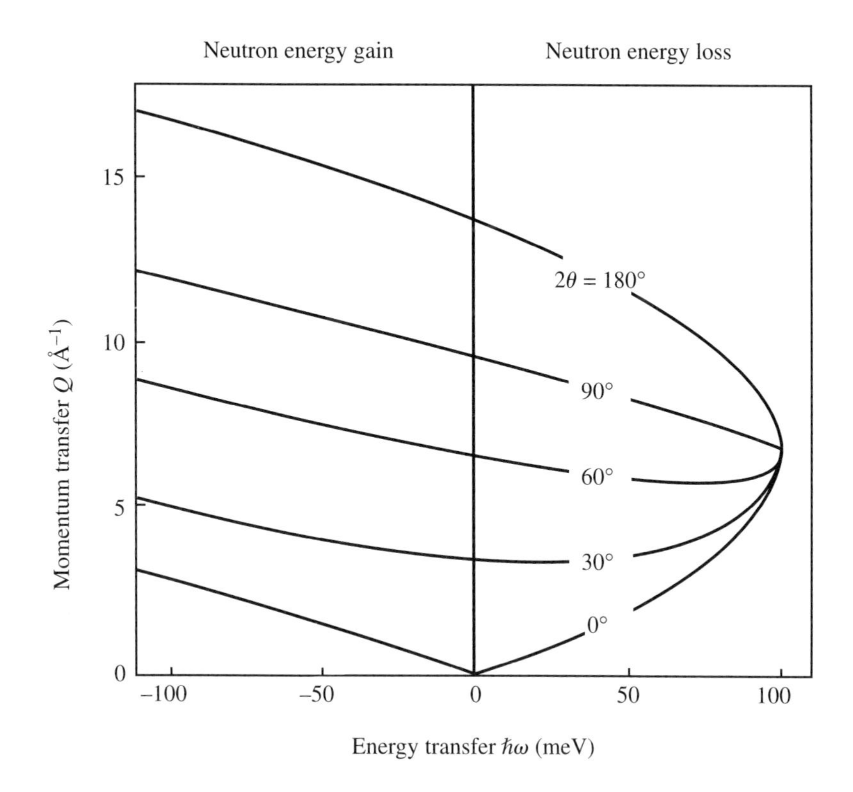

Fig. 2.9 The region of $(\mathbf{Q}, \omega)$ space accessible with a direct-geometry spectrometer using a fixed incident energy E_i = 100 meV and various scattering angles.

2.7 Magnetic scattering

The neutron is uncharged but possesses a magnetic dipole moment. This moment can interact with the magnetic field from unpaired electrons in a sample through (i) the magnetic field associated with orbital motion of the electron and (ii) the intrinsic dipole moment of the electron itself. The strength of the magnetic interaction is of the order of r_0^2 or 10^{-29} m^2, where r_0 is the classical electron radius $(=e^2/m_\mathrm{e}c^2)$. The nuclear interaction is of the order of b^2 or 10^{-28} m^2 where b is the nuclear scattering length. Thus, the nuclear and magnetic interactions are roughly of the same magnitude.

There are two significant differences between the theory of magnetic scattering of neutrons and the theory of nuclear scattering. The first difference is that the cross-section for nuclear scattering is a scalar quantity, whereas the magnetic cross-section has a vectorial component depending on the direction of the applied magnetic field. If the neutron spin is at an angle to the field, its state can change by precessing in the field. The second difference concerns the form factor for scattering. For magnetic scattering, the combined amplitudes in the forward direction $(2\theta \rightarrow 0)$ from all the unpaired electrons are in phase; however, as 2θ increases phase differences occur, because the dimension of the electronic cloud is comparable with the neutron wavelength. This form-factor effect is well

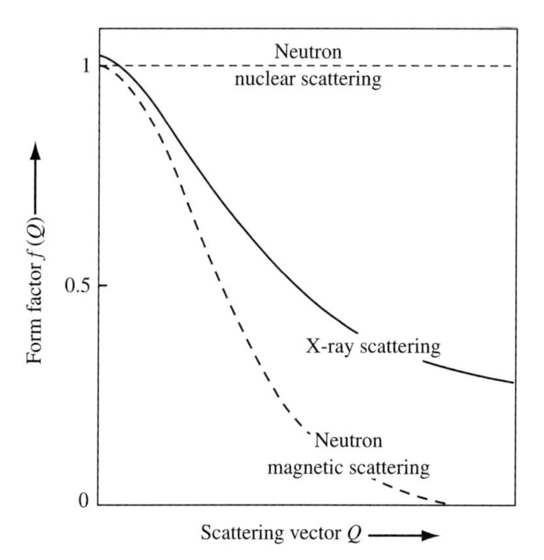

Fig. 2.10 The form factor $f(Q)$ for the scattering of X-rays and for the nuclear and magnetic scattering of neutrons.

known in X-ray scattering, but is even more marked for the magnetic scattering of neutrons where the scattering is from the outermost electronic orbitals (Fig. 2.10). In the case of nuclear scattering there is no fall off of scattering intensity with angle because the neutron wavelength is at least four orders of magnitude larger than the nuclear radius. We shall defer until Chapter 8 the treatment of magnetic scattering taking these additional features into account.

References

Arzumanov, S., Bondarenko, L., Chernyavsky, S., Drexel, W., Fomin, A., Geltenbort, P., Morosov, V., Panin, Yu, Pendlebury, M. J. and Schreckenbach, K. (2000) *Phys. Lett. B* **483** 15–22. "*Neutron lifetime value measured by storing ultracold neutrons.*"

Feynman, R. P., Leighton, R. B. and Sands, M. (1965) "*The Feynman lectures on physics.*" Volume 3. Chapter 1. Addison-Wesley, New Jersey, USA.

Lovesey, S. W. (1984) "*Theory of neutron scattering from condensed matter.*" Volumes 1 & 2. Clarendon Press, Oxford.

Squires, G. L. (1996) "*Introduction to the theory of thermal neutron scattering.*" Dover Publications, New York, USA.

Van Hove, L. (1954) *Phys. Rev.* **95** 249–262. "*Correlations in space and time and Born approximation in systems of interacting particles.*"

Zeilinger, A., Gähler, R., Shull, C. G., Treimer, W. and Mampe, W. (1988) *Rev. Modern Phys.* **60** 1067–1073. "*Single- and double-slit diffraction of neutrons.*"

The production of neutrons

3

3.1 Introduction

Much of the matter in the world around us consist of bound neutrons. The free neutron is an unstable particle, undergoing a three-body decay to a proton, an electron and an anti-neutrino, with a half-life of about 10 minutes. Neutrons are produced in a variety of nuclear reactions which includes fission, photofission, spallation and fusion. For scattering experiments the most significant of these are fission and spallation.

Fission in a nuclear research reactor provides the most common source of neutrons. Present-day neutron scatterers must acknowledge their debt to the strategic work done during the Second World War on the isolation of fissile materials from raw minerals and from spent fuel elements. Without this work, leading to the subsequent nuclear power programme, it is questionable whether powerful research reactors would have been available today.

The spallation of heavy atoms takes place using *accelerator-based* sources. The construction of these sources started almost four decades after instruments for neutron scattering had been first installed on research reactors. Reactors are now approaching the maximum neutron fluxes that are technologically achievable; pulsed accelerator sources have not yet attained this limit and offer the most promising method of realizing even higher fluxes. Nevertheless, it seems likely that all types of neutron sources for scattering experiments will remain relatively weak. At best, a monochromatic beam of around 10^8 neutrons/cm^2/sec strikes the sample in a neutron experiment: this compares with a photon flux at least three orders of magnitude higher, as produced by the characteristic spectral lines of a sealed-off X-ray tube, and more than 10 orders of magnitude higher, given by third-generation X-ray synchrotron sources.

Before the advent of fission reactors, the earliest scattering experiments were conducted using neutrons from radioactive sources. These will be described first. Such sources are still in use today as portable devices for installing and calibrating detectors.

3.2 Radioactive sources

The (α,n) reaction was employed by Chadwick in 1932 to isolate and identify the neutron. ^{210}Po is a naturally occurring alpha emitter which decays to ^{206}Pb with the emission of a 5.3 MeV alpha particle. (The alpha particle is the nucleus of the helium atom ^{4}He.) Chadwick bombarded beryllium with these alpha particles, producing neutrons by the exothermic reaction:

$$^{9}\text{Be} + {}^{4}\text{He} \rightarrow {}^{12}\text{C} + {}^{1}\text{n} + 5.7\ \text{MeV} \tag{3.1}$$

The reaction yields a weak source of neutrons with an energy spectrum resembling the fast neutron spectrum of a fission source.

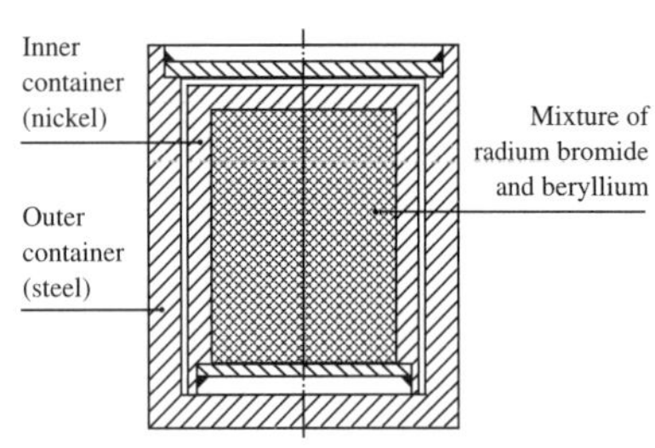

Fig. 3.1 Cross-section through a typical radium–beryllium (α,n) source.

The alpha emitter in a modern source of this kind is usually radium, americium or even plutonium. A typical design is illustrated in Fig. 3.1. The inner container of nickel prevents the release of radon from radium bromide. The outer container of steel is several mm thick. Because of the short range of alpha particles, the source is made from an intimate mixture of radium bromide and beryllium.

Photonuclear reactions of the (γ,n) type are also used. A convenient source consists of a rod of radioactive antimony surrounded by a cylinder of beryllium metal, as shown in Fig. 3.2. The rod contains the artificial radioisotope ^{124}Sb which is a gamma emitter. Neutrons are produced by the endothermic reaction:

$$\gamma + {}^{9}\text{Be} \rightarrow {}^{8}\text{Be} + {}^{1}\text{n} - 1.66\ \text{MeV} \tag{3.2}$$

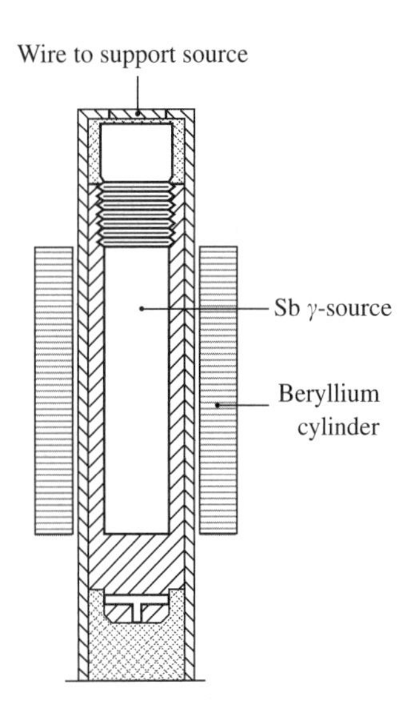

Fig. 3.2 Cross-section through an antimony–beryllium (γ,n) source. The central wire is used to withdraw the antimony rod and to 'switch off' the source.

When newly prepared, the source emits neutrons of approximately uniform energy at a rate of a few millions per second. ^{124}Sb has a half-life of 60 days, but after the activity has decayed it can be regenerated by inserting the antimony rod into a nuclear reactor. On account of the greater range of gamma rays compared with alpha particles, a (γ,n) source has the advantage that the two physical components of the source can be separated, making it possible to 'switch off' the reaction by separating the radioactive source from the beryllium.

3.3 Neutrons from nuclear reactors

Nuclear reactors consist of four basic components: fuel, moderator, coolant and shielding. The fuel contains a fissile material, such as uranium, thorium or plutonium, either in its elemental form or as a compound, such as uranium dioxide. The fissile material can be in its naturally occurring form (uranium or thorium) or else it can be enriched isotopically in a fissionable isotope (^{235}U or ^{239}Pu). The moderator is a material of low atomic mass, such as light water, heavy water and graphite. The coolant is usually water but it is sometimes also a gas, such as carbon dioxide or even helium. Fast reactors (in which fast neutrons alone maintain the chain reaction) do not have a moderator and they employ liquid metals, sodium or a sodium–potassium alloy, as coolant without moderating the neutrons.

In a nuclear reactor fuelled with uranium, thermal fission of ^{235}U takes place with the production, on average, of 2.7 fast neutrons and the simultaneous disintegration of the uranium nucleus into two fission fragments

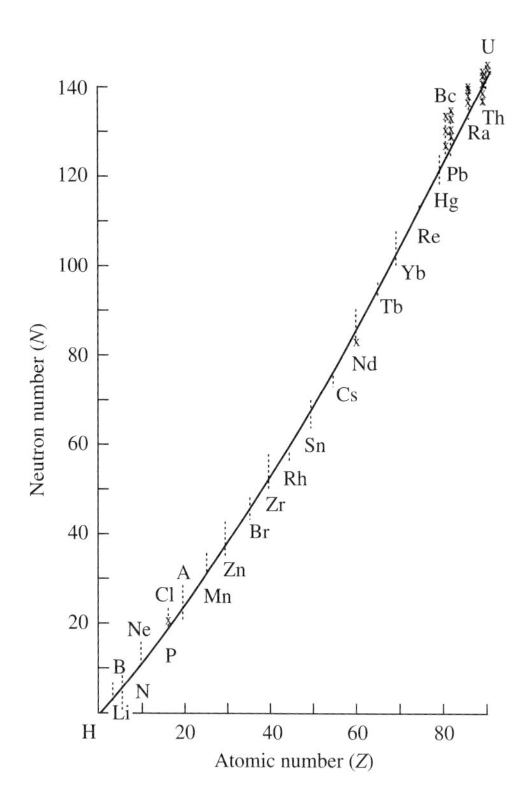

Fig. 3.3 Number of neutrons *versus* atomic number for stable nuclides. The full line is known as the line of beta stability. The dots refer to different isotopes. (*After* Hunt, 1987.)

of unequal mass. These fragments are predominantly rich in neutrons and are highly radioactive: they decay via a cascade of beta emissions (*i.e.* expulsion of electrons) to what is called the line of beta stability, which is obtained by plotting the number of neutrons *versus* the atomic number for the naturally occurring stable isotopes, as shown in Fig. 3.3.

The fission reaction becomes self-supporting provided that, from this yield of 2.7 neutrons, one neutron causes further fission in another ^{235}U nucleus, producing a self-sustaining chain reaction. The remaining 1.7 neutrons are either absorbed or escape from the surface of the reactor. It is these latter neutrons that form our source of slow neutrons for scattering experiments.

The purpose of the moderator is to slow down the source neutrons from their initial MeV (10^6 eV) energies to meV (10^{-3} eV) energies where the likelihood of fission is much higher. The neutron population is then approaching thermal equilibrium with the moderator. ^{235}U is a *fissile* material, which will undergo fission at all neutron energies down to zero energy, with the neutron yield for thermal fission exceeding that for intermediate or fast fission. In contrast, ^{238}U is a *fertile* material which does not fission with slow neutrons but produces a new fissile isotope ^{239}Pu by slow neutron capture and subsequent alpha decay. Because of the presence of the moderator, the core of a thermal fission reactor can be several metres across, whereas a fast reactor core may measure only 20 cm.

In nuclear fission, the incident neutron is absorbed into the nucleus of the fissile atom to produce a *compound nucleus*, and within 10^{-14} seconds the compound nucleus breaks up into two fission fragments, releasing fission neutrons. These are referred to as *prompt neutrons.* A small but significant fraction ($\approx$0.7%) of fission neutrons is released at longer times, up to a few minutes after fission: these are called *delayed neutrons* and are of the greatest importance in the control of reactors. The multiplication factor k from one generation of neutrons to the next originates from both the prompt neutrons and the delayed neutrons: a sustainable critical reaction occurs when $k = 1$. If $k > 1$ a *supercritical* reaction occurs and for $k < 1$ the reaction is *subcritical.* By ensuring that k for prompt neutrons is just less than unity, the rate of increase of neutron density in the reactor is determined essentially by the time constant of the delayed neutrons. Thus, the power output can be increased relatively slowly and controlled by the physical movement of neutron absorbers—the control rods—in and out of the reactor core.

The intense beta emission of fission fragments has two important practical consequences. The first is that the fuel elements in nuclear reactors rapidly become highly radioactive and a thick shield around the core is required, not only for the safety of personnel but also to reduce the background radiation recorded by the suite of instruments immediately around the reactor. The second consequence is that many of the fission fragments and their decay products are highly absorbing to neutrons and can stop the chain reaction. In particular, this is true of the isotope ^{135}Xe which has an absorption cross-section for thermal neutrons of 3.0×10^6 barns, one of the highest known cross-sections for any nuclear reaction. ^{135}Xe is known as a *reactor poison.* It is produced both directly as a fission product of ^{235}U, with a relatively low yield of 0.2% per fission, and indirectly as the granddaughter product of ^{135}Te, itself produced as a fission product with a much higher yield of 6.1%. During steady operation of a reactor, the concentration of ^{135}Xe reaches a constant value determined by the balance among its rate of production, its rate of radioactive decay and its rate of removal by neutron absorption. When the reactor is shut down, the removal process by neutron absorption stops and the ^{135}Xe concentration increases, as the half-life for its formation (6.7 hours) is less than its half-life for decay (9.2 hours). The xenon concentration reaches a maximum value about 10 hours after the reactor has shutdown, at which point the concentration begins to fall but does not reach its pre-shutdown value until much later.

Figure 3.4 illustrates the variation of xenon concentration as a function of time after shut down, following a period of operation at steady power. The higher the neutron flux in the reactor, the higher the concentration of xenon and the faster the rise in the concentration in the period immediately after shutdown. For low-flux reactors (flux $\sim 10^{13}$ n/cm^2/sec) the effect of xenon poisoning is not particularly important, except perhaps towards the end of a reactor cycle when the fuel has been burnt

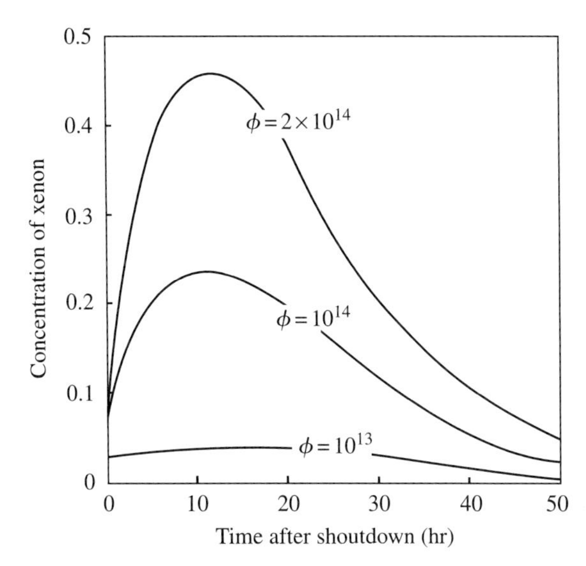

Fig. 3.4 Rise in concentration of xenon after the shutdown. ϕ is the neutron flux in $n/cm^2/sec$.

and the excess reactivity in the core is small. For a high-flux reactor (HFR) such as that at the *Institut Laue-Langevin* (ILL) in Grenoble, if a shutdown occurs at any time during the operating cycle, apart from the first few hours of operation, then within a few minutes the concentration of xenon rises to a level that exceeds the excess reactivity in the core. The reactor has then 'poisoned out' and cannot be restarted until the xenon concentration has fallen sufficiently, and typically this requires two more days of shutdown. A nuclear reactor can be designed with sufficient excess reactivity such that, at any time in the cycle, it can be brought back to criticality after a shutdown; this is an expensive contingency which is usually reserved for nuclear-powered submarines where poisoning out would have more severe consequences. This problem would have been even more serious for the Advanced Neutron Source (ANS) at Oak Ridge, USA, which was first proposed in 1985 but later cancelled. The ANS was designed to have a reactor power of 350 MW and a flux three or four times higher than the ILL flux. Fast reactors do not suffer from xenon poisoning.

High-flux beam reactors are specifically designed to provide the highest neutron density just outside the edge of the reactor core where the beam tubes terminate. Earlier research reactors built in the 1950s and 1960s were designed to undertake a range of activities including isotope production, gamma-ray activation analysis, basic reactor physics and neutron cross-section measurements. Therefore, the development of neutron beam reactors can be considered in three phases: the early experimental assemblies; the later muti-purpose research reactors; and the modern dedicated high-flux neutron beam reactors, such as the HFR reactor at the ILL or the FRM-II reactor at Munich, Germany.

The interested reader is referred to the classic texts on nuclear reactor physics [Glasstone *et al.* 1957 and 1994, Megreblian and Holmes 1960].

3.3.1 The ILL reactor

Since 1973 this reactor has been the world's most powerful source of neutrons for scattering experiments. The thermal flux at the beam tubes is $1.5 \times 10^{15}\,\mathrm{cm}^{-2}\ \mathrm{sec}^{-1}$, supplying neutron beams to instruments in the main reactor hall and in the two *neutron guide halls.* The thermal neutron flux, in equilibrium with the D_2O moderator at 300 K, has a peak in the Maxwellian distribution of wavelengths at 1.2 Å. For some of the beams, this distribution is modified by a hot source of graphite at 2,400 K, which enhances the flux at wavelengths below 0.8 Å, and two cold sources of liquid hydrogen at 25 K, which enhance the flux at wavelengths exceeding 3 Å. A vertical guide supplies neutrons to an upper laboratory for storage as *ultracold neutrons* in neutron bottles; their wavelengths are around 1,000 Å (see Chapter 11) and they are used for studying the fundamental properties of the neutron itself.

The reactor core of HFR is a single fuel element of uranium, 93% enriched in ^{235}U. This fuel element has a central cylindrical cavity containing the neutron-absorbing control rod made from a silver–indium–cadmium alloy. The coolant is water, which is pumped through the internal fins of the core at a velocity of 15.5 m/sec to remove the 58 MW of heat generated by the fission process. The coolant acts as the primary moderator but encircling the core is a heavy-water moderator, which also serves to reflect neutrons back into the core for further moderation and continuation of the chain reaction. The presence of the heavy water acts as a neutron reflector and causes a peaking of the thermal neutron flux just outside the reactor core where the front ends of the beam tubes are located.

Thus, the neutron beam tubes penetrate the biological shield of the reactor to the position of maximum flux surrounding the core. A hollow beam tube causes a local depression in the flux distribution around the core and the resultant neutron density gradient is the source of the high neutron current passing down the beam tube to the instruments at the other end of the tube. Most of the tubes do not point radially towards the reactor core but are tangential to it. This ensures that the background effects of the high fluxes of gamma rays, generated by both the fission reaction and the decaying fission fragments, are minimized, and also that the ratio of thermalized neutrons to fast neutrons (sometimes called the cadmium ratio) passing along the beam tube is as high as possible.

There are several arrays of neutron guide tubes, which transport neutrons to the two low-background experimental halls without the loss in neutron intensity experienced with non-reflecting beam tubes. These guides are up to 120 m long, allowing the installation of many additional instruments, which can make measurements with a high thermal or cold neutron flux and a low background. The guides feed more than twice as many instruments as the beam tubes in the reactor hall. (Guide tubes are discussed further in Chapter 5.) Figure 3.5 illustrates the layout of the instruments around the reactor and in the two guide tube halls.

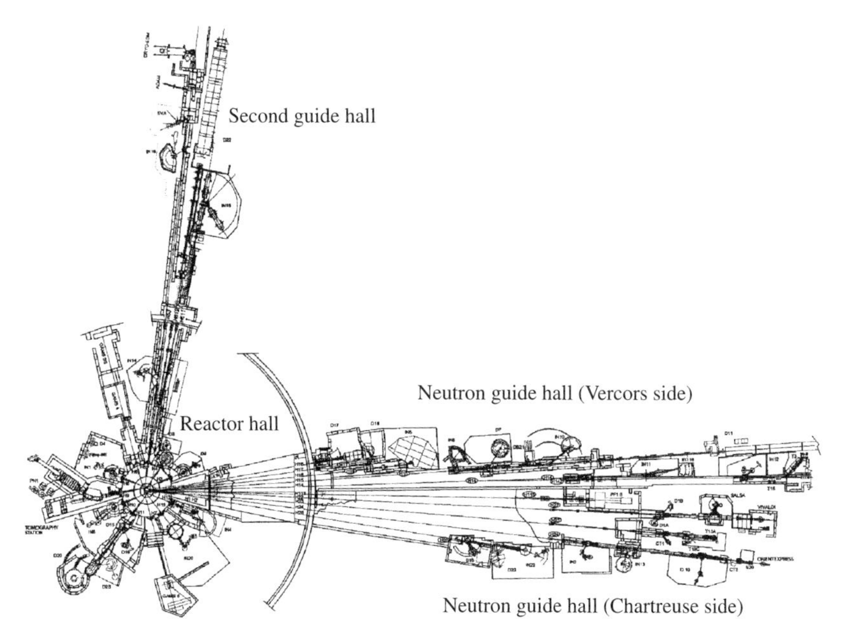

Fig. 3.5 Layout of instruments in the main reactor hall and in the two guide halls at the ILL.

3.3.2 Pulsed reactors

Time-of-flight methods operating with pulsed reactor beams offer an alternative procedure for carrying out neutron scattering experiments. Pulsed reactors are more compact than accelerator-based pulsed sources. Since 1960 they have been developed, in particular, at the *Joint Institute for Nuclear Research* in Dubna, Russia, with the construction of the IBR series of nuclear reactors.

A fission reactor can be pulsed in a number of ways:

(i) The reactivity for prompt neutrons in a subcritical assembly can be momentarily increased to become supercritical ($k > 1$), causing the assembly to emit an intense burst of neutrons. This can be achieved mechanically by designing either a piece of fissile material or a piece of reflector material, which needs to be brought close to the core of the subcritical assembly in a periodic manner. The neutron pulse length is determined by the speed with which the moving component can pass by the core. Because of the practical limitation on this speed, pulse lengths tend to be rather long ($100 - 500\,\mu$sec).

(ii) The reactivity of the assembly can be pulsed to a supercritical level by the instantaneous generation of a burst of neutrons in the core, which is created by an electron pulse hitting a target in the centre of the reactor and generating neutrons by the photofission process. These neutrons then multiply in the subcritical assembly. This is called a *static booster*. It has the disadvantage that *delayed neutrons* from the fission reaction undergo the same multiplication as the pulsed neutrons but are uncorrelated in time with respect

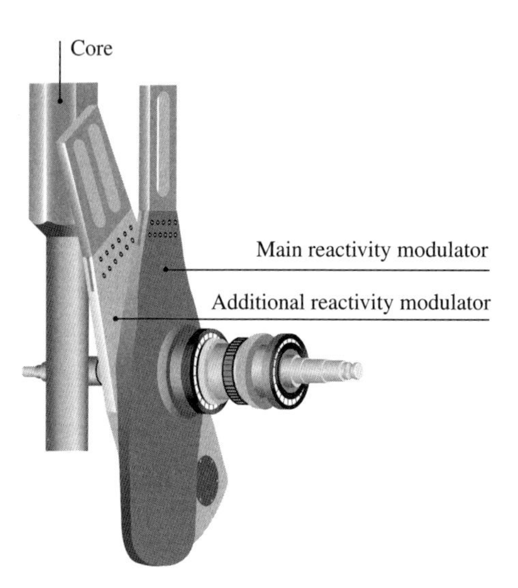

Fig. 3.6 The core of the pulsed reactor IBR-2 at Dubna, Russia. The reactivity is modulated by the two counterrotating neutron reflectors. At the centre of the core is an electron target for possible use in the booster mode.

to the original pulse, so that scattering measurements are made on a relatively high time-independent background.

(iii) The combination of a pulsed particle beam and a rotating wheel overcomes, to some extent, the deleterious effects of multiplication of the delayed neutrons between pulses, as the reactivity of the subcritical assembly can be designed to be well below criticality. This combination is called a *dynamic booster* and can give pulse widths as short as 5 μsec.

It is possible to make a useful distinction between sources of long and short pulses. If the time required to generate the pulse is significantly shorter than the moderation time of the neutrons, then the source can be classified as a short-pulse source; when this time is comparable to or greater than the moderation time, the source is a long-pulse source. For neutrons with energies less than 1 eV, accelerator-based spallation sources normally give short pulses, whereas reactor-based pulsed sources give long pulses. However, Long Pulse Spallation Sources (LPSS) offer some advantages over Short Pulse Spallation Sources (SPSS) for cold neutron experiments with relaxed resolution, for example, in experiments using small-angle scattering or neutron spin-echo. [The design of the European Spallation Source (ESS) incorporates a long-pulse target only, and the SNS source at Oak Ridge will have a second long-pulse target station in addition to the short-pulse target.]

A diagram of the pulsed reactor IBR-2 at Dubna, about 100 km north of Moscow, is shown in Fig. 3.6. The average thermal power of the reactor is 2 MW but the peak power in the pulse is 1,500 MW. The core is a compact fast reactor using highly enriched plutonium dioxide as fuel and liquid sodium (giving negligible moderation) as coolant. Moderation of the neutron beams is provided by light water viewed by the beam tubes. The peak thermal neutron flux at the surface of the moderator is 1.5×10^{16} n/cm^2/sec. The reactivity is pulsed by two rotating reflectors with an effective frequency of 5 Hz, providing a pulse length of 215 μsec for thermal neutrons. Particular attention is paid to control circuitry and safety aspects, as fluctuations in reflector frequency and transverse vibrations of one component with respect to another can cause undesirable reactivity excursions. The physical gap between different components is only 10 μm and fluctuations in power level, pulse-to-pulse, can be of the order of 20–30%. The use of two reflectors rotating at different frequencies keeps the pulses relatively short and the period between pulses long, as significant multiplication occurs only when both reflectors simultaneously pass close to the core. A consequence of this is the presence of a series of satellite pulses that occur between the main pulses when only one reflector passes through the core. These satellite pulses are 10^4 times weaker than the main pulse.

There is a possibility that the IBR-2 reactor can be converted into a dynamic booster mode by firing a short burst of electrons on to a target in the reactor core to initiate the neutron generation. In this way the peak flux would be maintained whilst the length of the pulse would be reduced to 5 μsec, bringing the pulsed reactor into the short-pulsed-source regime for both thermal and cold neutrons. It seems unlikely, however, that pulsed reactors will ever be mainstream sources of neutrons for condensed matter research.

3.4 Neutrons from pulsed accelerators

Since the 1950s work has been carried out at a number of laboratories using neutron beams provided by electron linear accelerators. This work provided neutron cross-section data in the eV to keV energy range for nuclear power programmes. The experiments were carried out by time-of-flight techniques where pulses of neutrons were allowed to disperse their energy over very long flight paths, leading to high precision in the measurements. It is remarkable that Enrico Fermi was able to assemble the first critical nuclear pile in 1942 without the aid of a comprehensive catalogue of neutron cross-sections. Such data are available today (*e.g.* published in the 'Barn Book' *BNL Report 325*) after being assembled from measurements made principally with pulsed sources.

The demand for more intense sources has led to the consideration of neutron generation by means other than fission reactors. This is because the power density in the core of high-flux fission reactors (~7 MW/litre) is limited by the achievable rate of cooling of the core. The first-ever pulsed spallation source for neutron scattering, ZING-P, was built and operated by John M. Carpenter and colleagues at the Argonne National laboratory (ANL). This led to the construction of the *Intense Pulsed Neutron Source*, IPNS, at the ANL in 1981. The attraction of pulsed sources is obvious: an intense burst of neutrons is produced in a time which is short compared with the period between pulses, and so the heat generated in the target during the pulses can be removed in the time between the pulses. Provided that scattering measurements using the full spectrum of energies in the white beam are carried out during the time between pulses, it is the peak neutron flux which determines the data collection rate and not the relatively low time-averaged neutron flux. Despite the apparent advantage of target cooling in pulsed sources, this still remains the most significant technological problem restricting the performance of high-power spallation sources such as SNS (in the United States) and J-PARC (in Japan). The step from the 1.4 MW liquid mercury target of the SNS to the previously proposed 5 MW short-pulse option for the ESS would pose severe challenges to engineers and physicists alike. The reference design for the ESS is a long-pulse option where the demands on the target are much diminished.

The development of pulsed sources has been assisted serendipitously by a number of factors. The run down of a series of accelerators for experiments in high-energy particle physics has provided an opportunity to reuse these discarded facilities as neutron sources. All the four high-intensity second generation pulsed source facilities (ISIS in the United Kingdom, IPNS and LANSCE in the USA and KENS in Japan) fall into this category. IPNS and KENS are no longer operational. Another significant factor is that, because of the environmental lobby, the necessary authority to build nuclear reactors is much more difficult to obtain than was the case in the 1960s. The events at Three Mile Island in the United States and at Chernobyl in the Ukraine have tipped the balance in favour of pulsed sources, as has the concern over nuclear proliferation which discourages the use of highly enriched uranium and the reprocessing of the used fuel elements.

The radioactive inventory of the SNS target at the end of an operational period of many years will equal, however, that in the fuel element of the ILL reactor at the end of its fifty day cycle. Data collection at pulsed sources relies crucially on time-sorting multidetectors for every instrument, leading to the subsequent processing and visualization of huge data sets. This places heavy demands on the availability of fast computing power and cheap data storage media. Kinetic experiments on instruments with large detector banks and operating on pulsed sources suffer from the relatively long time required to start and end the data runs. This will certainly be solved in future by advances in computer technology.

We now turn to two methods of generating pulsed neutrons using accelerators: the neutrons are produced by photofission with high-energy electrons and by spallation with high-energy protons.

3.4.1 Photofission sources using electron accelerators

High-energy electrons have been used to generate neutrons in a two-stage process via an initial production of gamma radiation. When electrons are incident on a heavy metal target, they suffer rapid deceleration: this slowdown of charged particles produces gamma radiation which is known as *Bremsstrahlung* or braking radiation. As the range of the electrons in the target is quite short—no more than a few millimetres—the dissipation of heat is concentrated at the front of the target, whereas the gamma rays illuminate the whole target, producing fast neutrons by the (γ,n) photofission reaction. The reaction is not very efficient, and the energy dissipated for each neutron produced is about 3,000 MeV. The integrated power generated is much greater, neutron for neutron, than in conventional fission or spallation sources, and its removal constitutes a major drawback of (γ,n) sources. In addition, the neutron beams are heavily contaminated by intense gamma fluxes which cause background problems in the neutron detectors, particularly when observing higher-energy neutrons at short times-of-flight.

Nevertheless, photofission sources using electron linear accelerators ('linacs') are relatively cheap and they have been the torch-bearers for the development of experimental techniques in pulsed neutron scattering. Linacs were used at Harwell, UK, from 1957 until 1990, and in Dubna from 1965 to the present day, and they have also been employed more recently at Toronto, Hokkaido, Kyoto and Frascati in neutron scattering experiments. The Harwell facility operated at an electron current of 1 A and at a maximum electron energy of 136 MeV, generating 90 kW of heat in a train of targets operating at a frequency of 300 Hz. The accelerator output was switched between several targets to produce neutrons in different energy ranges. The fast neutron target received electron pulses of a few nanoseconds duration for measuring cross-sections in the 10 MeV range. The lead target for neutron scattering experiments received pulses of length 5 μsec, which matches the moderation time for thermal neutrons. Many prototype time-of-flight instruments were built to operate at linacs such as these.

3.4.2 Spallation sources using proton accelerators

The *spallation* process is a nuclear reaction which occurs when high-energy particles bombard the nuclei of heavy atoms. Spallation only takes place above a certain threshold energy for the incident particle, typically 5–15 MeV. The reaction itself, involving the momentary incorporation of the incident particle by a target nucleus, is a sequential process. Initially there is an internal nucleon cascade within the excited target nucleus, followed by an internuclear cascade when high-energy particles including neutrons are ejected and absorbed by other nuclei, and finally there is de-excitation of the various target nuclei followed by the emission ('evaporation') of many lower-energy neutrons and a variety of nucleons, photons and neutrinos (see Fig. 3.7). The spallation reaction is often likened to a cannon ball careering through a wooden-hulled ship producing splinters as it passes through. The term *spallation* is borrowed from geology, where to spall a rock is to chip splinters off it with a hammer.

The cascade processes account for only about 3% of the neutrons generated in spallation but as these neutrons have very high energies, up

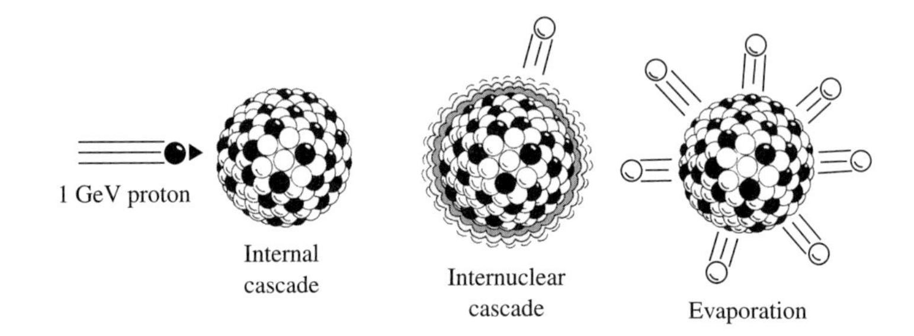

Fig. 3.7 The spallation process in which a high-energy proton interacts with a heavy nucleus. Typically 20–40 neutrons are evaporated with energies of many MeV in a single process which involves many nuclei. (*After* International Atomic Energy Agency, 1985.)

to the energy of the incident protons, they dominate the shielding requirements of the source. In the spallation reaction itself about 20–40 neutrons are generated per incident particle, with the actual yield being dependent on the target material and the energy of the incident particle. The energy released therefore for each neutron produced is quite low, typically about 55 MeV. The neutron source strength can be significantly increased by using a *fertile* target, such as depleted uranium (*i.e.* natural uranium depleted in the fissile isotope ^{235}U) which is fissile at incident neutron energies exceeding 1 MeV: the consequent disadvantage is that a delayed neutron background from fast fission processes in ^{238}U is produced and the target itself becomes contaminated with fission products.

Spallation sources for neutron scattering use a high-energy proton beam, which can be produced in three kinds of accelerators:

(i) *Linear Accelerators*, such as LAMPF (the Los Alamos Meson Production Facility), are high current, high repetition-rate accelerators. Linacs give pulses which are either too long or have frequencies that are too high to be used effectively for 'traditional' short-pulse neutron scattering, as pioneered on synchrotron spallation sources and electron linacs. Storage rings are therefore required at the end of the linear accelerator in order to compress the long pulses into the μsec range and extend the time period between pulses. The other option is to use the long pulse ($\approx$2 msec) from the linac directly into the target and develop instrumentation that can benefit from the extra neutrons generated by this technique. The SNS at Oak Ridge, USA, similar to the Los Alamos source, uses a combination of linear accelerator and proton storage ring to generate short pulses of neutrons. The European Spallation Source will not use a compressor ring and will generate long pulses of neutrons.

(ii) *Cyclotrons*, such as the SINQ source at the *Paul Scherrer Institute* in Zürich, produce a continuous beam of energetic protons and generate a steady current of neutrons via the spallation reaction.

(iii) *Synchrotrons*, such as ISIS at the Rutherford Appleton Laboratory in the United Kingdom, IPNS at the ANL in the United States and KENS at the KEK Tsukuba Laboratory in Japan, have operated with low currents until relatively recent times, when the implementation of rapid cycling techniques and weak focusing magnets allowed the acceleration of much higher currents. Narrow pulses ($< 1\,\mu$sec) can

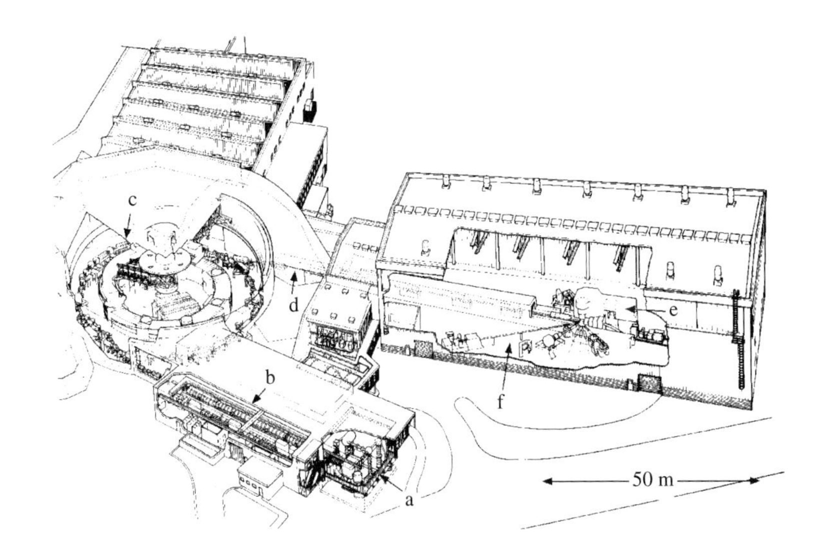

Fig. 3.8 Layout of the ISIS Laboratory (before the addition of a second target station). a: H^- ion source and pre-injector; b: linac; c: 800 MeV proton synchrotron; d: extracted proton beam; e: target station and shielding; f: experimental hall.

be produced at a modest repetition rate (20–60 Hz) which is well suited to neutron scattering measurements. Multi-turn injection of the beam into the closed orbit of the synchrotron is achieved by using a low-energy (<100 MeV) linear accelerator for H^- ions. At the point of injection, the beam passes through a thin alumina foil; the foil strips the two electrons from the ions, producing protons that enter the synchrotron for further acceleration (see Fig. 3.9). A neutron source of strength $\sim 5 \times 10^{16}$ fast neutrons/sec can be produced in a pulsed manner by this method. Protons are wound into the synchrotron ring rather like winding cotton on to a reel, and the intensity limit is reached when Coulomb repulsion causes the beam to expand too close to the walls of the vacuum chamber.

The ISIS pulsed neutron source

A cut-away view of the pulsed neutron source ISIS is shown in Fig. 3.8. It comprises a series of accelerators that produce an intense pulsed beam of 800 MeV protons, with 2.5×10^{13} protons per pulse and a current of 200 μA. This beam strikes a tantalum target, producing fast neutrons by the spallation process.

The ISIS facility can be subdivided into the following eight sections:

(i) A *Pre-injector*. This was originally a high-voltage generator of the Cockcroft–Walton potential-divider type. It produces pulsed voltages at 665 kV and 50 Hz and delivers a current of 40 mA in 500 μsec bursts to ...

(ii) An *H^- Ion Source*. The ion source produces negative hydrogen ions H^- in a discharge of caesium vapour. The vapour contains atomic hydrogen, which is generated by passing hydrogen gas through a heated nickel tube. The H^- ions of the pre-injector drift towards the first element of ...

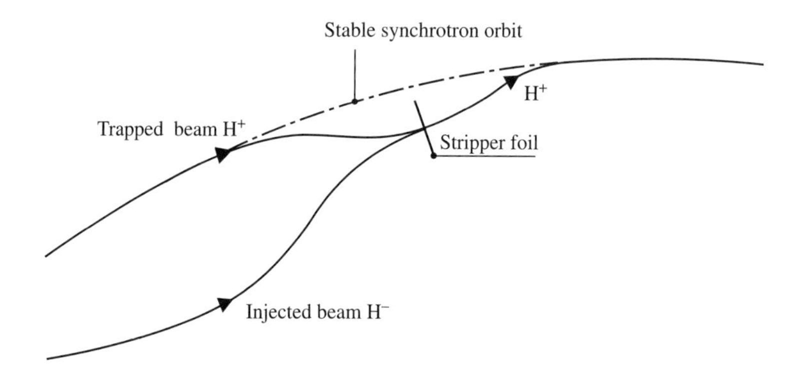

Fig. 3.9 Action of the stripper foil in facilitating injection of a beam into a closed orbit of the synchrotron.

(iii) A *Linear Accelerator Injector* of the Alvarez type. This consists of a series of potential gaps made up of copper electrodes and increases in length as the H^- ions gain energy in the accelerator. The linac accelerates the H^- ions from 665 keV to 70 MeV, producing a current of 20 mA in bursts of 500 μsec width every 20 msec (*i.e.* a frequency of 50 Hz). The H^- ion beam enters the synchrotron via ...

(iv) A *Straight Injection Section*, where a 0.25 μm thick foil of alumina—the *stripper foil*—removes the two electrons from the H^- ions. The resulting protons are trapped in the magnetic field of the synchrotron and join the acceleration process. The use of H^- ions and multi-turn injection allows the current in the synchrotron to be increased up to the space-charge limit. Thus the H^- ions incident upon the stripper foil, together with the protons from an earlier injection process but belonging to the same pulse, pass through the same magnetic field at the stripper foil and have equal radii of trajectory but opposite curvature. This is illustrated in the diagram in Fig. 3.9.

The *synchrotron* itself is 52 m in diameter and in the shape of a 10-sided polygon. Each of the 10 straight sections has common magnet elements for focusing the proton beam and for bending it by 36° into the next straight section. Each straight section serves a specific purpose. The first section is used for injection. The following six sections are used for acceleration, each containing a radio frequency (RF) ferrite-core cavity which feeds energy into the proton beam, appropriately phased. As the beam gains energy, the frequency of the RF cavities is increased to give the proton beam a slight increase in energy on each circuit. The six RF cavities accelerate the beam to 800 MeV within a 10 msec period, at which stage the protons are circulating in two narrow bunches on opposite sides of the synchrotron. This is known as second harmonic acceleration. Each bunch is 90 nsec wide and separated from its fellow bunch by 210 nsec. There are 2.5×10^{13} protons per pulse in the accelerated beam, which is equivalent to an average current of 200 μA. Two further straight sections of the synchrotron are used for the ports of the vacuum pumps: the ring must be held at a vacuum of 5×10^{-11} bar to avoid losing the beam by residual air scattering during the acceleration process. The vacuum vessel is made up of a specially developed ceramic to avoid eddy current heating. The beam area is approximately 150 cm^2 and the final straight section is ...

(v) *The Extraction Straight.* This contains a set of fast-acting kicker magnets, which are activated at the completion of the acceleration process. The electric current in the magnets is reversed in direction, thereby lifting the protons upwards by 1° and out of the closed orbit of the synchrotron.

The two bunches of protons in the ring are extracted from the synchrotron into ...

(vi) *The Extracted Proton Beamline.* The 800 MeV proton pulse is guided by quadrupole magnets along an evacuated beamline, 80 m in length, to ...

(vii) *The Target Station,* where it strikes a heavy metal target to produce a burst of fast neutrons by the spallation process and depositing 160 kW of energy. The target is 30 cm long and is made up of a series of tantalum plates, which are 9 cm in diameter. The thickness of the plates increases towards the end of the target, in order to ensure that comparable amounts of heat are deposited along the whole length of the target. The target is surrounded by three moderators. One is water at ambient temperature, the second is liquid methane at 105 K and the third is supercritical liquid hydrogen at 25 K. A beryllium reflector, which concentrates the fast neutron flux into the area occupied by the moderators, surrounds the whole target. Cadmium or boron decouplers are wrapped around those walls of the moderator facing the reflector: this restricts the incoming neutrons to the high-energy part of the spectrum and prevents the pulse width being degraded by slow neutrons which have been moderated in the beryllium reflector itself. The slow neutrons emitted by the front face of the moderator pass through collimators in steel and concrete shutters, 2 m long, to reach the beamlines and on to ...

(viii) *The Neutron Scattering Instruments.* There are 18 beam holes on the first Target Station of ISIS, nine on either side of the target. In a later extension to the ISIS facility, a second experimental hall was constructed to house a Second Target Station and its accompanying suite of instruments. This is achieved by diverting one in every five pulses from the synchrotron to this second station, whilst allowing the remaining pulses to continue on to the first target shown in Fig. 3.8. With the synchrotron continuing to operate at 50 Hz but with a higher current of 300 μA, the proton beam enters the second station at 10 Hz and gives a beam of long-wavelength neutrons. The layout of instruments at both target stations is shown in Fig. 3.10.

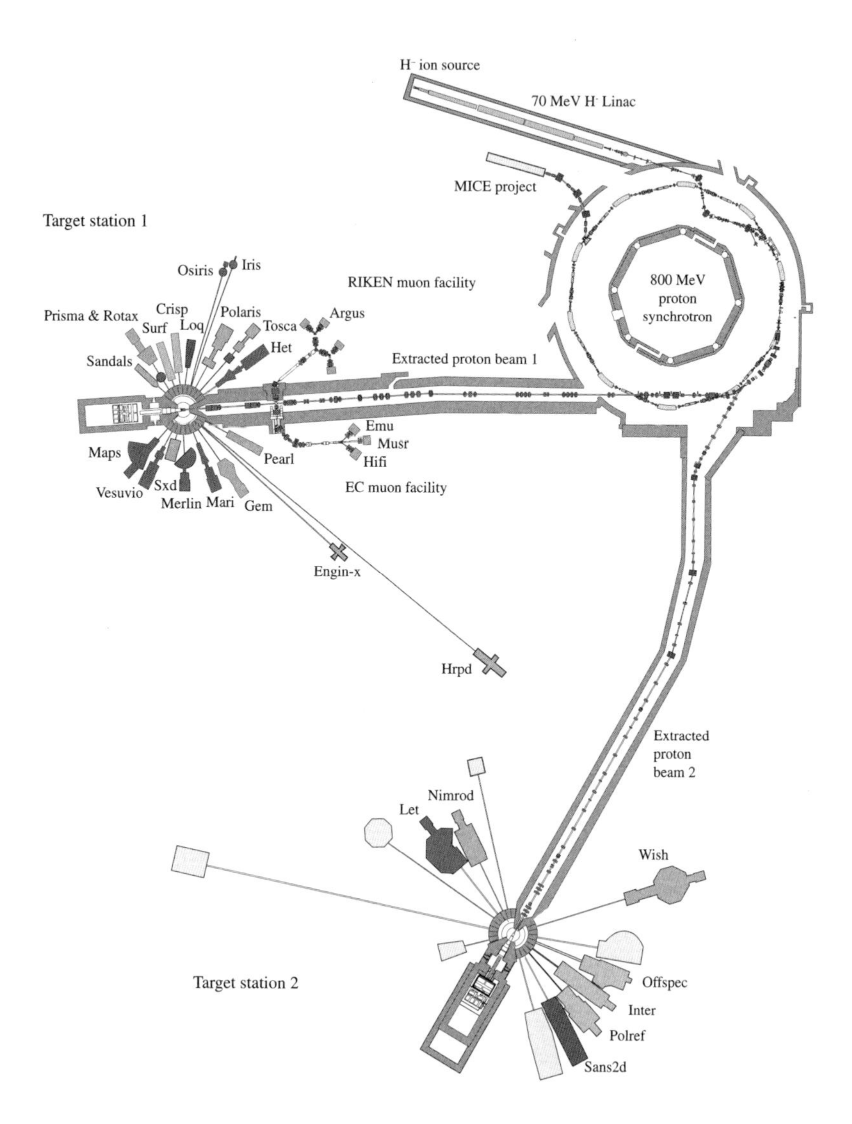

Fig. 3.10 A plan view of the components of the ISIS pulsed neutron source. The neutrons generated in the tantalum targets are directed along beam lines to the instruments in the two experimental halls.

3.5 Neutron moderation: hot and cold sources

The energy spectrum of neutrons emitted from a reactor or from an accelerator source depends on the temperature of the moderators surrounding the source, as the neutron 'gas' is in thermal equilibrium with the materials of the moderators. Figure 3.11 shows the Maxwellian flux for moderator temperatures of 25, 300 and 2,000 K. Most reactors are designed for a moderator temperature lying in the ambient range 300–350 K, corresponding to the middle curve of Fig. 3.11. However, for some experiments, such as high-resolution inelastic scattering, reflectometry and small-angle scattering, there is a considerable advantage in using long-wavelength neutrons which require a cold source giving a spectrum similar to that shown in the left-hand curve. On the other hand, to

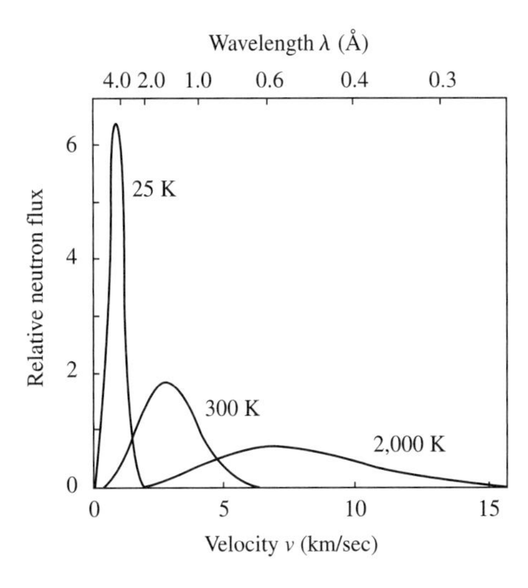

Fig. 3.11 The Maxwell–Boltzmann flux distribution for infinitely large moderators at three different temperatures.

extend the spectrum to shorter wavelengths, or higher energies, a hot source can be employed: the hot source of the HFR at the ILL consists of a graphite block heated to around 2,400 K in the radiation environment of the reactor. The epithermal tail of the under-moderated spectrum from a pulsed accelerator source already provides intense beams of hot neutrons without the need to employ a heated moderator. To achieve good coupling to the fast neutrons in the target and to maintain a narrow pulse and high neutron brightness, moderators on pulsed sources are considerably smaller than on reactors where there is no time-structure to preserve. The effect of this limited size is that the slow neutron spectra are appreciably under-moderated, and the characteristic temperature of the neutron spectrum is higher than the actual temperature of the moderator. Thus the neutrons are of higher energy and there is an epithermal component to the spectrum, which is absent from a reactor spectrum. A second consequence of the small size of moderators on pulsed sources is the greater difficulty of coupling to neutron guides efficiently, to achieve full illumination, particularly at longer neutron wavelengths.

The energy and wavelength ranges covered by moderators at temperatures of 25, 300 and 2,000 K are given in Table 3.1.

The material used for a low-temperature moderator must satisfy several conditions. It must have a low absorption cross-section and a high scattering cross-section. In addition, the material must be of low atomic number in order to ensure that the number of collisions required for reaching thermal energies is relatively low: in this respect hydrogen-containing moderators are the most effective. In addition, the moderator material must possess a suitable inelastic scattering mechanism for lowering the neutron energy.

Liquid hydrogen appears to be the most practical material for a cold source, in spite of its relatively high absorption cross-section and its low density. Hydrogen has the advantage that it does not suffer from radiation damage. Methane, which has a proton density in the liquid phase twice that of liquid hydrogen and exhibits inelastic processes below

Table 3.1 Energy and wavelength ranges of neutron moderators.

Moderator	Energy (meV)	Wavelength (Å)
Cold (20 K)	0.1–10	3–30
Thermal (300 K)	5–80	1–4
Hot (2,000 K)	80–500	0.4–1

100 μeV, is neutronically very favourable but is less practical than hydrogen as it suffers severely from *radiolysis*. Radiolysis (*i.e.* the production of free radicals by the breaking of chemical bonds) results in chemical recombination reactions, and these produce tars leading to eventual blockage of the moderator.

3.6 The relative merits of continuous and pulsed sources

An acceptable comparison of pulsed sources and continuous sources is fraught with difficulties and, whilst there have been several attempts to do this analytically, none has succeeded in demonstrating that there is a simple answer. Some general guidelines can only be drawn and these are changing constantly, as pulsed source technology is relatively new whereas reactor technology is more mature. In the end it comes down to where the best science can be done—on which instruments, at which sources and with which people. The body of knowledge accumulated by scientific endeavour is often subjective and so we shall confine ourselves to some general comments.

(a) *Politics.* Enriched uranium, *i.e.* uranium highly concentrated in the fissile isotope ^{235}U, is the raw material of atomic weapons as well as the fuel for research reactors. Its production, availability and use are the business of governments, which seek to control not only the supply of ^{235}U but also the information surrounding it. The U.S. government is currently committed to converting the use of highly enriched uranium (HEU with >20% content of ^{235}U) in research reactors to low enriched uranium (LEU with <20% content of ^{235}U). Fuel supplies, even for existing reactors, are therefore uncertain and the prospect of lower fluxes, following core reconfigurations adapted to the use of LEU, must be faced. Those reactors with the most compact cores—the HFRs—will suffer most.

The reprocessing of used fuel elements—a source of plutonium, which is a raw material for weapons—is another complex political and environmental issue. Even the transport of used fuel—let alone its reprocessing—is the focus of public dissent. Pulsed sources do not use fissile materials as targets, do not present the perceived risks of reactor sources and, although their targets become highly radioactive after many years of operation, they contain no fissile material as a result of the spallation process.

On the other hand, governments committed to a dependence on electricity generated by nuclear power plants or in maintaining independent nuclear weapons need to maintain a trained workforce in reactor technology in all its forms. There is the future prospect of converting the radioactive waste from nuclear reactors into more benign materials using high-power accelerators.

Environmentally, both power reactors and research reactors are perceived by the general public as being malign. Risks posed by supercriticality excursions, as in the Chernobyl and Three Mile Island incidents, are relatively low when set against other man-made risk factors in our daily lives. Death on the roads or from smoking tobacco are high risks but are well understood and are apparently acceptable to the general public, whereas the low risks from reactors are mysterious and unacceptable. It is unlikely that these perceptions will change significantly in the foreseeable future in favour of reactors. On the positive side the generation of electricity by nuclear power stations does not produce 'greenhouse gases', and the nuclear fuel itself, unlike oil, is not concentrated in politically unstable regions of the world. There will almost certainly be a revival of the fortunes of nuclear power, as governments struggle to meet the targets on carbon emissions set by the 1997 Kyoto Protocol on global warming.

(b) *Neutron production.* Reactors generate neutrons continuously. The moderator is an intrinsic component of the reactor, helping to maintain criticality by thermalizing the fast neutrons. This results in a well-moderated spectrum for neutron scattering measurements with a rich flux of low-energy neutrons (further enhanced by cold sources) and a relatively low flux of neutrons at energies exceeding 100 meV. In contrast, the moderators on a pulsed source serve no purpose in the neutron generation process and can therefore be tailored to suit the neutron instruments. In order to maximize the intensity and to retain the narrow pulse structure of the incident proton beam, there is a need to maintain good coupling to the target, and so the moderators are small and the resultant spectrum is under-moderated. This results in a significant flux of epithermal neutrons and a relatively low flux of cold neutrons. This apparent complementarity—using reactors for low-energy neutrons and pulsed sources for high-energy neutrons—is oversimplistic and is violated by the use of hot moderators on reactors and cold moderators on pulsed sources.

Reactors are simpler to operate than pulsed sources and are intrinsically more reliable. In a reactor the control of a single set of control rods to maintain reactor power requires the monitoring of a relatively small number of parameters, such as the temperature and the flow rate and the radioactivity levels of the coolant. This contrasts with the large number of components to be controlled and monitored on a pulsed source, with perhaps up to 500 identifiable components on the accelerator itself. The reliability of pulsed sources is, therefore, noticeably poorer than for reactor sources. This difference is further accentuated by the demand for the utmost safety on reactors. Components and systems are built to the highest nuclear standards and are often installed with triple redundancy, thereby enhancing overall reliability.

Because of its continuous operation the power density levels generated in the core of a research reactor can be very significant, up to 10 MW/litre, which must be removed by forced cooling to maintain steady temperatures. In contrast, on a pulsed source, power density levels in the target are much lower, around 5 kW/litre in present-day sources. This is partly a result of the pulsed nature of the neutron generation process and partly because the heat generated for each useful neutron ($\sim$30 neutrons generated per incident proton in the spallation process) is significantly lower than that in the fission process ($\sim$0.5 useful neutrons per fission reaction). Thus in fission the heat released is $\sim$200 MeV/n whereas in spallation it is $\sim$50 MeV/n. However, given that there is an 'instantaneous' production of heat in a small volume when the proton beam strikes the target of a pulsed source, effective power densities are significantly higher than the time-averaged heat loads quoted here. Reactors are now at the limits of material technology and so higher power densities, and hence higher neutron fluxes, are not foreseen. Pulsed sources, on the other hand, are operating within cooling limits and sources are under consideration which are 30 times more intense than those currently operating (*e.g.* the proposed 5 MW ESS). There are other problems to be considered with pulsed sources, such as the mechanical integrity of the target containment vessel as it is repeatedly struck by intense pulses of high-energy protons.

The possibility of designing a competitive pulsed source, which generates perhaps two orders of magnitude fewer fast neutrons than an equivalent reactor, requires the utilization of the peak neutron flux for a large part of the time between pulses. Nearly all the neutrons generated in fission in a reactor are removed by using a monochromator, and measurements are then made continuously with a monochromatic beam. On a pulsed source the well-defined time origin of the neutron burst allows the dispersion of neutrons of different energies before the beam strikes the sample, and if the instrument is properly designed then these neutrons fill the whole measuring time frame with polychromatic (white) neutrons at peak intensity. The peak flux on a pulsed source can be compared to the average flux on a reactor to give one measure of comparability of source strength. This is illustrated in Fig. 3.12 where the published values of the fluxes at ILL and ISIS are compared. It can be seen that pulsed sources compare particularly well at higher neutron energies. However, this simple comparison overlooks the fact that guides can be better coupled to larger moderators on reactor sources, that large-area focusing monochromators can be used on reactor sources to enhance the flux at the sample, and that on a white-beam pulsed source whole frames are not filled and the intensity within a frame can vary appreciably.

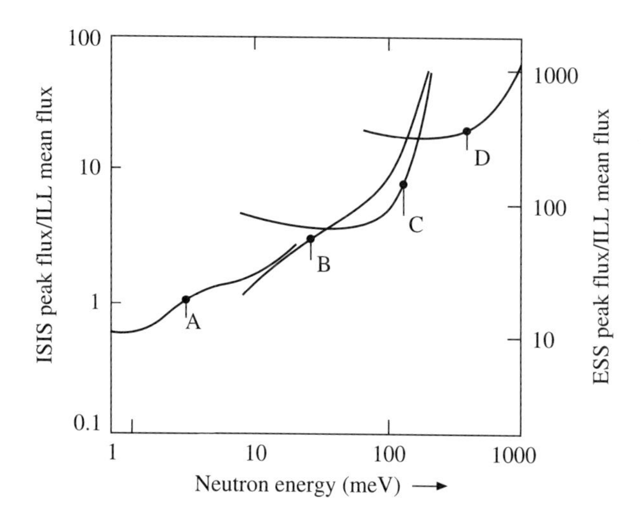

Fig. 3.12 Variation with energy of the ratio of the peak flux at the ISIS source and the mean flux at the ILL reactor source (*idealized*). A, B, C and D refer to a comparison of different moderators on the two sources.

(c) *Neutron utilization.* Whatever kind of source is used to record a diffraction pattern or an inelastic spectrum, the measurements are carried out in quite different ways. With a continuous source we can select a particular wavelength with a crystal monochromator and carry out measurements as a function of scattering angle, whereas with a pulsed source we use a wide band of wavelengths and analyse the neutrons by time-of-flight techniques. Because both the pulse width and the time-of-flight over a fixed distance are proportional to wavelength, pulsed source instruments have resolutions that are relatively constant over a wide dynamic range. Coupled to the rich flux of short wavelength neutrons from a pulsed source, a short-pulse source is of particular advantage in powder diffraction when high resolution over a wide range of Q ($=4\pi \sin\theta/\lambda$) is needed. This advantage is lost on a long-pulse source.

In general, pulsed source instruments are static during their operation whereas reactor instruments are mobile. This feature of pulsed sources can have advantages when using complex sample environment equipment where, for example, complete scattering patterns from a sample under pressure in a fixed window cell can be obtained at a single scattering angle. On the other hand, the resolution function of a reactor instrument is invariably symmetric in Q or in ω, whereas that from a pulsed source instrument is almost always asymmetric as a result of the sharp rise time and slow decay of the neutron pulse in the moderator. This can give rise to problems in data correction which are peculiar to pulsed source instruments. Correction factors are more straightforward for reactor data when the measurements are made at fixed wavelength. When using beams from neutron guides the distance of a pulsed source instrument from the moderator determines its resolution, whereas on a reactor it does not. There is therefore less flexibility in the siting of instruments on pulsed sources than there is on reactors.

In principle, because a pulsed source produces two orders of magnitude fewer neutrons than an equivalent reactor, instrumental backgrounds should be lower in the former case. In practice, this is only true for chopper spectrometers where very low signals can be sensed. The very fast neutrons generated by a pulsed source and the need to illuminate the sample with an intense white beam often militate against this apparent advantage. One further manifestation of this is the very large amount of steel and concrete shielding surrounding all pulsed source instruments, as well as the huge beam stops that are required. Combined with the need for long incident flight paths on white-beam instruments, the instruments themselves on pulsed sources are appreciably more expensive than their reactor counterparts.

A single-crystal diffraction study with a continuous source allows more flexibility than with a pulsed source. The crystal and detector can move independently of one another and so there is no restriction on the choice of scan in reciprocal space. An analogous situation arises in comparing the triple-axis spectrometer on a steady source with a time-of-flight spectrometer on a pulsed source. With a triple-axis spectrometer we can chose any type of scan in $(\mathbf{Q}, \omega)$ space whereas this is not so for the pulsed source analogue. On the other hand, the completeness of the pulsed source data set, whether from elastic or inelastic instruments, can reveal subtle effects which, if not expected, can remain unobserved on a reactor instrument.

In the field of liquid diffraction the benefits of high fluxes of high-energy neutrons mean that inelasticity corrections (so-called Placzek corrections) are minimized on a pulsed source instrument where measurements to very high momentum transfers can be made.

With reflectivity measurements there is a similar story. The fixed geometry of a pulsed source instrument yields great benefits in the study of surfaces (particularly so for liquids), whereas reactor instruments, in general, must scan in a $\theta/2\theta$ mode. However the scanning process, which can be disadvantageous from a stability point of view, allows the experimentalist on a reactor instrument to increase the dwell time continuously, point-by-point, to match the fall in signal between the maximum and minimum intensities displayed in the reflectivity pattern.

The comments above are not meant to be a definitive survey of pros and cons of the various instruments available on continuous and pulsed sources, but rather to give an insight into the differences in the two techniques. An advantage in one type of experiment can prove to be a disadvantage in another. There is also no doubt that the area of overlap between what can be done on pulsed source instruments and what can be done on reactor-based instruments has increased considerably with time. This process will continue as more opportunities are exploited on pulsed sources, but the original, tidy concept of complementarity of the two kinds of source has given way to a more realistic complementarity of instruments.

References

Beckurts, K. H. and Wirtz, K. (1964) "*Neutron Physics*" Springer-Verlag, Berlin.

Glasstone, S. and Edlund, M. C. (1957) "*The Elements of Nuclear Reactor Theory*" Van Nostrand, New York.

Glasstone, S. and Sesonske, A. (1994) "*Nuclear Reactor Engineering: Reactor Design Basics*" Kluwer Academic Publishers, Dordrecht, The Netherlands.

Hunt, S. E. (1987) "*Nuclear physics for engineers and scientists.*" Ellis Horwood, Chichester.

International Atomic Energy Agency (1985) "*Neutron scattering in the nineties.*" Conference Proceedings, Julich, 14–18 January. IAEA, Vienna.

Megreblian, R. V. and Holmes, D. K. (1960) "*Reactor Analysis*" McGraw-Hill, New York.

Neutron detection

4

4.1 Introduction

The neutron is a very weak probe, and so neutron scattering does not unduly disturb the properties of the sample under investigation. From the point of view of scattering theory this is a significant advantage, as it allows us to use the *Born approximation* (whereby the scattering process is treated by first-order perturbation theory) in describing the neutron–nucleus interaction. However, the weakness of its interaction with matter also renders neutrons difficult to detect directly. Thermal neutrons produce negligible ionization and they can only be detected by the production of secondary ionization arising after neutron absorption. The absorption process leads to the formation of a compound nucleus, which decays either by the emission of a gamma ray or by a splitting of the compound nucleus into two charged nuclei. In neither case the energy of the incident neutron can be measured directly, as the absorption reactions are strongly exothermic with the release of several MeV of energy. (If a method were devised both to detect thermal neutrons and to measure their energies to a reasonable accuracy, then it would transform the field of neutron scattering!)

The most common nuclear reactions used in neutron detection involve the nuclei of ^{3}He, ^{10}B, ^{6}Li and ^{235}U. The nuclear capture reactions proceed as follows:

$$\begin{aligned}
&^{3}\mathrm{He} + \mathrm{n} \rightarrow {}^{3}\mathrm{H} + \mathrm{p} + 0.8\ \mathrm{MeV}\\
&^{10}\mathrm{B} + \mathrm{n} \rightarrow {}^{7}\mathrm{Li} + {}^{4}\mathrm{He} + 2.3\ \mathrm{MeV}\\
&^{6}\mathrm{Li} + \mathrm{n} \rightarrow {}^{4}\mathrm{He} + {}^{3}\mathrm{He} + 4.8\ \mathrm{MeV}\\
&^{235}\mathrm{U} + \mathrm{n} \rightarrow \text{two fission fragments} + 190\ \mathrm{MeV}
\end{aligned} \tag{4.1}$$

The first two reactions are employed in gas proportional counters, the third in scintillation detectors and the last one in fission chambers for monitoring the incident beam. The absorption cross-sections σ_{a} for the first three reactions in eqn. (4.1) are shown as a function of neutron energy in Fig. 4.1. The ^{3}He cross-section is particularly high and is 3,000 barns ($3{,}000 \times 10^{-28}$ m^{2}) at a wavelength of 1.0 Å. Over the energy range considered, all these cross-sections are inversely proportional to velocity, $\sigma_{\mathrm{a}} \propto 1/v$. A simple physical interpretation of this so-called *one over v* behaviour is that the slower the neutron, the longer the dwell time near the nucleus and the higher the probability of absorption.

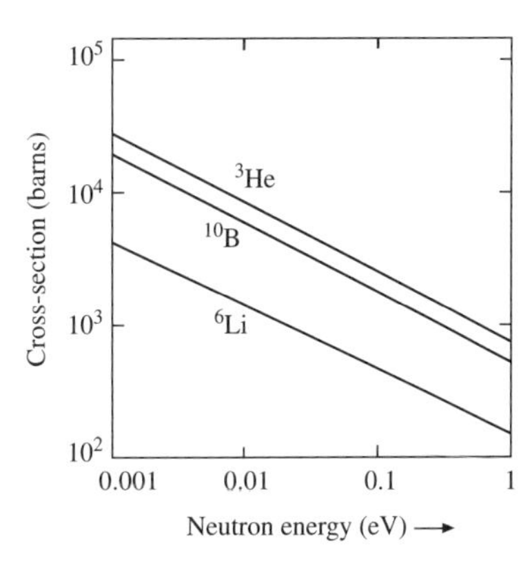

Fig. 4.1 Absorption cross-sections as a function of neutron energy.

4.2 Beam monitors

A beam monitor is necessary on all instruments in order to measure the intensity of the main beam; allowance can then be made automatically for any variations in beam strength during a measurement. A monitor is also required for the measurement of the spectrum of an incident white beam to be used in subsequent data correction, especially in time-of-flight experiments employing the whole spectrum. A typical beam monitor is a fission chamber consisting of a rectangular box, about 1 cm thick, containing gaseous uranium hexafluoride, UF_6. Fission fragments, produced by the absorption of neutrons as they pass through the chamber, give rise to ionization in the gas. This ionization creates an intense pulse which is readily detected electronically. The monitor is placed in the direct beam, and so the attenuation of the beam passing to the instrument must be low and is normally much less than 1%.

Instead of distributing the absorber uniformly, as in the case of a gas monitor, the absorption and ionization elements of the monitor can be separated. The absorber is then in the form of a foil or else is evaporated as a very thin layer on the inside of the cathode of a gas detector or ionization chamber. A few micrograms of ^{235}U evaporated on to the inside surface of a flat gas counter (see Section 4.3) gives an efficiency of about 0.1%. Foil detectors are commonly used for beam monitoring because they combine low efficiency and high transmission.

A novel beam monitor developed by Guerard (2002, unpublished) consists of a proportional counter using nitrogen as the neutron-absorbing gas. Nitrogen has a low absorption for neutrons and is accurately $1/v$. An extremely low-efficiency beam monitor can be constructed which allows the correction for incident spectral distribution and is particularly effective for time-of-flight instruments. The attenuation of the incident beam is very low, thus preserving the intensity falling on the sample.

4.3 Gas detectors

In this type of detector the absorbing material is a gas, such as ^{3}He and boron trifluoride, BF_3, enriched in the boron isotope ^{10}B. A nuclear reaction in the gas releases energy exothermically, and the reaction products fly apart in opposite directions since the kinetic energy of the total system (neutron plus absorbing nucleus) prior to absorption is negligible. The reaction products share the energy of the reaction in inverse proportion to their masses, so that in the absorption of neutrons by ^{3}He gas, for example, the energies of the resulting proton and triton [see eqn. (4.1)] are in the ratio 3:1. The particles cause ionization in the gas and this ionization is collected by an anode wire at high voltage. If the voltage is kept low, then only the electrons released in the initial ionization are collected and the current pulse is weak. As the voltage increases, the electrons are accelerated by an electric field which diverges as $1/r$, where r is the distance from the wire, and eventually the electrons produce their own ionization in an electron avalanche. The pulse is now much larger but it is still proportional to the energy deposited in the gas; therefore, the detector is operating in the so-called *proportional region*. This does not imply that we are able to determine the energy of the incident neutron itself, as each capture reaction releases the same large amount of energy, irrespective of the neutron energy. It does mean, however, that the detector can discriminate against gamma rays and that its efficiency is stable to changes in the electronic detecting chain arising, for example, from fluctuations in temperature. At still higher voltages the pulse is produced by a cascade effect, proportionality breaks down, and the detector operates in the *Geiger counter* mode.

Scattering experiments are often carried out in the presence of backgrounds from gamma radiation and fast neutrons. The fast neutron flux is readily reduced by surrounding the instrument with an outer shield of hydrogenous material such as polythene, which serves to slow down the fast and epithermal neutrons; a thin inner shield of boron is then sufficient to absorb the moderated neutrons. The background gamma radiation is more difficult to attenuate, even though the range of the electrons produced by a gamma ray is large compared with the detector size. A gamma ray releases much less energy than that dissipated by the reactions in eqn. (4.1), and so discrimination between neutrons and gamma rays is possible with a proportional counter. Geiger counters operating at very high voltages are rarely used in neutron scattering experiments, because they fail to discriminate against gamma rays. Also the dead time between the detection of successive ionization pulses is longer than the dead time ($\sim 1\,\mu$sec) in a proportional counter.

A gas detector is usually constructed with the neutron-sensitive gas enclosed in a cylindrical metal tube which forms the cathode of the detector and is held at earth potential. The anode is a fine central wire stretched along the axis of the tube and is insulated from the cathode with ceramic seals. For use in time-of-flight instruments, detectors are used in the 'side-on' configuration and are often flattened (Fig. 4.2), in order to reduce uncertainties in the path length. In a ^{3}He detector the ^{3}He gas is pressurized to increase the detection efficiency, and detectors are made with pressures varying from 1 to 40 atmospheres. An additive gas is included, such as krypton or methane, to produce further ionization and also to quench the discharge in the shortest possible time. BF_3 detectors are rarely held at pressures above 1 atmosphere in order to minimize the risk of releasing the toxic gas BF_3, which generates white hazardous fumes of hydrofluoric acid when in contact with the humidity in air.

The distribution of output charge per detected neutron can be measured using a pulse-height analyser. A typical pulse-height distribution for a ^{3}He detector is shown in Fig. 4.3. The peak at 1,070 eV corresponds to the total reaction energy being collected. The edges $E_{\rm p}$ and $E_{\rm t}$ are due to the so-called *wall effect*: the range of the reaction products in the gas is a few millimetres and so there is a finite probability, if the absorption of the neutron takes place close to the wall, that one of the

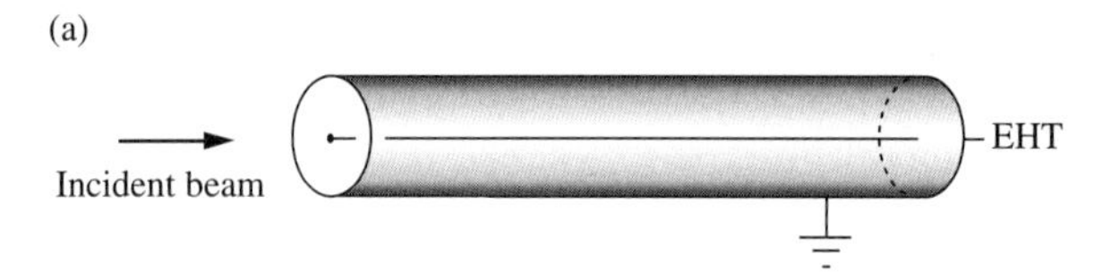

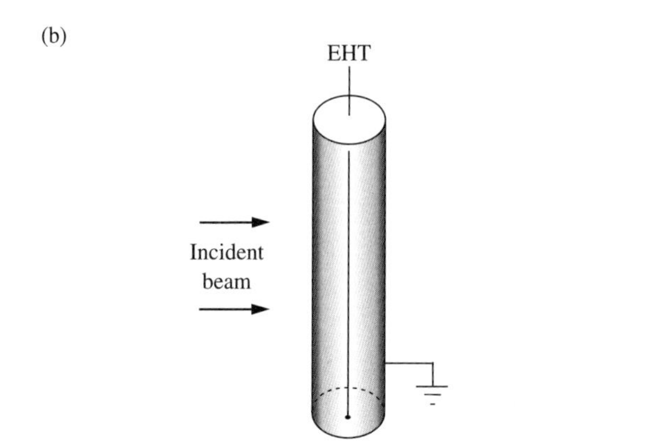

Fig. 4.2 Gas counter with extra high tension (EHT) along the anode wire. (a) End-on configuration. (b) Side-on configuration. Pulsed sources employ the side-on configuration in order to minimize the uncertainty in deriving the path length of the neutrons. (*After* Windsor, 1981.)

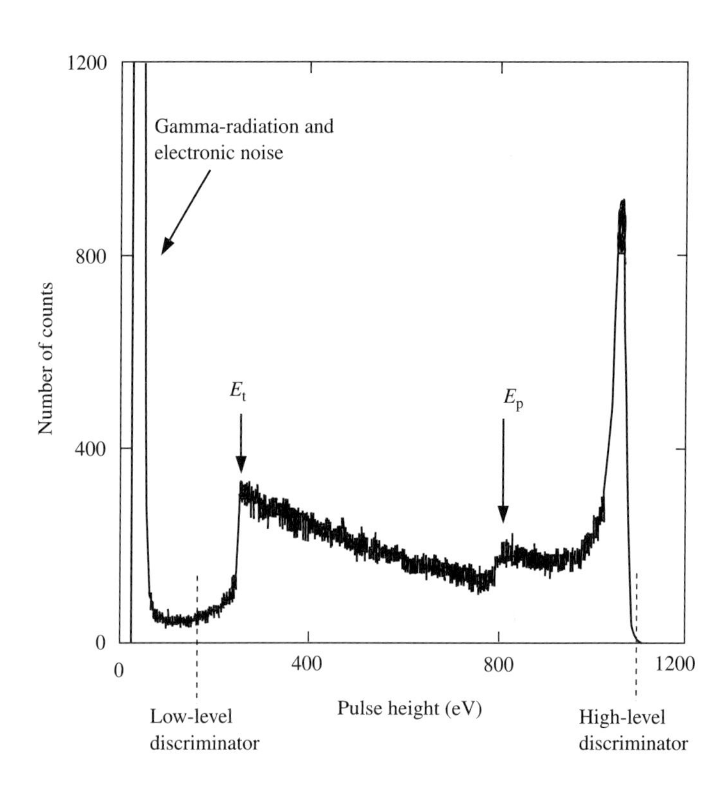

Fig. 4.3 Pulse-height spectrum of a ^{3}He detector.

products will deposit its energy into the wall material and not into the gas. As the two reaction products travel in opposite directions, the minimum energy deposited will be that of the less energetic particle. Thus, the two edges at E_{p} and E_{t} correspond, respectively, to full absorption of the ionization from the proton only and from the triton only. At low pulse heights, there is a long tail caused by electronic noise and the detection of gamma rays. This tail can be minimized by fabricating the wall (the cathode) from a low-Z material such as beryllium. To ensure that the detection efficiency is as close as possible to the theoretical maximum (determined by the neutron cross-section of the filling gas), an electronic discriminator is set just below the triton edge, and the voltage across the counter is set to ensure that this edge is well separated from the electronic noise. The function of the low-level discriminator is to suppress the smaller pulses from gamma rays. A quenching circuit lowers the voltage momentarily after discharge for a period slightly in excess of the recovery time, or dead time, of the detector.

Boron exists in the gaseous phase at room temperature as BF_3, a white corrosive gas, and this is the form used in detectors. Natural boron consists of the isotopes ^{10}B and ^{11}B, in the approximate proportion 1:4. ^{10}B is the absorbing isotope and, to ensure the highest efficiency, isotopically enriched BF_3 is used in detectors. ^{3}He has a higher absorption cross-section and is more efficient for the detection of thermal neutrons: BF_3 detectors, therefore, are less commonly used than in earlier times when ^{3}He was rare and expensive. ^{3}He is a by-product from oil wells; today it is no longer prohibitively expensive, and so ^{3}He has largely superseded BF_3 in detectors.

In the traditional way of carrying out a neutron scattering experiment, a single detector is used to measure the scattered intensity consecutively at selected points in $(\mathbf{Q}, \omega)$ space. However, neutrons are scattered from the sample along all directions in space, and so a more effective procedure is to measure the scattered neutrons using an extended or segmented area detector and then to select the required points in $(\mathbf{Q}, \omega)$ space after the measurements have been made. Improvements in data-handling electronics and computers have shifted the balance decisively towards this second procedure. Position-sensitive detectors were originally only employed in special instruments such as the powder diffractometer D1A at the Institut Laue-Langevin (ILL) or the MARX triple-axis spectrometer at Risø: today it is the norm to install the largest one-dimensional or two-dimensional position-sensitive detectors on new instruments. The various categories of position-sensitive detectors (PSDs) are described below. Either they are discrete-element digital devices, where the output from a single element determines the position of the detected neutron, or they are continuous neutron converters, where the output from a number of electrodes is analysed in an analogue fashion to give the neutron positions.

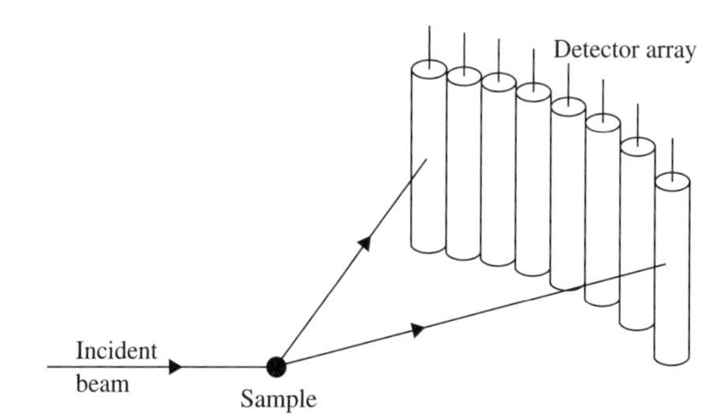

Fig. 4.4 Array of individual detectors.

4.3.1 Banks of individual detectors

This is the logical step following on from the single counter instrument. The detector consists of a close-packed array of separate gas detectors (see Fig. 4.4) whose diameters determine the spatial resolution of the extended detector system. The individual detectors are commercially available; setting up and maintenance is simple, efficiencies are high and backgrounds are low. In a recent development a matrix of such tubes is welded into a single manifold, adding rigidity to the structure and enabling large area detectors to be manufactured at modest costs.

4.3.2 Resistive-wire gas proportional counters

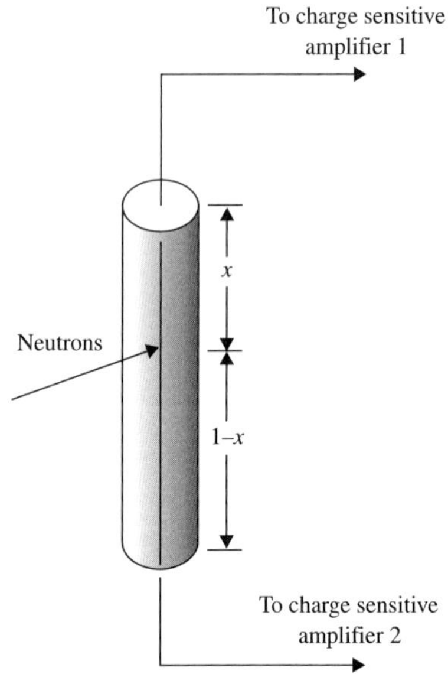

Fig. 4.5 The resistive-wire, position-sensitive gas counter. The charges measured by the two amplifiers are q_1, q_2 and x is given by $x = q_1/(q_1 + q_2)$.

This detector is an extension of the single gas counter, in which the anode wire is converted into a linear PSD. This is done by making the wire resistive, replacing it with a glass fibre coated with a thin metallic layer. A neutron captured at a position x along the wire produces a charge whose amplitude is measured at both ends of the detector. On account of the resistance of the wire, the charge is divided in the ratio of the distances x and $1 - x$ from the ends (see Fig. 4.5) and the position of the neutron is determined from the ratio q_1/q_2 of the two charge measurements. The positional resolution of the detector can be as low as a few millimetres and is determined by the diameter of the charge cloud surrounding the neutron absorption event. By stacking a number of these linear PSDs together, a two-dimensional position-sensitive detector array can be assembled.

This technique was employed in early time-of-flight spectrometers and extended later to the time-of-flight instruments at ISIS. Position-sensitive ^{3}He gas tubes up to 3 m in length and assembled in space-filling arrays to give a spatial resolution in two dimensions of 1 cm are used with IN5 at ILL and MAPS at ISIS.

4.3.3 Multi-wire proportional counter (MWPC)

The original MWPC was invented by Charpak *et al.* (1968), for which he was awarded the Nobel Prize. It consists of a large number of very thin anode wires which are mounted between two plates acting as cathodes in a gaseous volume. Each plate consists of independent cathode strips and the strips on the separate plates are mutually perpendicular, allowing the position of the neutron to be determined to a spatial resolution of about 1 mm. Thus this is an extension of the one-dimensional resistive-wire detector. An early example of such a detector is the large circular detector, filled with ^{3}He gas, which is used primarily on small-angle scattering instruments.

The electrons produced in the gas drift towards the anode wires where they undergo an avalanche amplification, as with Geiger counters, in the $1/r$ electric field surrounding the wires. At the same time the positively charged ions drift towards the cathodes, producing a high space charge around the cathode strips, which limits the field strength and reduces the gas gain. This limitation can be mitigated by reducing the drift path of the positive ions, leading to their faster evacuation. Such a possibility has been achieved with the micropattern gas counters described below.

4.3.4 Micropattern gas counters (MPGC)

To increase the maximum counting rate, the distance to the cathodes must be reduced, leading to a faster evacuation of the avalanche ions. In the MPGC detectors there are no wires at all, but instead they are made up of very narrowly spaced conductor strips, which may have a width of only a few microns. The fine electrode structures are produced by photolithography onto a glass substrate, a well-established technique for fabricating integrated circuits.

The first type of MPGC was described by Oed (1988). The anodes are very narrow strips located between two wider strips, acting as the cathodes, and the spacing between strips can be as small as 200 μm, which is five times finer than the spacing of the wires in MWPCs. The applied electric potential alternates between each strip, producing gas avalanche amplification between adjacent strips. An electron in the gas drifts towards the nearest positive strip, where it undergoes avalanche amplification, and the ions produced are then neutralized on the cathode strips close by. This leads to a significant reduction in the size of the space charge, so that the spatial resolution can be as small as 1 mm.

There are many other variations in the design of MPGCs, and great efforts have been made to overcome some of their weak points, such as the charging up of the surface of the insulator between the strips. A survey of these detector types and their characteristics is given in a review by Oed (2004).

4.4 Scintillation detectors

In this type of detector the neutron absorber, typically ^{6}Li in the form of a lithium salt, is mixed homogeneously with a scintillator material such as ZnS; the burst of light, which is emitted when a neutron is absorbed and ionizing secondary particles are formed, is amplified by a photomultiplier. High-intensity pulsed neutron sources require position-sensitive detectors capable of rapidly recording high and instantaneous count rates. Scintillation counters are, in general, much better for this purpose than gas detectors, as their detection mechanism (which does not depend on the drifting of ions) is intrinsically 100 times faster. The requirements for an efficient scintillation medium are that scintillation must occur immediately, the conversion efficiency to light must be high, and the material must be transparent to its own radiation. This last point is a significant restriction, as emission and absorption spectra are normally quite similar. Another advantage of a scintillator detector is that the range of the ionizing particles is reduced to a few microns, and so the spatial resolution is better than for gas detectors.

Scintillator detectors, being solid, are particularly sensitive to gamma rays and special precautions must be taken to minimize their gamma sensitivity: discrimination against a background of gamma rays is much easier with a gas counter. With a continuous neutron beam, a scintillator is used only if the gamma sensitivity is of little importance; discrimination against the gamma background is possible with a pulsed neutron beam, by taking advantage of the time-of-flight difference between neutrons and gamma rays. ^{6}Li is chosen as the neutron absorber because its reaction products have the highest energy and the longest range. The scintillator itself may be an *intrinsic* scintillator (*i.e.* self-scintillating) or, by the addition of a suitable activation impurity, an *extrinsic* scintillator. Intrinsic scintillators such as lithium iodide, when irradiated with energetic ionizing radiation, create electron-hole pairs, which de-excite with the emission of a photon. These materials are transparent to their own emitted radiation and have a short decay time ($\sim$75 nsec). However, they must be cooled to assist the de-excitation process via phonon creation, and special optical couplings are necessary to transmit the photons, which are normally emitted in the ultraviolet region of the electromagnetic spectrum.

A heavy-metal impurity such as Ag and Eu is incorporated in extrinsic scintillators. In this case, the presence of such activators produces additional energy levels in the band gap between the valence band and the conduction band. Just as for intrinsic scintillators, ionizing radiation produces electron-hole pairs, some of which decay with the emission of an ultraviolet photon. An activator atom is excited to its first excited state which then decays to the ground state with the emission of a light photon. The material will operate at room temperature but has the disadvantage, caused by the additional conversion process, of a longer decay time ($\sim$1 μsec) than an intrinsic scintillator.

There has been much development work on scintillation detectors, largely as a result of the advent of high-intensity pulsed neutron sources and the availability of faster computers to handle the increase in data rates obtainable with large position-sensitive detectors.

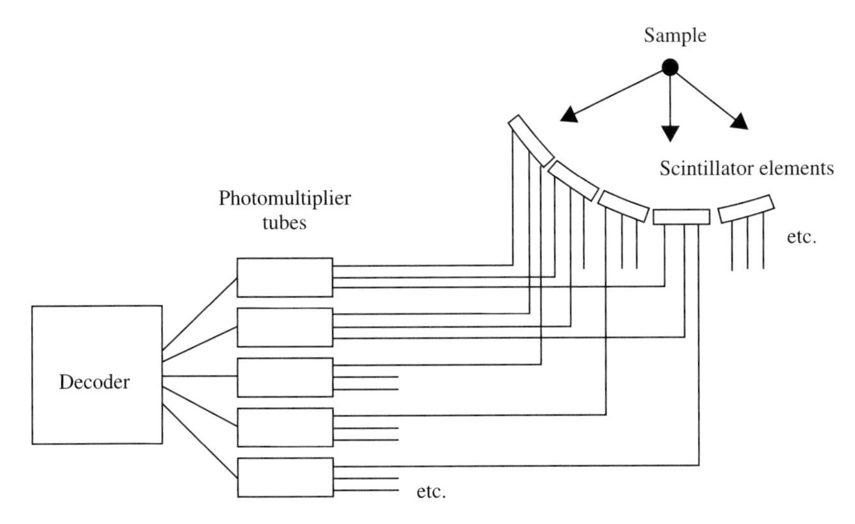

Fig. 4.6 Fibre-optic coded discrete scintillator detectors. By using triple coincidence of events, the number of photomultiplier tubes is reduced.

4.4.1 Discrete element scintillators

This type of detector is analogous to an array of linear position-sensitive gas detectors. Each scintillator element is linked to several photomultipliers. In a development by Davidson and Wroe (1976, unpublished), which utilizes fibre-optic encoding of the signals from individual elements, light from a single scintillator element excites three photomultipliers simultaneously via fibre-optic linkages (see Fig. 4.6). This leads to a reduction in the number of photomultipliers and, in addition, allows discrimination against background events by anticoincidence counting. In general, the number of elements n_e which can be encoded by n_p photomultipliers with three linkages per element is

$$n_\mathrm{e} = \frac{n_\mathrm{p}!}{3!\,(n_\mathrm{p}-3)!} \tag{4.2}$$

For example, 8 photomultipliers can encode 56 detector elements. Clearly, there is a considerable saving in space and expense by using this ingenious encoding technique, but manufacture is a lengthy process.

4.4.2 The Anger camera

This camera gets its name from its inventor (Anger, 1958), who first applied it to the detection of X-rays for medical purposes. The principle of the camera is illustrated in Fig. 4.7.

Neutrons are detected in a large area of flat glass scintillator which is directly viewed by a close-packed array of photomultipliers. By triangulation of the amplitudes of the signals received by a number of photomultipliers, the origin of the neutron absorption event can be obtained within a few millimetres. The Anger camera has the advantage of possessing a continuously sensitive surface for detecting neutrons, thus overcoming the problem presented by discrete element scintillators

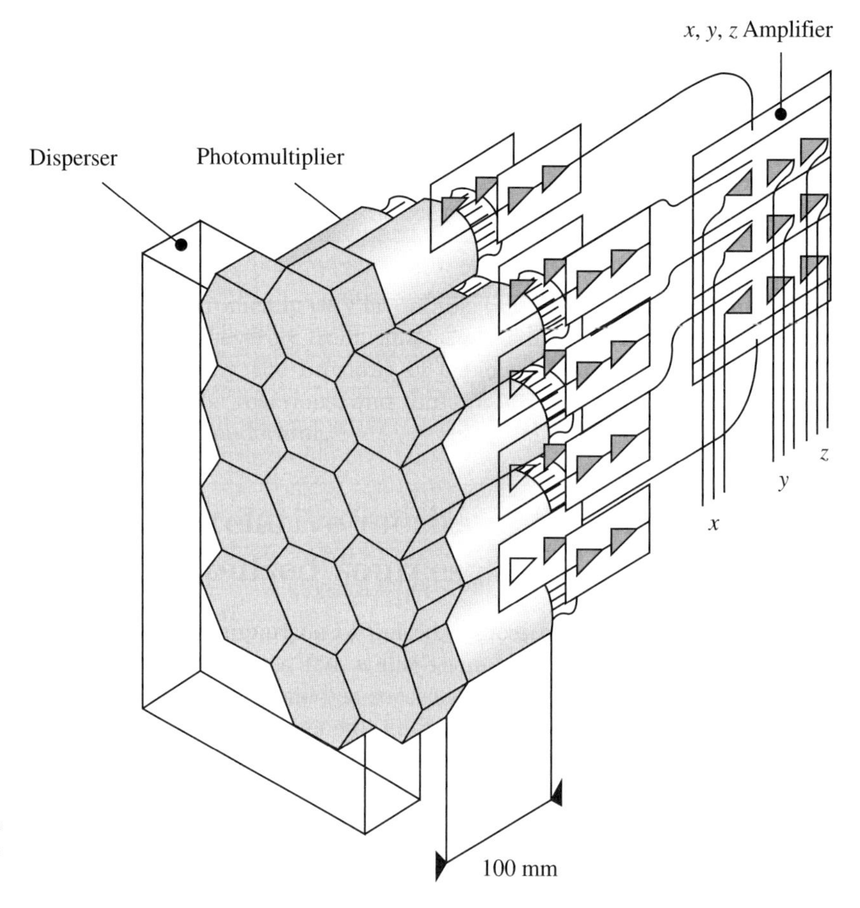

Fig. 4.7 Parallel encoding for the Anger camera. In front of the light disperser is a flat glass scintillator.

when measuring, for example, a Bragg reflection falling on the boundary between two elements.

4.5 Plate detectors

4.5.1 Photographic plates

Photographic techniques for neutron scattering have largely been restricted to determining the position of the sample in the incident beam, rather than quantitative measurement of scattered intensities. This is surprising because photographic techniques overcome many of the disadvantages exhibited by other types of PSD. For example, the spatial resolution is very good ($\sim 100\,\mu\text{m}$), the film is continuous, the method is extremely cheap and no electronics are required. The principal drawback is the lack of uniformity in the photographic film and the non-linear response to intensity, so that absolute intensities cannot be determined to better than 8%. As the film is an integrating device, it also records the light of the gamma background.

A diffraction instrument using photographic plates can give complete coverage in reciprocal space, where this feature is more important than accurate determination of intensities. Neutron diffraction topography has been successfully applied to the study of crystal defects (Baruchel *et al.*, 1977) using photographic plate detectors. The 'photographic' method has undergone a revival following the development of image plates at synchrotron sources.

4.5.2 Image plates

Image-plate detector systems are routinely used in X-ray protein crystallography where huge numbers of Bragg spots are to be measured. They are made from thin plastic sheets coated with finely powdered phosphor crystals doped with Eu^{2+} ions. After irradiation with X-rays or gamma rays, electrons are liberated by conversion of Eu^{2+} into Eu^{+} and these electrons are then trapped in metastable states with a decay period of many hours. Thus the electrons form a stored image which can be read by scanning with a laser beam of visible radiation. The beam releases the trapped electrons by photostimulation, emitting light lying in the blue region of the spectrum. With the help of an online scanner the inconvenience of chemical development of film is avoided.

To use these image plates as neutron detectors, it is necessary to convert the neutrons into electromagnetic radiation. Niimura *et al.* (1994) developed an image plate for detecting neutrons by mixing the luminescent material with ^{6}Li or Gd. [Gadolinium has a very strong (n,γ) cross-section for thermal neutrons.] This neutron image plate has been used successfully for the collection of single-crystal neutron diffraction data from lysozyme (Niimura *et al.*, 1997). The plate of cylindrical geometry encircles the protein sample, giving a total collection area of $400 \times 800\,\text{mm}^2$. For a pixel size of $0.2 \times 0.2\,\text{mm}^2$ the total number of pixels available for recording the Laue diffraction pattern is 8×10^6. In 10 days of instrument time 38,000 reflections were collected to 2 Å resolution from a tetragonal lysozyme crystal. Image plates do not suffer the linearity problems of photographic methods, and the dynamic intensity range covers six orders of magnitude. Like all integrating detectors, and particularly because of the presence of Gd with its high atomic number, neutron image plates are rather gamma sensitive.

Image-plate detectors are employed on the LADI and VIVALDI instruments at the ILL as well as on instruments at JRR-3 in Japan.

References

Anger, H. O. (1958) *Rev. Sci. Instrum.* **29** 27–33. "*Scintillation camera.*"

Baruchel, J., Schlenker, M. and Roth, W. L. (1977) *J. Appl. Phys.* **48** 5–8. "*Observation of antiferromagnetic domains in nickel oxide by neutron diffraction topography.*"

Charpak, G., Bouclier, R., Bressani, T., Favier, J. and Zupancic, C. (1968) *Nucl. Instrum. Methods* **62** 262–268. "*The use of multiwire proportional counters to select and localise charged particles.*"

Niimura, N., Karasawa, Y.,Tanaka, I., Miyahara, J., Takahashi, K., Saito, H., Koizumi, S. and Hidaka, M. (1994) *Nucl. Instrum. Methods Phys. Res. A* **349** 521–525. "*An imaging plate neutron detector.*"

Niimura, N., Minezaki, Y., Nonaka, T., Castagna, J. C., Cipriani, F., Hoghoj, P., Lehmann, M. S. and Wilkinson, C. (1997) *Nat. Struct. Biol.* **4** 909–914. "*Neutron Laue diffractometry with an imaging plate for neutron protein crystallography.*"

Oed, A. (1988) *Nucl. Instrum. Methods Phys. Res. A* **263** 351–359. "*Position-sensitive detector with microstrip anode for electron multiplication with gases.*"

Oed, A. (2004) *Nucl. Instrum. Methods Phys. Res. A* **525** 62–68. "*Detectors for thermal neutrons.*"

Windsor, C. G. (1981) "*Pulsed neutron scattering.*" Taylor & Francis, London.

Instrument components

5

In this chapter we shall describe some of the components that are used to tailor the neutron beam to satisfy the requirements of the experimentalist. These components perform such functions as limiting the wavelength range or the degree of collimation of the beam striking the sample. We shall be concerned with *unpolarized* beams only: the devices for producing polarized beams are described later in Chapter 9.

5.1 Neutron guide tubes

We begin by considering neutron-conducting tubes or guides, which allow the measuring instrument to be situated well away from the source itself and into more spacious regions of low background. The slow neutron beam is transported away from the neutron source without losing intensity by the inverse square law, and if the guide tube is slightly curved into an arc, it acts as a filter discriminating against higher-energy neutrons and gamma rays. A schematic illustration of a neutron guide is shown in Fig. 5.1. The immense value of neutron guide tubes has been universally recognized since 1971, when they were successfully installed on the high-flux reactor at the Institut Laue-Langevin (ILL). The initial development of neutron guide tubes is due to Maier-Leibnitz and Springer (1963).

Neutron guide tubes exploit the fact that the refractive index n for neutrons is less than unity for most materials, so that the phenomenon of total external reflection can occur at the boundary between the material and a vacuum. The refractive index is defined by the relationship

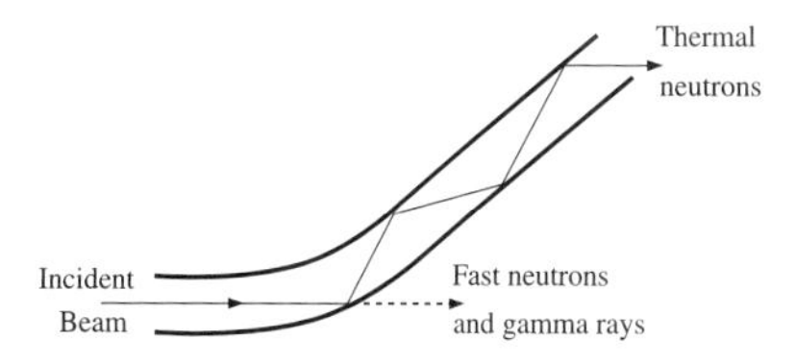

Fig. 5.1 Schematic diagram of a neutron guide tube. The incident thermal neutrons are repeatedly reflected by the curved guide and emerge without contamination by fast neutrons and gamma rays.

$$n = \left(\frac{\cos\theta}{\cos\theta'}\right) \tag{5.1}$$

where θ is the glancing angle of incidence and θ' is the glancing angle of reflection (see Fig. 5.2).

When $\theta' = 0$, then $n = \cos\theta_c$, where θ_c is defined as the critical angle for total external reflection. The refractive index for neutrons is related to the wavelength λ of the neutron by the expression

$$n^2 = 1 - \lambda^2\left(\frac{Nb}{\pi}\right) \tag{5.2}$$

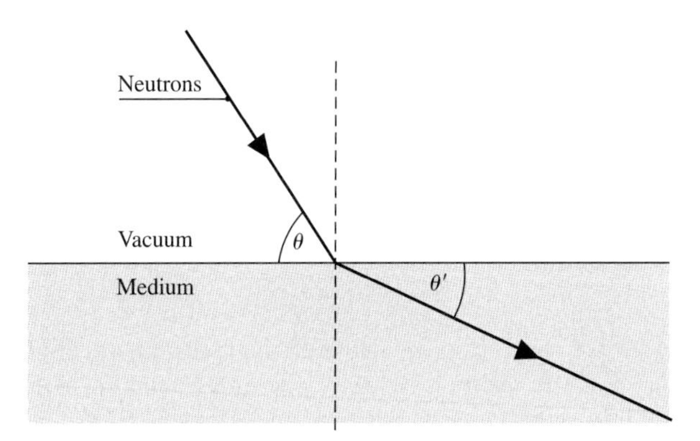

Fig. 5.2 Refraction of neutrons by a medium with $n < 1$.

(see Chapter 11) where N is the number of atoms per cm^3 and b is the coherent scattering amplitude of the bound atom, as listed in Appendix B. Nb is referred to as the scattering length density. b is positive for most materials and for the wavelengths used in scattering experiments N is less than but very close to unity. Hence $\theta_c \ll 1$ and we may approximate $\cos^2\theta_c$ to $(1-\theta_c^2)$. Substituting $n^2 = \cos^2\theta_c = 1-\theta_c^2$ into eqn. (5.2) gives

$$\theta_c = \lambda_c\sqrt{\frac{Nb}{\pi}} \tag{5.3}$$

where λ_c is the critical wavelength for total reflection.

The critical angle θ_c, as calculated from eqn. (5.3) for a few elements at a wavelength of 1 Å, are given in Table 5.1. Natural nickel is the material most commonly used as a reflecting coating in the construction of neutron guides, as it has a high scattering length density Nb and so gives a relatively large value of θ_c. Of the commonly available materials, nickel is followed closely by iron and carbon. Beryllium is marginally better than nickel but is ruled out because of its toxicity. The ^{58}Ni isotope is included in the table since its scattering length density is 20% higher than Nb for natural nickel. It is 40% more effective than natural nickel, as the solid angle transported by a neutron guide is proportional to the square of the critical angle.

The practical realization of neutron guides depends on the reflecting surface being extremely flat. The reflection process cannot be efficient if the fluctuation in the flatness of the surface approaches the critical reflecting angle, so that surface undulations must not exceed 0.1 mrad (or <1 minute of arc). This flatness can be achieved by polishing nickel metal, but this is an expensive procedure. Plate glass and modern high-quality float glass have the finish necessary for neutron guides and can be used as a substrate on to which a thin layer of nickel is evaporated. Neutron guides are produced by assembling rectangular cross-sectional tubes of glass, 1 m long, whose inner surfaces have been coated with nickel. Many 1 m long sections are then butted together inside a vacuum vessel to fabricate the guide.

Table 5.1 Critical angles of scattering of the elements for $\lambda = 1\,\text{Å}$.

	$N\,(\times 10^{29}/\text{m}^3)$	$b\,(\times 10^{-14}\text{ m})$	$Nb\,(\times 10^{38}/\text{m}^2)$	θ_c (mrad)[a]
^{58}Ni	9.0	1.44	13.0	2.03
Beryllium	12.3	0.77	9.5	1.73
Nickel	9.0	1.03	9.3	1.70
Iron	8.5	0.96	8.2	1.62
Carbon	11.1	0.66	7.3	1.61
Copper	8.5	0.79	6.7	1.39
Cobalt	8.9	0.25	2.2	0.86
Aluminium	6.1	0.35	2.1	0.81
Supermirror (Ni/Ti)	—	—	—	1.7–6.1

[a] 1 mrad = 3.44 minutes of arc.

It is necessary to distinguish between *surface flatness* and *surface finish* to understand the processes which diminish the effectiveness of a surface in reflecting neutrons. Flatness refers to a macroscopic variation in the angle of the surface over several centimetres. Finish represents microscopic variations over fractions of a millimetre, and these variations can be reduced by lapping or polishing. A parabolic telescope mirror, for example, has a high degree of finish but is not flat; on the other hand, a ground-glass screen has a poor finish but is quite flat.

This difference in scale between surface flatness (macroscopic surface defects) and finish (microscopic surface defects) is related to the penetration depth of a neutron wave packet which is incident on the surface. The amplitude $a(x)$ of the wave decreases with depth x according to the relationship

$$a\,(x) = a\,(0)\exp\left(-\frac{x}{d}\right) \tag{5.4}$$

where $a(0)$ is the amplitude at the surface and d is the $1/e$ penetration depth for a particular glancing angle θ. If the glancing angle is small, then d is given by

$$d = \left[4\pi Nb\left(1 - \frac{\theta^2}{\theta_c^2}\right)\right]^{-1/2} \tag{5.5}$$

At $\theta = 0$ the penetration depth for nickel is 95 Å, and it is not until the glancing angle is within 5% of the critical angle that the penetration depth reaches 1,000 Å. In practice, a 2,000 Å (or 0.2 μm) nickel layer ensures a reflectivity, averaged over all incident angles, which is close to unity. Thus we may set our arbitrary boundary between flatness and finish at a length scale of 2,000 Å, and this corresponds roughly to the finish obtained by polishing or mechanical lapping.

For straight neutron guides the gain in intensity at a distance L from the neutron source is proportional to the solid angle transmitted by the guide divided by the solid angle subtended by the source at the sample position. The solid angle subtended by a flat source of side s at a distance L is

$$\Omega_s = \frac{s^2}{L^2} \tag{5.6}$$

whereas the solid angle transmitted by a fully illuminated guide is $4\theta_c^2$. Hence the intensity gain factor g achieved by using a perfect guide instead of a non-reflecting collimator is

$$g = \left(\frac{2\theta_c L}{s}\right)^2 \tag{5.7}$$

For a nickel guide (see Table 5.1) $\theta_c = 0.0017$ rads at 1 Å, and so

$$g = 1.16 \times 10^{-5}\left[\frac{\theta L}{s}\right]^2 \tag{5.8}$$

For a guide of length 100 m viewing a square moderator of side 10 cm, g is 11.6 at a wavelength of 1 Å. Note, however, that this gain in intensity is only obtained at the expense of an increase in the beam divergence at the sample position.

A straight guide views the neutron source directly, but by curving the neutron guide slightly, the direct paths of fast neutrons and gamma rays can be blocked. Provided there is no line-of-sight along the guide, neutrons arriving at the sample position must have undergone at least one reflection. The minimum guide length L_0, which is required to ensure that there is no line-of-sight through the guide, is given by

$$L_0 = \sqrt{8aR} \tag{5.9}$$

where a is the width of the guide in the plane of curvature and R is the radius of curvature of the guide. Such a curved guide is known as a 'line-of-sight guide' see Fig. 5.3). The guide exit for a line-of-sight guide is displaced from the direct line by just four times the guide width.

For a line-of-sight neutron guide, a characteristic angle γ^* can be defined as the angle between the line-of-sight and the tangent to the guide at its entrance or its exit. This angle is given by

$$\gamma^* = \frac{L_0}{2R} = \sqrt{\frac{2a}{R}} \tag{5.10}$$

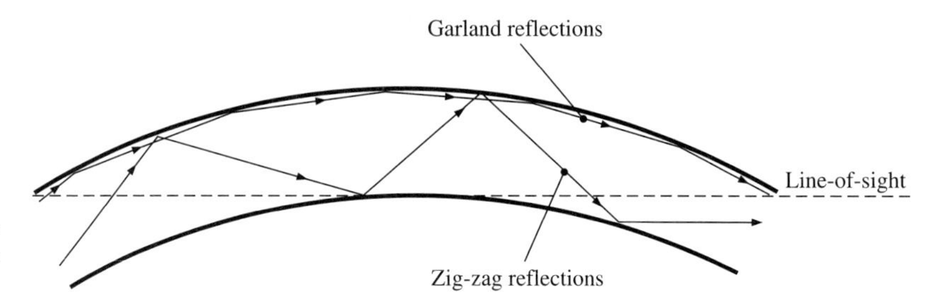

Fig. 5.3 Garland and zig-zag reflections in a line-of-sight guide. The broken line is the limiting 'line-of-sight'.

Fig. 5.4 Guide tube assembly at the ILL, as viewed from above.

From eqn. (5.3) we can define a characteristic wavelength λ^* for the guide corresponding to the angle γ^*. Thus $\lambda^* = \gamma^*/0.0017$ for a nickel-coated guide. λ^* has the following physical significance: for wavelengths less than λ^* transmission through the guide is by successive garland reflections along the outside surface of the neutron guide only, whereas for wavelengths longer than λ^* transmission can also take place by zig-zag reflections across the two walls (see Fig. 5.3). The characteristic wavelength λ^* is sometimes referred to as the guide cut-off, implying that no neutrons of shorter wavelength will be transmitted. Strictly speaking, this is not true but it is nevertheless useful to designate λ^* as the wavelength at which the guide begins to become ineffective.

The radius of curvature of neutron guides is several kilometres, and so the curvature of sections 100 m long is imperceptible. This makes it possible to approximate the arc of the curve with straight sections of guide, each 1 m long. The angle between successive 1 m sections is very much less than the critical glancing angle, and the transmission of the guide is only slightly less than that of a strictly circular curved guide.

Figure 5.4 shows the guide tubes leading from the reactor core to the experimental hall at the ILL. The guide tubes are in two bunches: the first bunch views the thermal moderator of the reactor and the second bunch views the liquid-hydrogen cold moderator.

5.1.1 Supermirrors

The search for more effective surface coatings for neutron guides has led to the development of multilayer surfaces for monochromatizing and polarizing long-wavelength neutrons. This work was pioneered by Schoenborn *et al.* (1974) and by Saxena and Schoenborn (1977). The basic idea is that an artificial one-dimensional crystal is formed by sequentially evaporating alternate layers of two materials on to a glass surface. Particularly suitable materials are ^{58}Ni and Ti which give rise to a large

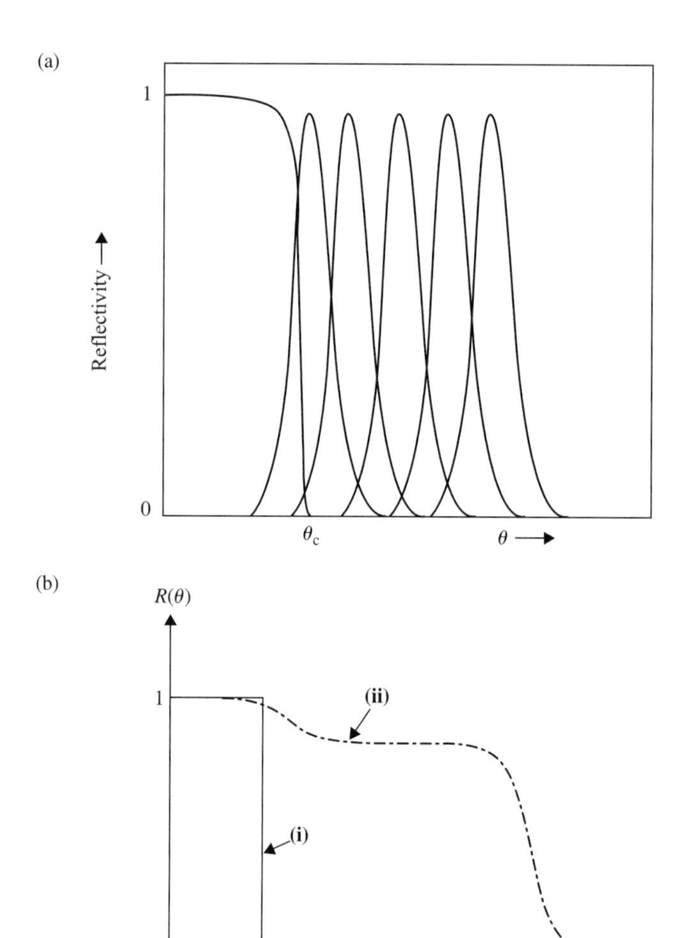

Fig. 5.5 (a) Reflectivity of bilayers with different layer spacings. (b) Reflectivity curves for (i) total reflecting mirror and (ii) supermirror. (*After* Mezei, 1976.)

contrast in scattering density between adjacent layers. The sputtering technique is a process whereby atoms in a solid target are ejected into the gas phase due to bombardment of the material by energetic ions. It has achieved particular importance in the manufacture of ultra-high density computer hard disks, exploiting the giant magnetic resistance (GMR) effect for which Peter Grünberg and Albert Fert were awarded the 2007 Nobel Prize in physics. Uniformly sputtered films of thickness up to a few hundred Ångstroms can be deposited over large areas: Bragg reflection from multilayers of large d-spacings takes place close to grazing incidence, and so long films are required to cover the beam width.

Mezei (1976) has pointed out that the performance of multilayer systems is improved by depositing the bilayers with a gradually increasing d-spacing. The overall effect is to cause the Bragg peaks from multilayers with different spacings and Bragg angles to overlap, as shown schematically in Fig. 5.5(a), where the angle θ for an individual peak is given by $\sin\theta/\lambda = 1/2d$. In this way the reflectivity is maintained at acceptable levels for an angle θ up to four times the critical angle θ_c for a simple mirror: Fig. 5.5(b) contrasts the reflectivity curves for a totally reflecting mirror and for a *supermirror*. Up to 600 layers are required to achieve this performance. Thus, substantial flux gains are achieved by replacing nickel guides with Ni/Ti supermirrors which accept a wider divergence of the primary incident beam. Supermirrors can also be used as neutron polarizers (see Chapter 9). Since the original suggestion of Schoenborn *et al.* (1974) the application of multilayers in neutron optics has increased immensely and this has led, in turn, to significant improvements in deposition techniques. The methods for depositing multilayers for neutron mirrors are reviewed by Anderson *et al.* (1996) and Böni (1997); Schärpf (1991) has described their wide range of applications.

5.1.2 Guides on reactors and pulsed sources

The construction of neutron guides on reactor sources and on accelerator sources is very similar. However, there are two obvious differences in performance. First, the moderator size on a reactor is quite large, as the neutron source comprises the whole moderator surrounding the reactor core. On accelerator-based neutron sources the moderators are relatively small, as their only purpose is to slow down the neutrons and not to maintain the production of neutrons via a critical reaction. In general, a large moderator on a pulsed source gives poorer coupling to the target and lower neutron brightness on the moderator surface. Therefore, guides on accelerator sources must be more accurately aligned to the smaller moderator and remain more stable in time, and more attention must be paid to ensuring that the entrance of the neutron guide is close enough to the moderator to ensure complete illumination. This difference is less apparent when we consider cold neutron sources, which are limited in size on both reactors and accelerators. Even so, the ILL cold source has a cross-sectional area of 600 cm^2 compared with the area of the cold moderator on ISIS of just 100 cm^2, and the entrance of the neutron guide needs to be less than 2 m away from the moderator face for full illumination with cold neutrons particularly at the longest wavelengths.

The second difference between the use of a neutron guide on a reactor and a pulsed source is related to the mode of operation of the source and of the instruments. A reactor is continuous in operation, so that the source-related parameters of interest are the neutron intensity at the guide exit and its spectral variation. To a first approximation, distance from the source is immaterial. For example, a powder diffractometer at the end of a guide tube a distance of 150 m from a reactor source will operate just as effectively as if it were only 30 m from the reactor, apart from a flux loss of perhaps a factor of two. Instrument positioning along neutron guides on a reactor is therefore quite flexible. On a pulsed source this is not so. At 150 m the resolution of a pulsed beam is five times better than at 30 m, and so a diffractometer at 150 m would have a considerably higher performance than that at 30 m. (Because of frame overlap the wavelength band available at 150 m is five times narrower however than that at 30 m.)

The siting of instruments along guides on pulsed sources requires, therefore, more careful consideration than it does on reactors. This does not just apply to white-beam instruments where the sample sits in the incident beam (and would require a much-prized end-of-guide position on a reactor), but also to instruments using a monochromatized beam. As the resolution on pulsed source instruments is measured in the time domain, the selection by a monochromator of a beam of wavelength λ and width $\Delta\lambda$ is observed as a time width Δt at a time of arrival t given by

$$\frac{\Delta t}{t} = \frac{\Delta\lambda}{\lambda} \tag{5.11}$$

Thus,

$$\Delta t\,(\mu\text{sec}) = 252.82L\ (\text{m})\ \Delta\lambda\ (\text{Å}) \tag{5.12}$$

and, whilst the wavelength width selected is independent of the distance from source to monochromator, its time width increases with distance L. Clearly, the siting of the instrument is very dependent on the required performance and on the resolution characteristics of the source.

5.2 Collimators

5.2.1 Soller slit collimators

Better instrumental resolution can be obtained by reducing the angular divergence of the primary or scattered neutron beam. This can be done by limiting the beam with defining apertures, but the method most frequently used is the one first used by Soller (1924) for the collimation of X-rays.

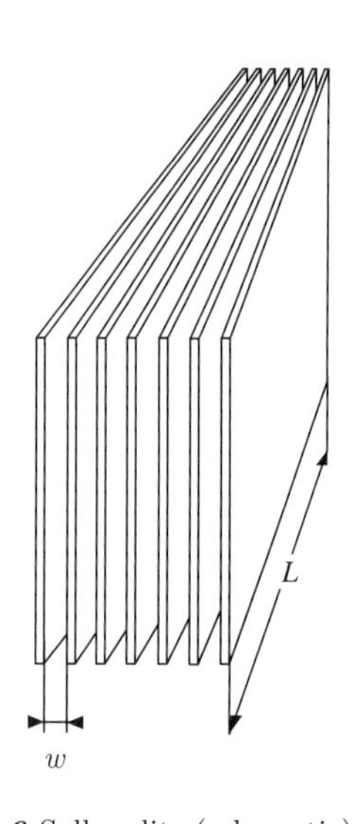

Fig. 5.6 Soller slits (schematic).

A Soller collimator consists of a series of narrow parallel channels separated by thin absorbing blades (see Fig. 5.6). The collimator thus consists of transparent slots and opaque slats sometimes referred to as septa. The divergence α of the collimator is defined as

$$\alpha = \frac{w}{L} \tag{5.13}$$

where w is the width and L the length of the slot. A Soller collimator effectively produces a stack of overlapping beams of narrow divergence. In this way the angular deviation of the overall beam is reduced, but the area of source viewed and the area of sample illuminated remain approximately the same, so that the intensity at the sample is unchanged.

The efficiency of Soller slit collimators depends upon the following factors: (i) the uniformity of the blade spacing throughout the length of the collimator; (ii) the thickness of the absorbing blades; (iii) the straightness of the blade edges at the entrance and exit to the collimators; and (iv) the macroscopic neutron capture cross-section of the septa material. These factors must be optimized for high transmission and a good signal-to-background ratio.

Early Soller collimators for neutrons used self-supporting steel plates as the septa, and some versions of these robust devices had transmissions less than 25% of the theoretical maximum. Greatly improved collimators were developed later by Meister and Weckermann (1973). They used septa of model-aeroplane paper coated with varnish and boron-loaded paint, which, on drying, caused the septa to contract. The septa were held between clamps and became quite taut and flat, giving transmissions up to 85% of the possible maximum. Further improvement was achieved by applying a technique first used in the production of spark chambers for high-energy particle detectors. Stretched mylar film (polyethylene terephthalate) of 25 μm thickness is glued on to aluminium frames and painted with epoxy resin loaded with gadolinium oxide, Gd_3O_4. These blades are stacked up to the desired beam area and bolted on to side plates.

When the whole collimator is assembled, the ends of the frames are cut off leaving a finished collimator. By substituting ^{10}B for the Gd_3O_4 in the absorber paint the useful energy range of the collimator is extended up to 1 eV. (^{10}B is a more efficient neutron absorber than gadolinium at energies greater than 100 meV.)

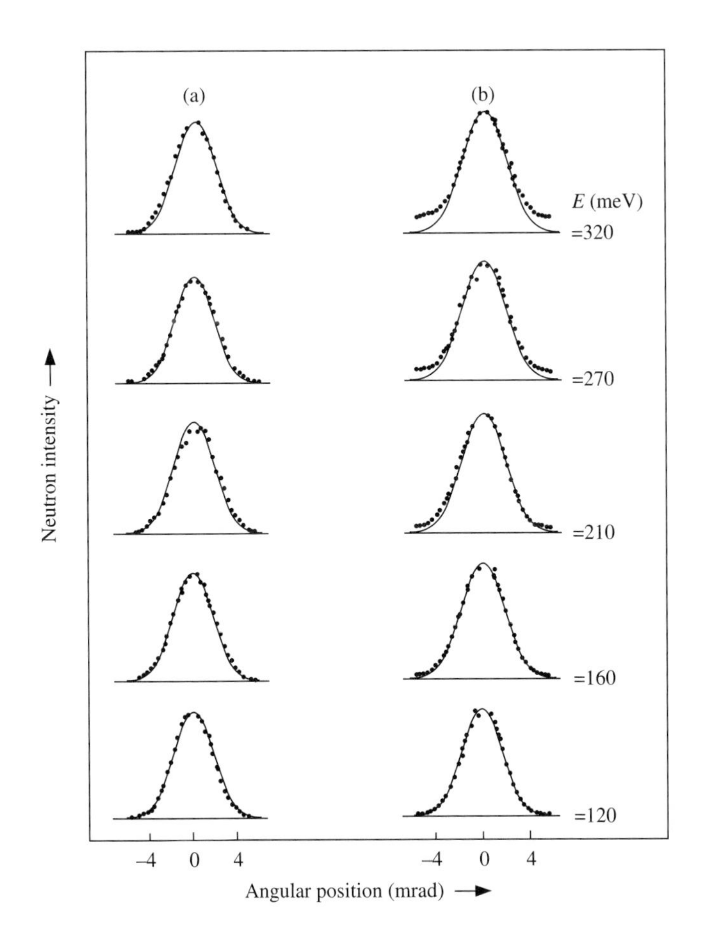

Fig. 5.7 Rocking curves showing the intensity transmitted by mylar collimators coated with (a) ^{10}B, and (b) Gd_3O_4. (*After* Carlile *et al.*, 1978.)

Measurements of the rocking curves of one collimator against another are used to determine the collimator transmission. The data shown in Fig. 5.7 illustrate the increase in transparency in the wings of the Gd_3O_4 collimators at higher energies. At thermal energies the transmission of these collimators is better than 95% of the theoretical maximum.

By applying the neutron guide principle to the Soller collimator, it is possible to produce a 'beam bender'. Layers of nickel 2,000 Å thick are evaporated on to thin sheets of flexible glass and stacked together like a Soller collimator into a curved array. Since the separation between adjacent blades is small, a line-of-sight is achievable in a relatively short distance and the beam may be bent by several degrees. This is particularly useful when it is necessary to shift the beam away from the straight-through position and there is not sufficient distance to install a long neutron guide. A bender has been used on the small-angle scattering instrument LOQ at ISIS. Similar devices, using multilayers of magnetic and non-magnetic materials, have also been used as polarizers and analysers on the D7 instrument at the ILL.

5.2.2 Radial collimators

The purpose of a radial collimator (see Fig. 5.8) is not to improve the instrumental resolution but instead to reduce the background in those experiments where large area detectors are used, such as a photographic plate or a position-sensitive detector. This device is based on a Venetian blind principle with the blades of the collimator radiating from the scattering centre towards the detector (Wright *et al.*, 1981). The volume of sample which is both illuminated by the incident beam and seen by the detector is limited to a region close to the scattering centre. Neutrons scattered from materials surrounding the sample cannot reach the detector. The collimator is particularly effective when the detector has a wide angular coverage, or the sample is contained in a substantial container such as a high-pressure cell or when small 'gauge volumes' are probed in studying microscopic strain in mechanical components (see Chapter 8). In this mode of operation the detector is shadowed by each blade of the

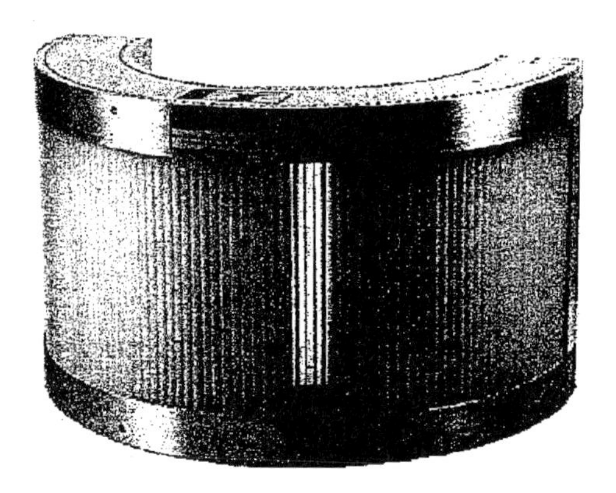

Fig. 5.8 Radial collimator (courtesy Euro Collimators).

collimator, but this problem can be overcome by oscillating the collimator slowly about a vertical axis centred on the sample. Provided that the amplitude of oscillation exceeds the angular divergence of the slots and also that the angular velocity of oscillation is constant, the effect of shadowing is completely masked.

5.3 Beam filters and resonance foils

It is sometimes possible to improve the signal-to-noise ratio on instruments by interposing filters in the incident or scattered beams. These filters operate either by absorbing unwanted components or by scattering them out of the direct path of the neutron beam. They can be designed to remove neutrons over a narrow or a wide energy range.

Let us first consider those filters which operate by scattering. For a polycrystalline filter, neutrons of a given wavelength will be scattered by the crystallites in the filter with orientations satisfying the Bragg condition. As all orientations are present in a polycrystalline material, all wavelengths will be Bragg reflected out of the beam provided that the filter is sufficiently long. However, there is a maximum wavelength λ_{max} beyond which there is no Bragg scattering, and this is given by

$$\lambda_{\text{max}} = 2d_{\text{max}} \tag{5.14}$$

where d_{max} is the maximum plane spacing in the material. This is known as the Bragg cut-off wavelength (or simply Bragg cut-off) for the filter. The cut-offs for various polycrystalline materials, of which the most commonly used is beryllium, are shown in Fig. 5.9. Beyond the Bragg cut-off the material can be nearly transparent to neutrons. Such a filter can be used, therefore, to remove thermal and fast neutrons from an incident beam of cold neutrons or to act as a coarse energy analyser in a scattered beam. It can also be used to remove higher orders of Bragg scattering from a cold-neutron beam which has been monochromated by a single crystal.

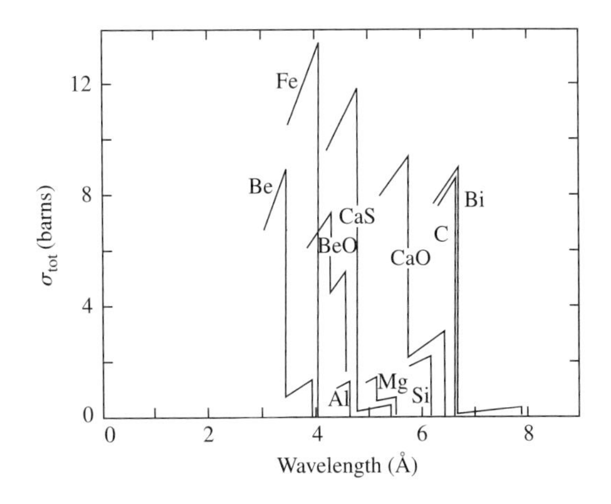

Fig. 5.9 Wavelength dependence of the total scattering cross-section of polycrystalline materials, showing their cut-off wavelengths. Beryllium cuts off at 3.9 Å and graphite at 6.7 Å. (*After* Iyengar, 1965.)

A suitable material for use as a filter must have a low absorption cross-section, a low incoherent scattering cross-section and a high coherent scattering cross-section. Also the operational temperature of the filter must be low compared with its *Debye temperature* so that thermal diffuse scattering (*i.e.* inelastic scattering by thermal vibrations) is reduced to a minimum. The materials most commonly used for polycrystalline filters are beryllium, beryllium oxide and graphite. The relevant properties of some filter materials are shown in Table 5.2.

In designing a filter to remove thermal neutrons, the length of the filter must be carefully optimized. If it is too short, the transmission of short-wavelength neutrons is too high; if it is too long, the transmission of long-wavelength neutrons is reduced too much by inelastic scattering. In practice, beryllium filters are between 15 and 30 cm long, and by cooling them to liquid-nitrogen temperature there is a reduction in thermal diffuse scattering and an increase in the transmission of neutrons beyond the Bragg cut-off. Cooling to temperatures lower than 77 K only marginally improves the transmission of beryllium filters, whereas graphite filters show enhanced performance even down to 4 K.

Next we consider single-crystal filters. The commonest one is pyrolytic graphite which has a layered structure with orientational disorder in the planes of the layers. The transmission for such a filter, in which the incident beam is normal to the layers, is shown in Fig. 5.10. (These measurements were made in 1974 using the Harwell Linac.) Long-wavelength neutrons are still removed by scattering over narrow

Table 5.2 Properties of common filter materials.

	Be	BeO	Fe	Bi	Si	C (graphite)
σ_{coh} (barns)	7.63	5.93	11.22	9.15	2.16	5.55
σ_{inc} (barns)	0.002	0.001	0.40	0.008	0.004	0.001
σ_{abs} (barns)	0.008	0.004	2.56	0.034	0.171	0.004
Debye temperature (K)	1,440	1,553	470	119	645	[a]
λ_{max}(Å)	3.96	4.38	5.08	8.44	6.27	6.70

[a] A single Debye temperature cannot describe the properties of graphite as it is a layered material with very different properties within and perpendicular to the layers.

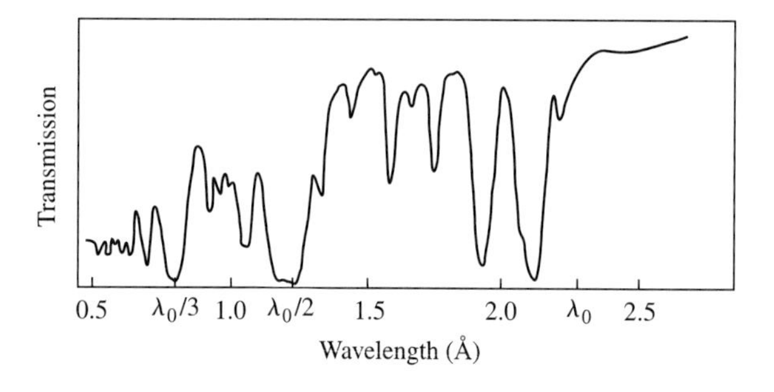

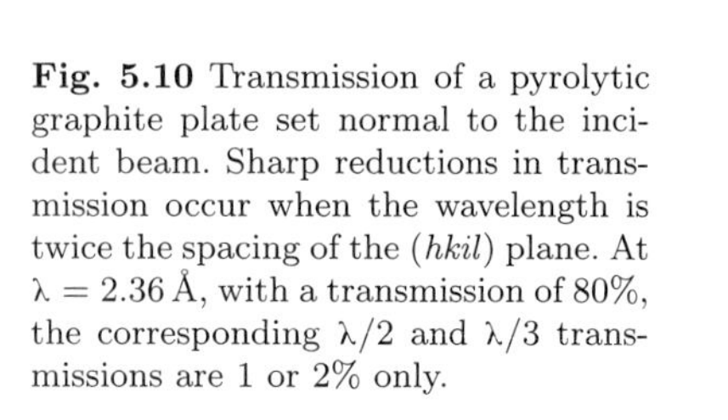

Fig. 5.10 Transmission of a pyrolytic graphite plate set normal to the incident beam. Sharp reductions in transmission occur when the wavelength is twice the spacing of the (*hkil*) plane. At $\lambda = 2.36$ Å, with a transmission of 80%, the corresponding $\lambda/2$ and $\lambda/3$ transmissions are 1 or 2% only.

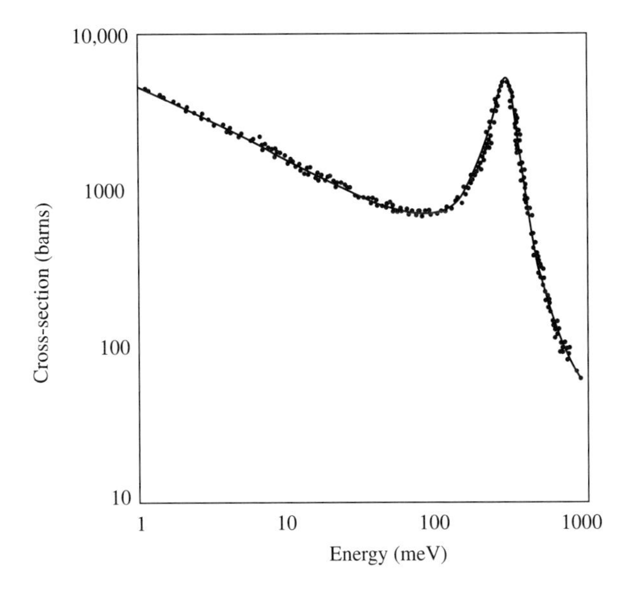

Fig. 5.11 Resonant absorption of ^{239}Pu. The cross-section rises to a sharp maximum at around 400 meV.

wavelength bands, but at short wavelengths these bands overlap considerably as more reflections are excited. In operating a crystal monochromator at $\lambda_0 = 2.36$ Å, the second-order ($\lambda_0/2$) and third-order ($\lambda_0/3$) neutrons are significantly suppressed. With a filter of thickness 8 cm this discrimination is achieved with a reduction of only 30% in the intensity of the primary beam.

Single-crystal silicon is a very effective bandpass filter whose cross-section is dominated at low energies by absorption and at high energies by thermal diffuse scattering. At high energies the attenuation is large whilst low-energy neutrons are still transmitted efficiently: the minimum in the cross-section curve occurs at increasingly higher energies as the temperature is lowered, because of the progressive reduction in thermal diffuse scattering. Sapphire or quartz are materials used for this purpose.

Another kind of filter is based on the resonant absorption process. Many materials exhibit sharp increases in their absorption cross-section at high energies due to the excitation of nuclear resonances, mainly in the eV to keV energy range. ^{239}Pu, however, has a strong broad resonance at 400 meV and can be used to eliminate the second-order reflection (wavelength 0.53 Å) from an incident beam employing a monochromator set to 1.06 Å. The cross-section of ^{239}Pu is shown in Fig. 5.11. Its resonant absorption cross-section is so high that a foil only 100 μm thick will reduce the intensity of the 0.53 Å neutrons by two orders of magnitude compared with the intensity of the 1.06 Å neutrons (of energy 72 meV).

Choppers and crystal monochromators are ineffective at electron volt energies, and so resonance foils provide the best method of selecting particular energies in the field of high-energy spectroscopy (see Chapter 15). As the foils used for this purpose remove a monochromatic band from the spectrum—in contrast to a monochromator which selects the monochromatic beam—at least two measurements are needed, one with the filter in place and one after removing the filter: the difference gives the response of the sample to the monochromatic neutrons at the resonant energy.

Cadmium is commonly used as a shielding material (particularly around samples, monochromators and detectors), as it exhibits a broad low-energy resonance at 178 meV which merges with a $1/v$ absorption cross-section at lower energies. Its use is so common that it is not always realized that at energies above 500 meV it is practically transparent to neutrons. Cadmium is therefore an extremely effective thermal neutron absorber but is ineffective for fast neutrons. Care must be taken in using it as a shielding material on pulsed-neutron sources because of the hard spectrum of the incident beam. When cadmium is used to stop cold neutrons, the radiological hazard from the hard gamma rays arising from the (n,γ) reaction can be remarkably severe. Where detectors are slightly gamma sensitive, as in photographic image plates, borated shielding must be used.

5.4 Selection and measurement of neutron energies

There are three principal methods of determining the energy of a neutron: from its velocity, its wavelength, or its Larmor precession frequency.

A **neutron's velocity** v is related to its energy by the kinetic energy equation $E = \frac{1}{2} \cdot m_\mathrm{n} v^2$. The velocity is derived by measuring the neutron flight time over a known distance. This flight time is accessed routinely on pulsed-neutron sources and also on continuous-neutron sources with mechanical choppers or rotating crystal monochromators. This method of determining the energy utilizes the particle nature of the neutron.

The **neutron wavelength** λ can be measured using a crystal monochromator. The energy of the neutron, which is derived from the de Broglie relationship between momentum and wavelength

$$\lambda = \frac{h}{m_\mathrm{n} v} \tag{5.15}$$

is given by

$$E = \frac{h^2}{2 m_\mathrm{n} \lambda^2} \tag{5.16}$$

or approximately by

$$E\ (\mathrm{meV}) = \left(\frac{9}{\lambda(\text{Å})} \right)^2 \tag{5.17}$$

This method of determining the energy utilizes the wave nature of the neutron.

The **Larmor precession rate** ω_{L} of the neutron in a magnetic field is given by

$$\omega_{\mathrm{L}} = 2.196 \text{ kHz/gauss} \tag{5.18}$$

By measuring the number of precessions of a neutron over a known distance in a constant field, its energy is very accurately determined. Such a measurement can be carried out using the method of *neutron spin echo*.

Of the three techniques, the measurement of neutron velocity and neutron wavelength can be undertaken with about the same precision. The measurement of the Larmor precession rate by neutron spin echo, on the other hand, is capable of providing 2–3 orders of magnitude higher precision (see Chapter 16). In this chapter we shall be concerned only with the first two techniques.

5.4.1 Choppers

Helical velocity selector

A neutron beam can be monochromated and pulsed with the aid of one or more mechanical choppers. Conceptually, a cylindrical drum of radius r and length l rotates about its axis with a frequency ω, and the incident neutron beam follows a path parallel to the axis of rotation. The drum is fabricated from a neutron-absorbing material and has a helical slot cut along its cylindrical surface. Only neutrons of one velocity will pass through the channel without striking either wall of the slot. This neutron velocity is given by

$$v = \frac{\omega l}{\phi} \tag{5.19}$$

where ϕ is the turn angle of the helix over the length of the velocity selector. Changing the speed of the chopper changes the neutron energy selected. For a parallel incident beam and a slot of width s, the uncertainty Δv in the transmitted velocity is

$$\frac{\Delta v}{v} = \frac{2s}{r\phi} \tag{5.20}$$

Typically, resolutions of 10% are obtainable using helical velocity selectors, which are coarse monochromators. A simpler construction is possible using a cylinder with a straight slot, but with the cylinder axis turned at an angle to the incident beam. This has the added advantage that the resolution of the monochromatic beam can be varied by changing the angle of the cylinder axis with respect to the beam axis.

For the helical velocity selector the speed required at the circumference to monochromate high-energy neutrons approaches the breaking limits of most materials. Such a device, therefore, tends to be used only for slow neutrons and finds particular application in the selection of long-wavelength beams for small-angle scattering instruments. In a practical velocity selector, the cylinder is not solid but is composed of a number

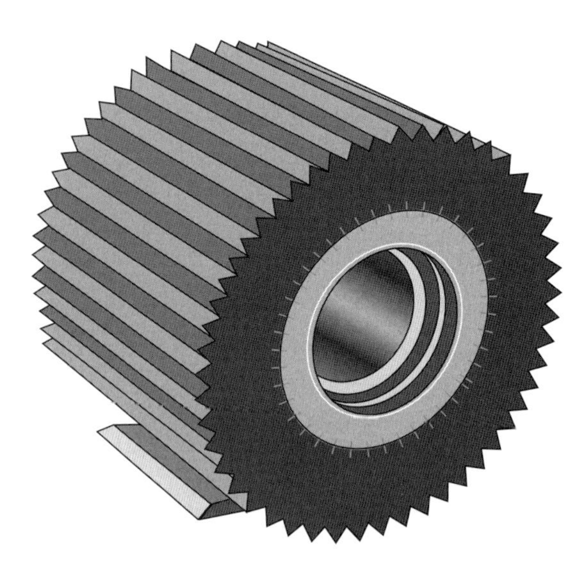

Fig. 5.12 Helical velocity selector.

of equidistant discs mounted on the same axle (see Fig. 5.12). Each disc contains a number of slots at regular intervals around the circumference, and each slot is in line with its equivalent slot in the next disc. In this way, a continuous monochromatic neutron beam can be selected with a relatively high efficiency. The transmission of such a chopper depends upon what is known as the 'mark-to-space ratio', that is, the ratio of open area to total area. Other designs employ carbon-fibre blades rather similar to the blades in jet turbines which combine lightness with strength.

A multiple-disc velocity selector will also transmit longer-wavelength harmonics of the fundamental neutron wavelength. This is caused by slower neutrons being transmitted by slots $1, 2, 3, \ldots$ in successive discs. This effect can be removed by siting one or more of the discs at a non-equidistant location along the axle and adjusting their orientation accordingly. Typically, a helical velocity selector runs at a rotational frequency of 100 Hz and can be up to 1 m long.

Disc choppers

In a disc chopper, the beam of neutrons passes through a slot cut parallel to the axis of rotation and near the periphery of the disc. The chopper can be considered to be a single blade of the helical selector. Operating singly, the disc chopper produces a polychromatic pulse of neutrons with a pulse width determined by the time taken by the slot in the chopper to traverse the neutron beam. This traverse time, or scan time, is usually not less than 40 μsec and so the single-disc chopper is not a high-resolution device. It has particular application in limiting the *frame overlap* on pulsed source instruments where the aperture on the circumference of the disc can be quite wide.

With a pulsed source of period τ, neutrons are generated in the moderator at times $0, \tau, 2\tau, 3\tau, \ldots$ and the neutrons of different energy disperse as they travel towards the instrument. Frame overlap occurs when fast neutrons from a later pulse catch up with slow neutrons from an earlier pulse. If τ_{min} and τ_{max} are the minimum and maximum flight-times to the detector for the fastest and slowest neutrons selected by the chopper, respectively, then the necessary condition to avoid overlap is

$$\tau_{\mathrm{max}} - \tau_{\mathrm{min}} \leq \tau \tag{5.21}$$

Time and wavelength are linearly related, and the widest wavelength band $\Delta\lambda$ accepted before the overlap occurs is

$$\Delta\lambda \leq \left(\frac{h}{m_{\mathrm{n}} L}\right) \tau \tag{5.22}$$

Note that this wavelength band is independent of the absolute value of the wavelength. Thus for a 50 Hz source and an instrument with a path length of 40 m the band width is 2 Å. The band can cover any 2 Å range (*e.g.* 4–6 Å or 0.5–2.5 Å) and it depends only on the phase of the opening of the disc chopper with respect to the neutron pulse.

Phased multiple disc choppers

When two disc choppers are phased together at a fixed distance apart, they monochromate the incident beam as well as pulsing it in time. By reducing the chopper aperture to the width of the incident neutron beam, the transmission function is precisely triangular. This principle is used in the long-wavelength time-of-flight quasielastic spectrometer IN5 at the ILL, where seven disc choppers are phased together. The separation of the first and final choppers determines the resolution of the incident beam, and the phase difference between these choppers determines the energy selected. The first pair of counter-rotating choppers removes slower harmonics from the beam (*cf.* the helical velocity selector) and the second pair of choppers, which run at a lower speed, removes pulses to avoid frame overlap. A sharp-edged resolution function is a great advantage when looking for line broadening effects, as in measurements of quasielastic scattering (see Chapter 16). This concept was first applied on the NEAT spectrometer at the *HMI*, Berlin, where seven disc choppers are also used to monochromate the beam. This design halves the time width of the neutron pulse for the same transmitted intensity. The new version of IN5 at ILL also adopts the same principle which has increased the luminosity of the instrument by more than an order of magnitude. It has been further increased by the addition of a large area detector array composed of multiple banks of ^{3}He detectors.

Fermi choppers

A Fermi chopper operates with its rotational axis perpendicular to the incident beam. Slots cut across (or close to) the diameter of the chopper, and perpendicular to the axis of rotation, select the pulsed beam (Fig. 5.13). Only neutrons which are fast enough to cross the chopper

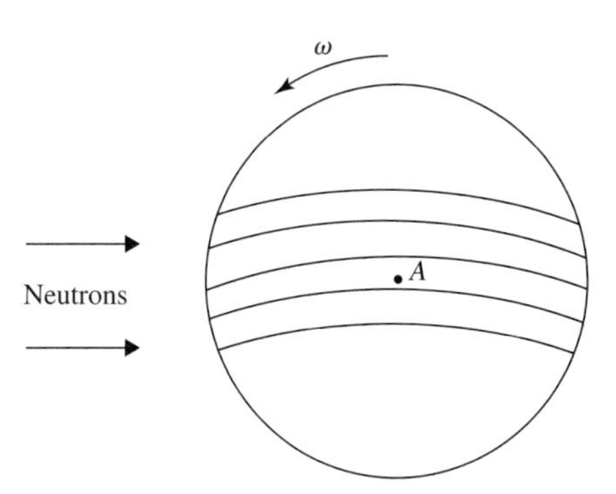

Fig. 5.13 Multi-slotted Fermi chopper. The rotation axis A is perpendicular to the disc.

diameter before the slot closes will be transmitted: the remaining neutrons are absorbed in the slot walls. When the chopper is closed, a large amount of material is placed in the beam, effectively suppressing the background. In the earliest designs of Fermi chopper the slot was simply an open hole through the body of the chopper. Modern designs employ a Soller-type slot package with alternating transmitting and absorbing materials.

Fermi choppers are more compact than disc choppers and consequently can be run faster and achieve shorter pulse widths than disc choppers. They are used for measurements at higher resolution than is possible with disc choppers, but they are technically more complex to operate. Some versions run suspended on magnetic bearings and are beautifully engineered devices, spinning at up to 600 Hz and producing neutron pulses as narrow as 1 μsec.

The transmission function of a Fermi chopper has been derived by Stone and Slovacek (1959) for a straight slot chopping a parallel neutron beam. Very fast neutrons are transmitted unattenuated, but there is a velocity cut-off at slow neutron speeds. This cut-off velocity is

$$v_{\mathrm{co}} = \frac{\omega R^2}{d} \tag{5.23}$$

where d is the slot width, R is the chopper radius and ω is its angular velocity. If we define β as

$$\beta = \frac{v_{\mathrm{co}}}{v} \tag{5.24}$$

then $\beta = 1$ at the cut-off velocity. The transmission function is

$$T(\beta) = 1 - \frac{8}{3}\beta^2, \quad 0 \le \beta \le \frac{1}{4} \tag{5.25a}$$

$$T(\beta) = \frac{16}{3}\sqrt{\beta} - 8\beta + \frac{8}{3}\beta^2, \quad \frac{1}{4} \le \beta \le 1 \tag{5.25b}$$

$$T(\beta) = 0, \quad 1 < \beta \tag{5.25c}$$

This transmission function is plotted in Fig. 5.14 (curve A).

The *burst time* δt_{b} of the chopper is the time for the slot to transmit neutrons arriving from a single point on the moderator surface and is given by

$$\delta t_{\mathrm{b}} = \frac{d}{2\omega R} \tag{5.26}$$

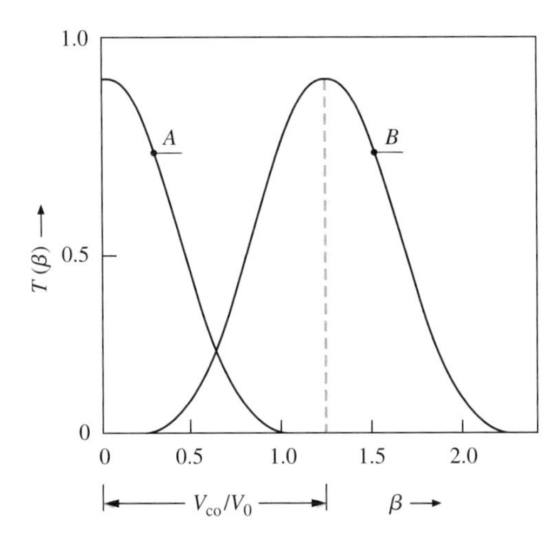

Fig. 5.14 Transmissions for straight slot (A) and curved slot (B) choppers.

If the chopper views a finite neutron source of width s at a distance L, then an additional time width δt_s must be folded with the burst time to give the overall pulse length which is transmitted by the chopper. This is called the *sweep time* or *scan time* and is given by

$$\delta t_s = \frac{s}{2\omega L} \tag{5.27}$$

The neutron traces out a curved path, which is an *Archimedean spiral*, in the frame of reference of the rotating chopper. A straight-slot chopper transmits fast neutrons, but is ineffective in transmitting slow neutrons. If we wish to increase the transmission for slow neutrons, the slot must be curved. Marseguerra and Pauli (1959) have extended the analysis of Stone and Slovacek to derive a transmission function identical to that for the straight slot chopper but with β redefined as

$$\beta = \frac{\omega R^2}{d}\left[\frac{1}{v} - \frac{1}{v_0}\right] \tag{5.28}$$

where v_0 is the velocity for which maximum transmission occurs. Such a chopper acts as a neutron monochromator (see curve B in Fig. 5.14).

Practical examples of Fermi choppers include multi-slotted rotors, converging-slotted rotors and banana-shaped slotted rotors. Choppers with burst times of 1 μsec for 1 eV neutrons have been developed for the spectrometers at *ISIS* using boron-fibre/aluminium laminates for the slots. Increasingly, choppers are built with magnetic suspensions rather than solid or air bearings, as this provides higher reliability and more precise definition of the pulse in high-resolution instruments.

5.4.2 Crystal monochromators

Diffraction from a single crystal provides the simplest method of selecting a monochromatic beam from a white beam. If the white neutron beam is incident upon the planes of a single crystal at a glancing angle θ, then a wavelength λ will be diffracted from the crystal at the same glancing angle. The value of the wavelength reflected depends on the plane spacing d of the crystal according to Bragg's equation

$$\lambda = 2d \sin \theta_B \tag{5.29}$$

The Bragg angle θ_B is equal to one half of the scattering angle. Bragg scattering from single crystals is commonly used on reactor instruments to produce monochromatic beams, and a single crystal after the sample can be used in a similar manner to analyse the scattered beam. Crystal analysers are used more extensively on pulsed source instruments, and on triple-axis and back-scattering instruments operating on reactors.

A crystal monochromator, set to reflect the wavelength λ from the (hkl) family of planes, will also reflect neutrons of wavelengths $\lambda/2, \lambda/3, \lambda/4, \ldots$ from the planes $(2h, 2k, 2l), (3h, 3k, 3l), (4h, 4k, 4l), \ldots$ at the same crystal orientation. These reflections are referred to as second-order, third-order, fourth-order, ... reflections, and the presence of multiple orders is referred to as higher-order harmonic contamination. Attempts to eliminate this contamination have occupied much time and effort by users of neutron instruments. The use of resonant absorbers, *e.g.* ^{239}Pu foils, in this context was mentioned earlier (Section 5.3).

The first materials to be used as neutron monochromators were ionic crystals such as sodium chloride and lithium fluoride, which were available as large, almost perfect, single-crystal ingots. However, the intensity reflected by such crystals is low precisely because of their perfection, so that the crystals give a reflected beam which is narrow in its angular divergence and in the width of its reflected band of wavelengths. The reflected intensity can be increased by mechanically straining the crystal, for instance, by bending or squashing it or by roughening its surface. The

Table 5.3 Properties of common monochromator crystals.

Material	Structure	Lattice parameters (Å)	$\sigma_{inc}/\sigma_{tot}$	σ_{abs} (barns at 1.8 Å)
Aluminium	Face-centred cubic	$a = 4.05$	0.0056	0.23
Beryllium	Hexagonal close-packed	$a = 2.28$ $c = 3.58$	0.0007	0.008
Copper	f.c.c.	$a = 3.62$	0.07	3.8
Lead	f.c.c.	$a = 4.95$	0.0003	0.17
Nickel	f.c.c.	$a = 3.52$	0.281	4.6
Pyrolytic graphite	Hexagonal (layered)	$a = 2.46$ $c = 6.71$	0.0002	0.003
Silicon	Cubic diamond	$a = 5.43$	0.0007	0.17
Zinc	h.c.p.	$a = 2.67$ $c = 4.95$	0.02	1.1

gain from surface roughening is greatest for those materials with a high absorption cross-section, such as LiF, as might be expected for a surface phenomenon. Metallic single crystals, such as aluminium, copper, zinc and lead, give intensities considerably higher (by factors of 2 or 3) than ionic crystals and are in common use today (see Table 5.3). Pyrolytic graphite, which is not a true single crystal but a two-dimensional crystal with a layered structure, is an extremely effective monochromator of longer-wavelength neutrons in the range 2–6 Å. Beryllium is generally accepted as being the ultimate monochromator material for high-energy neutrons but it is technically very difficult to grow in large enough crystals having the appropriate qualities. As Bacon (1975) puts it: 'continued attempts to reproduce the virtues of certain legendary crystals [of beryllium] have not proved successful'.

The inelastic scattering of neutrons by phonons reduces the crystal reflectivity. Thermal diffuse scattering is manifest as a broad background feature, which rises to a maximum under the Bragg reflection and causes a loss of intensity of the monochromatic beam. Beryllium has a Debye temperature θ_D of 1,440 K whereas copper has $\theta_D = 343\,\mathrm{K}$, so that at the same temperature copper generates much more TDS background than does beryllium. The Debye temperature corresponds to the upper limit of the energies of the acoustic phonons in the material, and gives a rough-and-ready way of expressing the range of the phonon spectrum and its ability to be populated at a given temperature. Thus in some cases it is favourable to cool the monochromator or analyser in order to reduce its physical temperature well below its Debye temperature. This is the case with copper for neutrons of high energy and with pyrolytic graphite for low energy, but it is less effective for beryllium because of its relatively high Debye temperature.

A monochromatic beam can be reflected from a single crystal in two different geometries. For *Bragg geometry* the beam is reflected from the crystal surface, and for *Laue geometry* the beam emerges from the monochromator after being transmitted through the surface (see Fig. 5.15). Bragg geometry is better for long wavelength neutrons (large scattering angle) and higher resolution, whereas Laue geometry is suitable for monochromation of short wavelengths (small scattering angles).

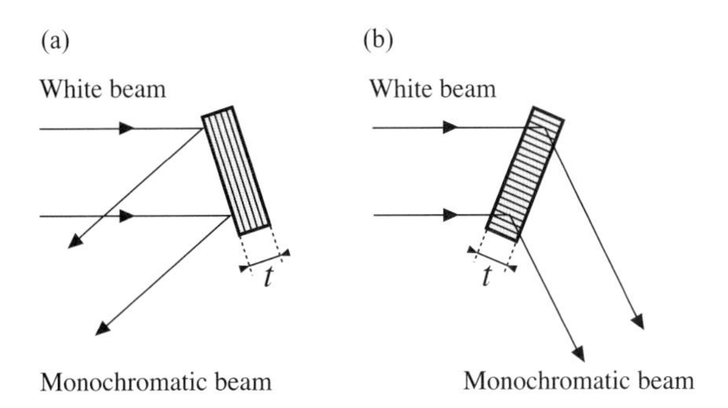

Fig. 5.15 (a) Bragg geometry and (b) Laue geometry.

In the Bragg case the reflectivity of the crystal as a function of thickness t can, in principle, rise to 100%. In practice, this limit is only approached with pyrolytic graphite. In the Laue case, as the thickness of the crystal is increased, the reflectivity rises to a maximum (theoretically 50% for zero absorption) and then falls, on account of absorption.

To summarize, an ideal material for a monochromator or an analyser must satisfy the following conditions:

- A high coherent scattering length density
- A low incoherent scattering cross-section
- A high Debye temperature
- A low absorption cross-section
- A suitable mosaic structure

Mosaic structure refers to the microcrystalline properties of the material. Some crystals, such as silicon produced for the semiconductor industry, are perfect, that is, the atoms in a single crystal are perfectly ordered over distances of centimetres or more. The 'rocking curve' of such a crystal, or the intensity recorded in the reflected beam as the crystal is rotated through the Bragg angle, has an angular width of only a few seconds of arc, and the integrated intensity under this curve is relatively low despite the fact that the peak reflectivity approaches 100%. Perfect crystals do not produce good monochromators but they are ideal for neutron interferometers (see Section 11.3).

Most crystals contain an appreciable density ($\sim 10^{18}/cm^3$) of dislocations and the crystal effectively consists of a large number of small perfect crystallites, each slightly misorientated with respect to one another. This mosaic structure gives rise to a rocking curve whose width is referred to as the *mosaic spread* of the crystal. Mosaic spreads, which range up to many minutes of arc, correspond to the typical angular divergences transmitted by neutron beam tubes and guide tubes. Mosaic crystals are well matched to these instrumental components and so are much better monochromators than perfect single crystals.

Focusing geometries

Neutron scattering is an intensity-limited technique, and the need for higher count rates quickly led to the development of focusing techniques with extended geometries. Single monochromator and analyser crystals can be replaced by extended arrays of monochromators and analysers which focus the neutron beam in space and wavelength. These focusing geometries can produce gains of more than an order of magnitude over a single monochromator.

For crystal arrays, there are two different kinds of focusing geometry, known as vertical focusing and horizontal focusing. Let us consider the triangle formed by the entrance point of the neutron beam A, the crystal B and the exit point of the neutron beam C, as shown in Fig. 5.16(a).

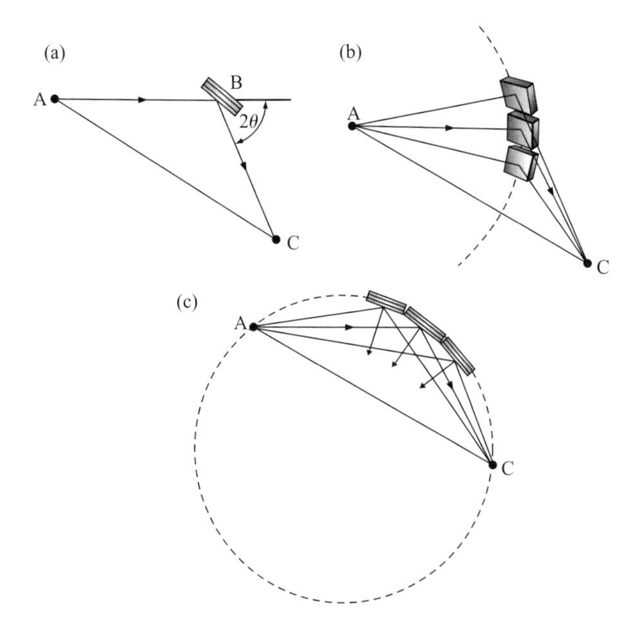

Fig. 5.16 Different focusing geometries for crystal arrays. Diagram (a) refers to one crystal. Diagram (b) shows vertical focusing and (c) shows horizontal focusing. The Rowland circle is indicated by broken lines.

Vertical focusing is achieved by rotating the triangle about the line AC and stacking crystals at different heights above and below the horizontal scattering plane: Fig. 5.16(b). The beam reflected from the crystals sitting on the arc of a vertical circle will still have the same degree of monochromaticity as for the single crystal case and will be focused spatially at the point C. In addition, all paths from A to C will be of the same length and therefore such a geometry would not defocus a time-of-flight instrument. A drawback is that the vertical divergence has increased and in the case of some instruments (*e.g.* the triple-axis spectrometer) spurious effects can be created which can be confused with the real signal (Shirane *et al.*, 2002).

In horizontal focusing the point B [Fig. 5.16(a)] of the crystal is kept in the horizontal plane but is moved along the circumference of the circle which passes through the points A, B, C. The angle subtended at point B remains constant ($180°–2\theta$) for crystals located on this circle. With the normal to the reflecting plane for each crystal set to bisect the angle between the incident and the reflected beam, the beam arriving at point C is still monochromatic and focused in space. This circle is the *Rowland circle*. A drawback of this geometry is the fact that the horizontal divergence of the beam will increase and the path lengths from A to C via B are no longer constant, which is disadvantageous for time-of-flight applications. Monochromator and analyser arrays employing both horizontal and vertical focusing are in use on many instruments; they are rarely operated at a fixed wavelength and so rather sophisticated mechanics to optimize the focus are required when changing the neutron wavelength.

Neutron guides, extended detectors and extended single-crystal arrays are frequently employed. However, increase in signal is not welcome if it brings with it disproportionate increases in background. Extended detectors and extended crystal arrays have open geometries with the potential to increase the background more than the signal. Also large active areas often imply increased beam divergence which can mean poorer resolution. Instruments measuring incoherent scattering processes, where the variation in the signal with scattering angle is less pronounced than for coherent scattering, can benefit particularly by these extended techniques.

References

Anderson, I. S., Schärpf, O. H., Høghøj, P. and Ageron, P. (1996) *J. Neutron Res.* **5** 51–61. "*Multilayers for neutron optics.*"

Bacon, G. E. (1975) "*Neutron diffraction, 3rd edition.*" p. 98. Oxford University Press, Oxford.

Böni, P. (1997) *J. Neutron Res.* **5** 63–70. "*Polarising supermirrors.*"

Carlile, C. J., Penfold, J. and Williams, W. G. (1978) *J. Phys. E: Sci. Instrum.* **11** 837–838. "*High-efficiency Soller slit collimators for 1 eV neutrons.*"

Iyengar, P. K. (1965) "*Thermal neutron scattering.*" Chapter 3 (edited by P. A. Egelstaff). Academic Press, London.

Maier-Leibnitz, H. and Springer, T. (1963) *J. Nucl. Energy* **17** 217–225. "*The use of neutron optical devices on beam-hole experiments.*"

Marseguerra, M. and Pauli, G. (1959) *Nucl. Instrum. Methods* **4** 140–150. "*Neutron transmission probability through a curved revolving slit.*"

Meister, H. and Weckermann, B. (1973) *Nucl. Instrum. Methods* **108** 107–111. "*Neutron collimators with plates of self-contracting foils.*"

Mezei, F. (1976) *Commun. Phys.* **1** 81–85. "*Novel polarised neutron devices: supermirror and spin component amplifier.*"

Saxena, A. M. and Schoenborn, B. P. (1977) *Acta Cryst. A* **33**, 805–813. "*Multilayer neutron monochromators.*"

Schärpf, O. (1991) *Physica B* **174** 514–527. "*Thin-film devices and their role in future neutron spectroscopic investigations.*"

Schoenborn, B. P., Caspar, D. L. D. and Kammerer, O. F. (1974) *J. Appl. Cryst.* **7** 508–510. "*A novel neutron monochromator.*"

Shirane, G., Shapiro, S. M. and Tranquada, J. M. (2002) "*Neutron scattering with a triple-axis spectrometer.*" Cambridge University Press, Cambridge.

Soller, W. (1924) *Phys. Rev.* **24** 158–167. "*A new precision X-ray spectrometer.*"

Stone, R. S. and Slovacek, R. E. (1959) *Nucl. Sci. Eng.* **6** 466–474. "*Neutron spectra measurements.*"

Wright, A. F., Berneron, M. and Heathman, S. P. (1981) *Nucl. Instrum. Methods* **180** 655–658. "*Radial collimator system for reducing background noise during neutron diffraction with area detectors.*"

Sample preparation, sample environment and radiological safety

6

6.1 General considerations

Because the neutron is electrically neutral, it penetrates most substances readily. The exceptions are hydrogenous materials and strongly absorbing materials, such as cadmium, gadolinium and boron. This penetrability not only means that the true internal microscopic behaviour of a solid or liquid can be probed but also that a sample under study can be held within bulky sample environments (*e.g.* cryostats, furnaces, pressure cells or reaction vessels) and, with properly collimated beam geometries, still give good quality data. For this reason neutrons are capable of exploring a wide range of thermodynamic variables. Temperatures from mK to thousands of degrees centigrade; pressures from mbars to Mbars; magnetic fields of 15 Tesla or higher; chemical reaction processes involving reaction vessels at elevated temperatures with gas overpressures; and kinetic processes such as extrusion, flow and shear: all these variables are accessible to neutron scattering.

However, in terms of absolute fluxes of particles, neutron sources are not intense when compared to other radiation sources such as electrons and photons. Not only that, but the source itself is large in extent meaning that brilliance, or brightness, that is intensity per unit source area, is even lower on a relative scale. For this reason, count rates have to be recouped by the use of large samples (often several cm^3 in volume) and large detectors. Together with the requirements for bulky radiation protection, this all adds up to large instruments representing an appreciable capital investment.

With the need for improvement in rates of data collection, the utilization factor of neutron instruments is very high. Much effort is expended in minimizing downtime on instruments. This is manifested in many ways: sample changers to automate the transition from one sample to another; 24 hours a day data collection; rapid changeover from one experiment to another; computer control of parameters to be varied (*e.g.* temperature); and careful experiment protocol. Neutron instruments are rarely left idle, in a similar way to their counterparts in particle physics and astronomy. This is in sharp contrast to other techniques used for the study of condensed matter, such as Raman spectrometry, electron microscopy, X-ray diffraction, NMR, or indeed techniques using synchrotron radiation where set-up time regularly exceeds measuring time.

It is fortunate that the true absorption of neutrons is usually quite small, and so every effort is made to increase the size of the sample, consistent with the problems of multiple scattering, extinction effects, cost or availability. For powder diffraction, a sample is normally a few cm^3 in volume, compared with a few mm^3 or less for X-ray diffraction using a laboratory X-ray source. In general, if a sample can be synthesized in the laboratory, then it can be produced in sufficient quantities for X-ray studies. This is not always true for neutron work: biological samples, isotopically enriched samples or newly discovered materials are often not available in large enough sizes. This is often considered to be a financial problem or a production problem, but there is a cultural aspect to this as well. Biologists, for example, are often not accustomed to handling large samples and the requirements for a successful neutron experiment are sometimes perceived as unattainable.

Financial questions of sample production also need to be put into context. A modern high-intensity neutron source is an expensive facility costing many millions of euros per year to operate. With perhaps 25 instruments the reader will realize that for 200 days of user operation a year, one day's measuring time on an instrument costs perhaps 10,000 euros. An average experiment lasts 4 days, making 40,000 euros for the direct cost of the experiment. In an ideal world, it would not therefore be disproportionate for the experimenter to spend 10% of this sum on the sample, say 4,000 euros, but this is probably much higher than what is actually expended in a large majority of cases. Furthermore, it is sometimes said that neutrons should only be used as a last resort, because they are so expensive when compared with X-rays. This is no longer true with the advent of synchrotron sources of radiation, where beam time is comparable in cost to neutron beam time.

6.2 Containment of sample and ideal sample sizes

There is a bewildering variety of samples encountered in neutron scattering, ranging from high-T_c superconductors to graphitic intercalation compounds that absorb molecular gases, from extruded polymer blends to biological polysaccharide gels. Each poses its own problem in considering appropriate sample support, containment and environment. Some samples are studied at high temperatures and some at high pressures. Others need to be orientated continuously through a solid angle of 4π or examined in a magnetic field. Still more, samples such as surfactants on liquid surfaces must be kept free from vibrations or maintained under a constant humidity atmosphere and in conditions of extreme cleanliness.

A large number are studied at low temperatures: neutron scattering experiments at liquid helium temperature (4.2 K) are quite routine, but evermore samples are being studied at dilution refrigerator temperatures of 30 mK where subtle magnetic ordering effects, for example, are observed.

Neutrons are ideal probes when these types of sample environment are required. It is more common that a sample is studied at 4.2 K, rather than at room temperature, and it is unusual nowadays to find a sample being investigated at ambient conditions. A neutron scattering institute will, therefore, have arrays of cryostats, furnaces, pressure cells, water baths and magnets. These devices will usually permit automatic scanning through the desired thermodynamic parameter. This parameter is maintained and controlled electronically, and its values are recorded by computer in experimental runs started and ended automatically according to a pre-programmed sequence.

The sample and the sample environment are the kernel of a good scattering experiment, and as much care must be devoted to them as to the instrumentation recording the response to the neutron beam. Very few samples are self-supporting solids, with single crystals, metallic alloys or polymer plates being the exception. The majority of samples must be placed in a container in order to define a precise geometry for the experiment, or to minimize multiple scattering, or to avoid chemical degradation by contact with air or water vapour. Indeed, containment may be necessary to protect the experimentalist from the toxic nature of the sample itself or its induced radioactivity after irradiation in the neutron beam.

In general, there are two kinds of materials that are used to contain samples: those which scatter predominantly incoherently such as vanadium, and those which scatter predominantly coherently such as aluminium. Strongly scattering materials (including those containing hydrogen), strongly absorbing materials and materials such as copper (which are activated in the neutron beam) are to be avoided. Equally, when magnetic fields or polarized neutrons are used, ferromagnetic materials are to be avoided and, in such cases, not only sample holders but also the instruments themselves are often constructed of totally non-magnetic materials.

Single-crystal samples are normally supported on small goniometer heads. Such goniometers have x and y linear translation slides and two orientational arcs whose centres of rotation coincide with the crystal centre. Increasingly, single-crystal samples are aligned offline on parasitic beams prior to the experiment. Typical crystal sizes range from 0.5 mm^3 to several cm^3. Crystals at the lower end of this range can be tolerated for diffraction studies, but larger crystals are needed for measurements of phonon dispersion relations where the inelastic signal is perhaps two orders of magnitude less than the elastic signal.

While self-supporting polycrystalline samples are often held in the beam by shielded clamps without any container, powder samples must be contained in a holder. For diffraction studies, where the Bragg reflections are sharply peaked across the whole measuring range, the sample holder is made from an isotropically scattering material giving rise to a flat background, which can be easily subtracted from the signal. Vanadium, almost totally spin incoherent, is often used. There are advantages in using so-called *null matrix alloys* such as TiZr where, by balancing negative and positive scattering lengths, coherent scattering can be eliminated. TiZr alloy shows a perceptible oscillation in $S(Q)$ due to short-range ordering, but this is overcome in VNb alloy (5.1% Nb) where the scattering pattern is almost featureless. For spectroscopic studies, where inelastic scattering signals are relatively weak, it is important not to flood the signal with high flat backgrounds. Accordingly, it is customary to use thin sample holders which scatter coherently and which are often made from aluminium.

For high temperature work, materials such as titanium or niobium, which have high melting points and are resistant to oxidation, are used. For small-angle scattering experiments the sample holder itself must not generate small-angle scattering, *i.e.* it should not contain pores or defects or have a texture on the scale of the order of 100 Å. In these circumstances the sample holder can be made from quartz.

The geometry of the sample holder is also important. In many experiments flat plates are used in order to minimize multiple scattering. Alternatively, where an instrument has detectors covering a wide range of scattering angles a cylindrical geometry may be necessary. Cylindrical holders are also employed with powder diffractometers, as the sample can then be rotated to reduce the influence of preferred orientation in the powder whilst at the same time maintaining a constant sample geometry. Annular holders are often used with spectrometers where the detectors cover a wide range of scattering angles, but this implies that thin samples are needed. In certain special cases, for example in the study of ^{3}He which is highly absorbing, it is necessary to devise a complex design of holder, which maximizes scattering from the sample surface and minimizes the effects of absorption.

Holders for powder samples can be adapted for liquids, filling being carried out by hypodermic syringes. One of the most effective sample holders for a powder sample is simply to wrap it in aluminium foil, roll it flat and tape it (using cryogenic aluminium tape) on to a cadmium backing sheet. Background is minimized in this way.

Choosing the optimum sample thickness is important. The sample must be as thick as possible without the effects of multiple scattering spoiling the data. A general rule of thumb is that a sample should scatter no more than 10% of the incident beam. In this case, no more than one tenth of once scattered neutrons should scatter a second time, meaning that multiple scattering accounts for less than 1% of the signal. In the absence of absorption, the ratio of the transmitted intensity to the incident intensity (which should exceed 0.9) is

$$\frac{I}{I_0} = \exp(-N\sigma t) \tag{6.1}$$

where N is the number of atoms per cm^3, σ is the atomic scattering cross-section and t the sample thickness.

Computer codes exist to correct data for self-shielding by the sample of the incident and scattered beams as they traverse the sample volume. Self-shielding factors are the first terms in multiple scattering corrections, for which Monte Carlo codes are used. Multiple scattering is an important effect since it can redistribute the scattered beam in both Q and $\hbar\omega$ giving rise to spurious effects in the data.

6.3 Cryostats

The attainment of low temperatures is essential for many studies, such as magnetic ordering processes and quantum tunnelling in molecular systems. The ability to measure at temperatures down to and below that of liquid helium has opened up a rich area of science for which neutrons, thanks to their penetrating power, are well suited.

Helium is used in cryogenic devices for attaining the lowest temperatures. Whilst compressed helium refrigerators can routinely operate down to 15 K, liquid helium cryostats will reach 4.2 K, or 1.2 K if pumped. ^{3}He sorption fridges will reach 300 mK, ^{3}He/^{4}He dilution refrigerators will reach 30 mK, and more specialized devices can even reach below 1 mK. At the lowest temperatures beam heating can be a problem.

The variable-temperature liquid-helium cryostat is a standard item of equipment. A sample can be cooled continuously from room temperature to 1.2 K (the temperature of pumped helium) in these cryostats. In the 'Orange Cryostat' [a licensed design from the Institut Laue-Langevin (ILL), Grenoble; see Fig. 6.1], the sample itself is located at the end of a 1 metre long centre stick which is inserted into the closed sample volume of the cryostat. The temperature within this volume varies from the set temperature (defined by the user) at the sample position to room temperature at the top. The sample volume is held at ~2 mbar pressure using a helium exchange gas, which allows sufficient heat transfer for attaining the set temperature of the sample whilst minimizing the heat transfer down the sample volume. Baffles along the centre stick minimize convection. The sample volume is surrounded along its whole length by an annular space through which the helium coolant flows. This is bled as a liquid from the annular liquid-helium reservoir through a capillary to a cylindrical heat exchanger wrapped around the sample volume, and vented into the surrounding annulus. The entry flow rate is controlled by a liquid valve (the cold valve) at the base of the helium reservoir and the exit flow rate is controlled by a gas valve (the warm valve) at the top of the annular volume. This mode of operation can take the sample down to 4.2 K and, by balancing the cooling power of the helium with electrical heat input to the sample centre stick or to the heat exchanger, any temperature up to room temperature can be maintained to an accuracy of ±0.1 K. By applying a high-power vacuum pump to the annular space,

Fig. 6.1 An Orange Cryostat shown with Garry McIntyre and the D10 diffractometer at the ILL.

the latent heat of the liquid helium can be utilized to take the sample down to 1.2 K. To limit radiation losses, a nitrogen shield, sitting at 77 K, surrounds the sample volume, and the reservoir for the liquid nitrogen nestles round the helium reservoir to extend the cryostat hold time. The cryostat itself is kept within a vacuum vessel. The hold time of such a cryostat can be up to 24 hours but as it can be filled online it can be kept cold continuously for over a month, with samples being changed online as necessary without disturbing the operation of the cryostat.

Orange cryostats use cryogenic liquids which are normally produced off-site and brought in by tanker to the facility. The only 'physics' involved in obtaining low temperatures is the pumping process to attain temperatures below 4.2 K. Otherwise it could be considered as brute force cooling. An orange cryostat is a delicate device which requires expert preparation and care in its use.

Closed cycle refrigerators (CCRs) using helium as the refrigerant have been developed relatively recently. Single stage devices can attain 12 K and top-loading double-stage devices can reach 3–4 K. Such CCRs are more reliable than cryostats and require much less operational attention. Accordingly, they are much more cost-effective to run.

The history of cryogenics is of interest. Towards the end of the 19th century much effort went into the liquefaction of the so-called *permanent gases*—hydrogen, oxygen, nitrogen, methane and carbon monoxide—which could not be liquefied by pressure alone. Success was finally attained in 1883 by Wroblewski and Olszewski in Krakow, who were the first to liquefy oxygen and a few days later nitrogen. These were produced as recognizable liquids with a meniscus, using the Joule–Thomson cooling effect (with forced expansion of a gas through a nozzle) combined with counter current heat exchangers.

In 1898 hydrogen was liquefied by Dewar in London, and helium itself was liquefied by Kamerlingh Onnes in Leiden in 1908 using liquid hydrogen as a pre-coolant. Helium had been added to the list of permanent gases following its isolation in 1895 by Ramsay. In 1908 Onnes had only 360 litres of gaseous helium at his disposal, entirely recovered by the processing of monazite sand.

The liquefaction of both hydrogen and helium had been held up for the want of improved insulation methods. The development of the evacuated insulated vessel, which we call the Dewar and which we take so much for granted, was a source of controversy since both Olszewski and d'Arsonval appear to have preceded Dewar in its development. Funding limitations were cited even then as the cause of delays in scientific progress. Onnes pointed out that Dewar had the use of 50 kg of ethylene (which can be cooled to 150 K under pressure) whereas he could only afford 1.5 kg!

Constraining a gas to do external work by adiabatic expansion against a piston in an expansion engine is thermodynamically more efficient than the Joule–Thomson effect which harnesses internal work. The combination of these two techniques was finally employed in the design of modern commercial air liquefiers.

6.4 Pressure cells

The traditional problems associated with work at high pressures are exacerbated when these techniques are applied to the samples examined in neutron scattering experiments. Thus, the materials used for the high-pressure cells and the pressure-transmitting media must be compatible with the properties of the neutron. The mechanical integrity of the cells may have to be compromised to some extent by the existence of windows for the incident and emerging neutrons. The sample dimensions, necessary for an adequate signal, influence the design of the cell and can prohibit the use of classical designs of pressure cell. In many experiments sample temperatures in the range 4–77 K are required. Finally, the cells and pressurizing systems have to work in a crowded experimental environment and must fit into the existing instruments used for the experiments.

6.4.1 Materials

The materials of the cell and of the pressure-transmitting medium can absorb neutrons from the beam and produce additional scattered neutrons (background), which compete with those from the sample (signal). Materials with high absorption and/or high scattering cross-sections must be avoided and the wall thicknesses of even acceptable materials must be carefully minimized. Data for absorption and scattering cross-sections are given in Appendix B. It will be noticed that in contrast to X-rays the neutron cross-sections vary irregularly with atomic number.

The requirements for good mechanical properties in the temperature range from 4 K to room temperature and for acceptable neutron properties leave a disappointingly small selection of materials for either the walls of the cell itself or the pressure-transmitting medium. Thin-walled steel and aluminium cells can be used at 3–4 kbar but at 10 kbar the choice is reduced to alumina, synthetic sapphire crystals, or alloys of beryllium, aluminium, titanium, zinc or zirconium.

Because of the high-scattering cross-section for hydrogen, hydrogenous materials cannot be tolerated. This has important ramifications in the choice of pressure-transmitting materials. The most commonly used liquid has been carbon disulphide, although at least one cell has successfully used hydrocarbon oils where the hydrogen atoms have been replaced by those of fluorine. Helium is the only acceptable gaseous medium if the cell is required to work down to 77 K. Below 77 K helium solidifies at $\sim$15 kbar pressure but some further compression can be achieved since solid helium appears to have rather spongy properties.

Even acceptable wall materials such as aluminium alloys may produce Bragg peaks at angles where the sample scattering is important. For Bragg scattering studies null-matrix pressure cells, working up to 10 kbar and made from TiZr, have been developed.

6.4.2 Windows

The low neutron fluxes that are expected make it important that the full area of the neutron beam illuminates the sample and that the number of scattered neutrons is maximized. This in turn requires that window areas, where the cell walls have to be as thin as possible, may have to be of the order of many square centimetres, thereby compromising the upper pressure limit of the cell. Windows for allowing the scattered neutrons to emerge present severe problems in instruments such as the twin-axis diffractometer, where the neutron wavelength λ is kept constant and the angle of scattering 2θ is varied. Here the windows have to cover a large angular range and ideally should be continuous around the cell. With time-of-flight experiments, where λ is varied and the scattered neutrons are detected at fixed angles, the output window or windows can be placed at specific angles covering only the area required by the detectors. In particular, studies at $2\theta = 90°$ are very effective as collimation of the incident and scattered beams can render much of the pressure cell invisible.

The provision of windows for high-pressure sample cells in neutron work, and the inevitable compromise with the mechanical design and safety demands, has called for much ingenuity. The restricted geometry of multi-anvil devices allows little scope for developing windows of the required area; the directly pressurized cell, or the piston-and-cylinder device and the system with opposed anvil cells have more favourable geometries.

6.4.3 Environment and safety

The cells and pressurizing systems have to be integrated into existing instruments in close proximity to a nuclear reactor or spallation neutron source. This imposes restrictions on the weight and size of the apparatus, further reduces the choice of which type of cell can be used, and clearly has important consequences for safety requirements.

Safety codes for high-pressure apparatus rightly advise that the equipment be installed in a completely shielded enclosure room or sited remotely from personnel and from other apparatus. Additional precautions can include remote control of valves, indirect viewing of gauges, and interlocks to prevent unauthorized access. Few of these recommendations can be met in neutron scattering experiments, particularly if the cell must be capable of working on many different types of instrument. Usually, the risks can be reduced to acceptable levels by careful design of the cells, provision of shielding very close to the cell, preliminary tests of all ancillary apparatus, and, in some cases, tests to destruction of the proposed cells with detailed studies of the mode of bursting and the resultant fragments.

Many of the conventional pressure cells discussed in the literature require pressure rams applied directly to the cell. As these are bulky devices there are usually problems in fitting them into the sample position of existing neutron scattering instruments. Cells, in which the pressure is transmitted indirectly to the sample via a pressure-transmitting fluid from an intensifier, offer a clear advantage, as the intensifier can be mounted some distance away from the instrument. There are additional problems, of course, if it is required to rotate the sample during the experiment.

6.4.4 Pressure cells used for neutron scattering

As indicated above, pressure cells used for neutron scattering can be separated into two categories: continuous-window devices and fixed-angle window devices. With a sample cell which has continuous windows it is possible, in principle, to perform any type of neutron scattering experiment with a spectrometer, and diffraction investigations may be performed in the same cell as one used for incoherent inelastic experiments. Attenuation of the neutron beam by the windows must be sufficiently low to allow any inelastic processes to be observed. Diffraction processes are, in general, of higher intensity and can tolerate thicker and stronger windows, allowing the high-pressure limit of the cell to be raised. The sample size for coherent scatterers can be quite large before multiple scattering becomes an embarrassment; incoherent scatterers, on the other hand, are normally hydrogenous materials and must be less than 1 mm in thickness.

6.5 Radiological safety

Hazards from using neutron radiation arise from the primary beam, from the secondary scattered beam, and from induced radioactivity in the samples and the parts of the apparatus exposed to the primary beam. All who work with neutron beams should understand these hazards (see Pochin, 1983). External exposure in operating close to a neutron scattering instrument is due to gamma rays. Atoms in the human body are displaced by direct collisions with fast neutrons, and chemical bonds are broken because the binding energy holding an atom in a molecule (less than 10 eV) is much less than the recoil energy. These collisions account for only a small part of the biological damage, most of which is caused by knock-on events. A carbon atom struck with a 1 MeV neutron recoils with an energy of about 150,000 electron volts, dislodging up to 60 other carbon atoms before it is slowed down to a harmless velocity.

6.5.1 Definitions

We start by introducing the units used for estimating radioactivity and radiation dose. These are the units recommended by the International Commission on Radiological Protection (ICRP).

The rate at which spontaneous transitions occur in a given amount of radioactive material resulting in decay of the nucleus is known as its *activity*. The unit of *activity* is the becquerel, Bq, which is an SI unit named after the French physicist Henri Becquerel and has superseded the original measure of the Curie, Ci. (The Curie was defined as the number of disintegrations per second in one gram of radium in equilibrium with the isotopes in its decay chain.) 1 Bq is equal to 1 disintegration per second, and 1 Ci $= 3.7\times10^{10}$ Bq. 1 Bq is so small that a much larger unit is frequently used, such as the megabequerel, MBq, which is one million becquerels. For example, 1 gm of ^{239}Pu has an activity of 2,000 MBq and it emits 2×10^9 alpha particles per second.

The *absorbed dose* is the amount of energy that ionizing radiation deposits in matter such as human tissue. The unit of absorbed dose is the gray (symbol Gy), named after the English physicist Harold Gray. A dose of one gray is equal to the deposition of an energy of one joule in one kilogram.

$$1 \text{ Gy} = 1 \text{ J/kg}(=10^4 \text{ erg/g}) \tag{6.2}$$

Further units are required because equal amounts of deposited energy do not necessarily have equal biological effects. Thus, 1 Gy from alpha radiation is more harmful than 1 Gy from beta radiation, because the alpha particle is slower and more heavily charged, and it loses its energy much more densely along its path. In order to put ionizing radiation on the same basis for dealing with harmful effects, we use another quantity, known as the *equivalent dose* and expressed in a unit called the sievert, Sv (after the Swedish physicist Rolf Sievert). The sievert is used to assess the effects of ionizing radiation on the cells of living tissue.

Equivalent dose is equal to the absorbed dose multiplied by a *radiation weighting factor* which takes into account the relative effectiveness of the radiation in causing biological harm. The factor is arbitrarily set at unity for X-rays and beta radiation and is 20 for alpha particles. Fast neutrons of around 1 MeV in energy produce recoil atoms in the energy band from 10^3 up to 10^5 eV which is the most damaging spectral range of recoil atoms. The radiation weighting factor is 5 for 1 keV neutrons, 20 at 1 MeV and again falls to 5 above 20 MeV. Finally, to calculate the *effective dose* we must weight the equivalent dose by a *tissue weighting factor* expressing the susceptibility to harm different tissues. The overall effective dose from exposure to radiation is then obtained by summing the effective dose to the major organs. Biological damage varies from burn-like lesions (somatic) to more severe effects (genetic), principally cancer or leukemia, and at the highest doses to rapid death. The lethal dose for 50% of the adult human population, known as the LD50, is about 5 Sv.

6.5.2 Limitation of effective dose

The natural background radiation level from all sources and in most parts of the world is within the range 0.1–1 μSv (microsieverts) per hour, ranging from a dose of 1mSv to 10mSv per year from natural causes. The dose limit for radiation workers recommended by the ICRP (1991; *ICRP Publication* 60) is 20 mSv per year over a five-year period with the dose in any year not exceeding 50 mSv. The annual exposure of a typical user working at a reactor source is likely to be much less than 10 mSv. Areas around an instrument with an equivalent dose rate of less than 0.5 mSv/hr are usually not classified as designated areas for radiation protection purposes. Further information on the effects of ionizing radiation and on radiological protection is to be found in the highly readable booklet *Living with radiation* published by the NRPB (1998) of the United Kingdom.

6.5.3 Activation of samples by neutron irradiation

In planning a neutron experiment it is important to know whether activation of the sample by the primary neutron beam will be significant. In calculating the degree of activation we need to know the time of exposure, the neutron spectrum, the mass of the sample, its isotopic content, *etc.* Table 6.1 is based on calculations for a 5 cm^3 solid sample of the

Table 6.1 Guidelines for the activation of certain elements in a neutron beam.

Element	Storage time	Prompt activation (1 unit = 37 Bq/g)
Antimony	520 days	800
Arsenic	18 days	8.4×10^4
Barium	6 days	80
Bromine	18 days	1.4×10^4
Cadmium	190 days	370
Chromium	61 days	40
Cobalt	24 years	5.2×10^4
Copper	7.4 days	1×10^4
Europium	50 years	2,200
Gadolinium	11 days	7,400
Germanium	6 days	1,100
Gold	29 days	3,000
Hafnium	1.6 years	620
Iridium	4.2 years	5×10^4
Mercury	24 days	700
Molybdenum	30 days	430
Platinum	20 days	230
Rubidium	56 days	1,800
Samarium	35 days	6,200
Silver	7.4 years	1.6×10^4
Tantalum	3 years	160
Tungsten	15 days	3.7×10^4
Zinc	5 days	1,600

pure element exposed to a pulsed neutron beam for 1 day. *Storage time* is the time required to decay to 74 Bq/g (note the archaic link with the definition of a Curie), which is a typical limit for shipping the sample as 'non-radioactive'. *Prompt activation* is the anticipated activation 2 minutes after exposure to the neutron beam has stopped. The calculations are for a pulsed source of 200 kW power, and they may be overestimates for reactor experiments where there are no epithermal neutrons. Elements giving very little activation are not included in the list. A more complete compilation can be found in the *Neutron Data Booklet* published by Dianoux and Lander (2003).

6.5.4 Shielding from neutron radiation

Shielding is required to protect the experimentalist from the harmful effects of radiation. It also is necessary to reduce the background received by the detector of the scattering instrument. The shielding absorbs both fast neutrons and gamma rays: a 10 MeV photon produced in neutron capture is attenuated by an order of magnitude in passing through 5 cm of lead or 50 cm of concrete.

Thermal neutrons are readily stopped by capture in materials with a high absorption cross-section σ_{abs} (see Appendix B) such as cadmium, gadolinium, boron and lithium. In the case of cadmium and gadolinium the absorption is accompanied by the strong emission of gamma rays, so that additional shielding is necessary to attenuate the gamma rays. This is not so for lithium and is much less pronounced for boron, and so these latter materials are preferable, especially acting as the outside layers of shielding in an experimental environment.

Fast neutrons of higher energy cannot be stopped directly by neutron capture and so they have to be slowed down first. In the MeV region three types of material are generally added together: a dense material for inelastic collisions, a hydrogenated material for moderation, and an absorbing material for capture. 'Heavy concrete', which contains iron, hydrogen and boron, combines all three features. Heavy concrete is used for primary shielding around the neutron source.

References

Dianoux, A.-J. and Lander, G. H. (editors) (2003) "*Neutron Data Booklet.*" ILL, Grenoble.

ICRP (1991) *Ann. ICRP* **21**(1–3) 67–77. "1990 *Recommendations of the International Commission on Radiological Protection.*"

NRPB (1998) "*Living with radiation,*" 5th edition. National Radiological Protection Board, London.

Pochin, E. (1983) "*Nuclear radiation: risks and benefits.*" Clarendon Press, Oxford.

Part II

Neutron diffraction

Single crystal diffraction

7

The differential cross-section $d\sigma/d\Omega$ is measured with a *diffractometer* by counting the number of neutrons scattered into a small solid angle $d\Omega$ at an angle 2θ to the direction of the incident beam. These measurements are classified as experiments in *neutron diffraction* and will be considered in Part II. Experiments in *neutron spectroscopy*, in which the neutrons are scattered inelastically and their change of energy is measured, are discussed in Part III.

In this chapter, we consider *nuclear* Bragg scattering from single crystals and in the next chapter the same type of scattering from polycrystalline materials. If the sample contains unpaired electrons, as in the elements of the first transition series with incomplete $3d$ electronic shells, then there is also *magnetic* scattering. We shall consider magnetic neutron diffraction in Chapter 9.

In Bragg scattering the energies (or wavelengths) of the incident and scattered neutrons are the same. Thus substituting $k_i = k_f$ into eqn. (2.54) gives

$$Q = 2k_i \sin\theta = \frac{(4\pi \sin\theta)}{\lambda} \tag{7.1}$$

where λ is the wavelength. The last equation indicates that there are two distinct ways of conducting an experiment in Bragg scattering. In the first method, neutrons of a fixed wavelength are incident upon the sample and the scattered intensity is measured as a function of the scattering angle 2θ; alternatively, by reversing the roles of 2θ and λ, the intensity is measured as a function of wavelength at a fixed value of the scattering angle. In this chapter, we shall consider these two procedures for measuring the Bragg diffraction from single crystals, and in the next chapter we shall describe the same type of measurement from polycrystalline samples. However, first we must refer to some general considerations relating to the diffraction process.

7.1 General

7.1.1 The unit cell and the crystal lattice

A single crystal consists of a basic structural unit, a parallelipiped-shaped volume known as the *unit cell*, which undergoes a translational repetition in three dimensions to generate the entire crystal. The contents of the cell may be as small as a single atom or as large as a biological molecule. The particular choice of cell is not necessarily unique, but it is conventional to choose the smallest unit cell displaying the highest possible symmetry. Any repeated pattern, such as the distribution of atoms in a crystal, can be related to an imaginary *lattice* of points: the lattice defines the way in which the repetition takes place without any reference to the nature or contents of the pattern itself. The most general lattice is the *triclinic* lattice, in which the unit cell is a parallelepiped with unequal edges a, b, c and unequal angles α, β, γ (see Fig. 7.1). Some lattices have points not only at the corners of the cell but also at the centre of the cell or at the centre of one or more of the faces of the cell. Lattices and cells of this type are *non-primitive*, whereas *primitive* cells are those associated with just one point of the crystal lattice.

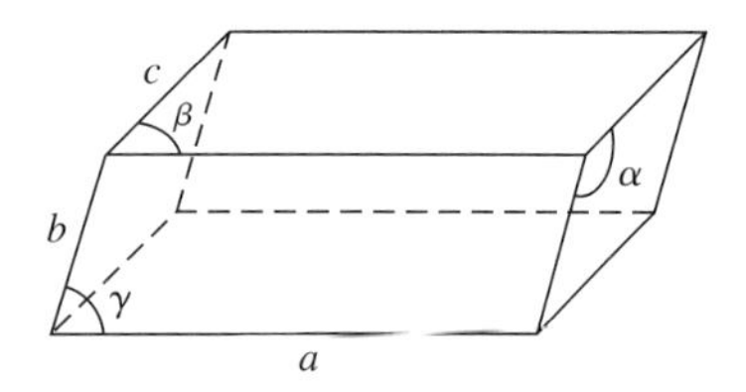

Fig. 7.1 The triclinic unit cell: $a \neq b \neq c$ and $\alpha \neq \beta \neq \gamma$.

There are 14 distinct crystal lattices or *Bravais lattices*, which are distinguished from one another by their symmetry. Seven of the lattices are primitive and seven non-primitive. In the cubic system there are three types of lattice: these are the primitive, body-centred and face-centred lattices, each possessing equal sides and equal interaxial angles of 90°.

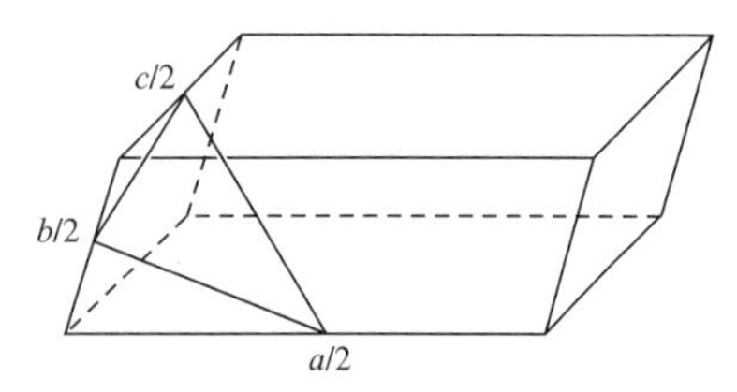

Fig. 7.2 The (222) plane in the triclinic system.

A family of parallel and equidistant planes in the crystal lattice is described by its *Miller indices.* If **a**, **b**, **c** are the vectors along the sides of the unit cell, the planes with Miller indices $(h\ k\ l)$ make intercepts on these axes which are in the ratio $a/h{:}b/k{:}c/l$. Thus (100) is the family of planes which is parallel to the plane containing the vectors **b** and **c**, whereas (321) is the family with fractional intercepts in the ratio 2:3:6. In Fig. 7.2, the plane with intercepts halfway along each axis is described by the Miller indices (222).

7.1.2 The reciprocal lattice

The *reciprocal lattice* is derived from the crystal lattice and was first introduced into diffraction theory by Ewald (1921). The translation vectors $\mathbf{a}^*$, $\mathbf{b}^*$, $\mathbf{c}^*$ of the reciprocal lattice are defined by the equations

$$\mathbf{a}^* = \frac{2\pi}{V_c}(\mathbf{b} \times \mathbf{c}),\ \mathbf{b}^* = \frac{2\pi}{V_c}(\mathbf{c} \times \mathbf{a}),\ \mathbf{c}^* = \frac{2\pi}{V_c}(\mathbf{a} \times \mathbf{b}) \qquad (7.2)$$

where $V_c = \mathbf{a} \cdot (\mathbf{b} \times \mathbf{c})$ is the volume of the unit cell in direct (or real) space, with the dot symbol $[\cdot]$ indicating scalar product and the cross symbol $[\times]$ indicating vector product. Thus $\mathbf{a}^*$ is perpendicular to the plane containing **b** and **c**, $\mathbf{b}^*$ is perpendicular to **c** and **a**, and $\mathbf{c}^*$ is perpendicular to **a** and **b**. The volume of the unit cell of the reciprocal lattice is $(2\pi)^3/V_c$.

There are two important properties of the reciprocal lattice, which we shall use later:

(i) The vector $\mathbf{H} = h\mathbf{a}^* + k\mathbf{b}^* + l\mathbf{c}^*$ in the reciprocal lattice is normal to the family of planes in the direct lattice with Miller indices (hkl).

(ii) The magnitude of **H** is $2\pi/d_{hkl}$, where d_{hkl} is the interplanar spacing of the (hkl) planes.

7.1.3 The Ewald sphere

The equation

$$\lambda = 2d_{hkl} \sin \theta \tag{7.3}$$

is the expression of Bragg's law in direct space. A formulation of the law in reciprocal space is obtained by introducing the so-called *Ewald sphere* (or *sphere of reflection*).

To define the Ewald sphere we refer to Fig. 7.3. O is the origin of reciprocal space, and the vector $\vec{CO}$, which terminates at O, represents the wave vector $\mathbf{k}_i$ of the incident radiation. The Ewald sphere is the sphere with centre C and radius k_i or $2\pi/\lambda$. [In X-ray crystallography, it is customary to define the reciprocal lattice by removing the factors 2π in eqn. (7.2), and the radius of the Ewald sphere is then taken as $1/\lambda$.]

AA' in Fig. 7.3 represents an axis of rotation of the crystal in a diffraction experiment. Suppose that during this rotation the end point of $\mathbf{H}$ touches the sphere at P. If 2θ is the angle between the vectors $\vec{CO}$ and $\vec{CP}$, we have $H = |\mathbf{H}| = 2k_i \sin\theta = (4\pi \sin\theta)/\lambda$. But $H = 2\pi/d_{hkl}$ and so $\lambda = 2d_{hkl} \sin\theta$, which is Bragg's law. Thus comparison with eqn. (7.1) shows that Bragg reflection takes place when the scattering vector $\mathbf{Q}$ coincides with the reciprocal lattice vector $\mathbf{H}$.

The vector $\vec{CP}$ in Fig. 7.3 lies in the plane containing the incident beam $\mathbf{k}_\mathrm{i}$ and the diffraction vector $\mathbf{H}$, and it makes angle of 2θ with $\mathbf{k}_\mathrm{i}$. Its magnitude is the same as the magnitude of $\mathbf{k}_\mathrm{i}$, and so $\vec{CP}$ is the wave vector $\mathbf{k}_\mathrm{f}$ of the scattered beam. The condition for the Bragg equation to be satisfied for the (hkl) family of planes is that the corresponding diffraction vector $\mathbf{H}$ terminates on the surface of the Ewald sphere at P;

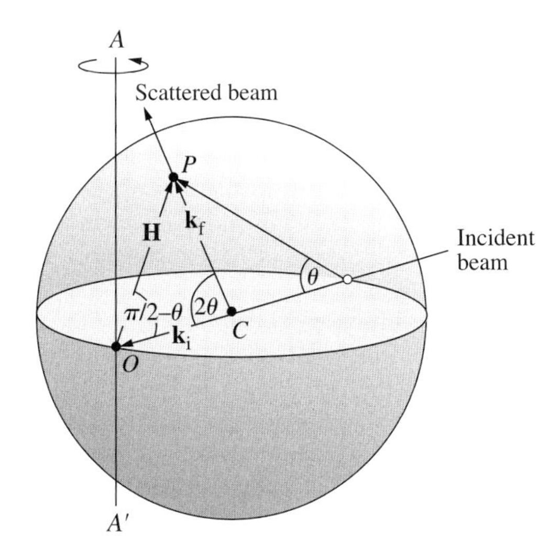

Fig. 7.3 Ewald sphere of reflection with centre C. O is the origin of reciprocal space and P is the hkl reciprocal lattice point. A diffracted beam occurs along the direction $\vec{CP}$ when the point P touches the surface of the sphere.

when this occurs, a Bragg reflection flashes out from the crystal in the direction joining the centre of the sphere with P.

7.1.4 Nuclear Bragg scattering

The *nuclear structure factor* F_{hkl} describes the amplitude of the wave scattered into the hkl Bragg reflection by the atomic nuclei in one unit cell. It is calculated by summing the nuclear scattering lengths of the atoms after multiplication by the appropriate phase factors. For an atom with fractional coordinates x_j, y_j, z_j along the cell edges a, b, c, the phase angle is $2\pi(hx_j + ky_j + lz_j)$ and so

$$F_{hkl} = \sum_j b_j \exp\left[2\pi i\left(hx_j + ky_j + lz_j\right)\right] \exp\left(-\frac{1}{2}H^2\overline{u_j^2}\right) \tag{7.4}$$

where b_j is the coherent scattering length of the jth atom in the unit cell and the summation is over all atoms in the cell. The Debye–Waller temperature factor $\exp\left(-\frac{1}{2}H^2\overline{u_j^2}\right)$ with $\overline{u_j^2}$ representing the mean-square atomic displacement accounts for the reduction in scattered amplitude arising from thermal vibrations.

Equation (7.4) can also be expressed in the form

$$F\left(H\right) = \sum_j b_j \exp\left\{i\boldsymbol{H}\cdot\boldsymbol{r}_j\right\} \exp\left(-\frac{1}{2}H^2\overline{u_j^2}\right) \tag{7.5}$$

where $\mathbf{r}_j$ is the vector to the jth atom in the unit cell.

The nuclear density $\rho(xyz)$ at any point xyz in the unit cell is given by the three-dimensional Fourier series

$$\rho\left(xyz\right) = \frac{1}{V_c} \sum_{h=-\infty}^{\infty} \sum_{k=-\infty}^{\infty} \sum_{l=-\infty}^{\infty} F_{hkl} \exp\left\{-2\pi i\left(hx + ky + lz\right)\right\} \tag{7.6}$$

The structure factor F_{hkl} in eqn. (7.6) is a complex quantity which can be represented in an amplitude-phase vector diagram by a single resultant vector obtained by adding the vectors of the individual nuclei. Thus, F_{hkl} is the product of its magnitude $|F_{hkl}|$ and its phase factor $\exp(\mathrm{i}\phi_{hkl})$, where ϕ_{hkl} is the phase of this resultant vector: or $F_{hkl} = |F_{hkl}| \exp(\mathrm{i}\phi_{hkl})$.

The experimental determination of the structure factor requires the measurement of the *integrated intensity* I_{hkl} of the Bragg reflection as it sweeps uniformly through the reflecting position. The majority of crystals have a mosaic structure (see Section 7.4) and for these the integrated intensity is proportional to the square of the structure factor. Hence the magnitudes $|F_{hkl}|$ can be obtained immediately from the measured intensities, but not the phase angles ϕ_{hkl}. The phases ϕ_{hkl} may have any value between 0° and 360° relative to the phase of a wave scattered by an atom at the origin of the unit cell; the determination of these phases constitutes the central 'phase problem' of crystallography. Once the phases are known the crystal structure can be derived by computing the Fourier synthesis in eqn. (7.6).

X-ray crystallographers have devised numerous ways of dealing with the phase problem (see Woolfson and Fan, 1995). Amongst the most important are those known as *direct methods*, whereby estimates of the phases are derived from theoretical relationships between the phases ϕ_{hkl} of different *hkl* reflections and their corresponding structure amplitudes $|F_{hkl}|$. Direct methods are rarely employed in neutron crystallography because neutron studies are usually confined to those systems for which an approximate structure is already known. This structure will give approximate phases, which will be sufficient to lead to structure refinement by standard iterative procedures. The phase problem is of fundamental importance in X-ray structure analysis, but it is less significant in neutron studies.

7.2 Diffraction at constant wavelength

7.2.1 The four-circle diffractometer

The four-circle diffractometer is the standard instrument used for single-crystal diffraction measurements at constant wavelength (see Arndt and Willis, 1966). It is designed to bring the *hkl* diffracted beams, one at a time, into the detector position; the *integrated intensity* I_{hkl} of each reflection is then measured by recording the total intensity as the crystal rotates at uniform angular velocity through the reflecting position. (I_{hkl} determines the magnitude of the structure factor $|F_{hkl}|$ and is defined in Section 7.4.) The rate of data collection is enhanced by replacing the single detector with an area detector (Section 4.5), but the single-detector diffractometer remains the instrument of choice for the most precise studies.

Figure 7.4 is a schematic diagram of a four-circle diffractometer, whose overall design has changed little since it was introduced in the early 1960s. Figure 7.5 is an illustration of the instrument used then at the Harwell Laboratory in the United Kingdom. The three circles, ϕ, χ and ω, constitute a Eulerian cradle for orienting the crystal, and the independent fourth circle (2θ) is employed to rotate the detector in the horizontal plane. The crystal is mounted on a goniometer head attached to the ϕ circle; the ϕ circle rotates about the χ-axis, and the combined $\phi - \chi$ assembly rotates about the vertical ω-axis which is coaxial with the 2θ-axis. The *hkl* reflections are measured by bringing the diffraction vectors $\mathbf{H}$ into the horizontal plane and at an angle $90° - \theta_B$ to the incident beam.

Once the crystal is centred on the diffractometer but in an arbitrary orientation, approximate values of the lattice parameters of the crystal are found by searching for reflections at fairly low values of 2θ

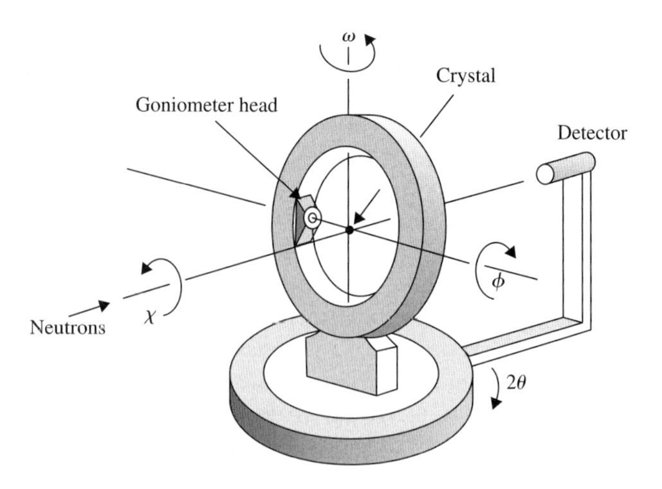

Fig. 7.4 Four-circle diffractometer (schematic).

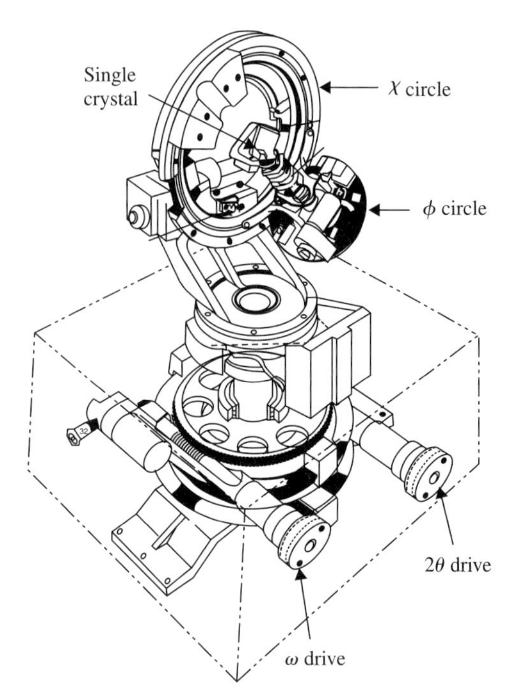

Fig. 7.5 Drawing of the original Harwell four-circle diffractometer.

(Giacovazzo, 2002). The approximate values of the setting angles for the remaining possible reflections can then be computed, and by carrying out a least-squares minimization of the differences between calculated and observed angles, precise setting angles and lattice parameters are obtained.

Each reflection is superimposed on a background which may include incoherent scattering (see Section 2.4) in addition to the thermal diffuse scattering from lattice vibrations. By scanning the diffraction vector **H** in small steps and at constant rate through the Ewald sphere, and by summing the neutron counts which are recorded at each step, the integrated intensity I_{hkl} is measured. This scan can be carried out by rotating the crystal about the ω-axis, either with the detector fixed (ω-scan) or with the detector rotating about the 2θ-axis at twice the angular velocity of the ω-axis ($\omega/2\theta$ or $\theta/2\theta$ scan). All these operations are carried out under computer control, and many thousands of reflections can be scanned without any intervention by the experimentalist.

Care must be taken in choosing the width of the scan across the Bragg peak. If the scan is too wide, that is, it contains too much background on either side of the peak, the estimated standard deviation $\sigma(I_{hkl})$ in determining I_{hkl} is too large, whereas if the scan width is too small a portion of the Bragg peak may be excluded from the measurement of I_{hkl}. According to Lehmann and Larsen (1974), the optimum width of the scan is that for which the ratio $\sigma(I_{hkl}) \div I_{hkl}$ is a minimum.

It is assumed in constructing Fig. 7.3 that the incident beam is strictly monochromatic and that it has no angular divergence. The Ewald construction is readily adapted to account for any departure from these ideal conditions by replacing the single sphere with a nest of spheres all passing through the origin of reciprocal space (see Fig. 7.6). Alternatively, we can retain the single Ewald sphere, but replace each reciprocal lattice point by a finite volume representing the *resolution function*. This resolution function accounts for the influence of many factors on the width of the Bragg reflection, such as the divergence of the incident beam, the wavelength dispersion of the monochromator and the sample size. The resolution function is measured or calculated as an ellipsoid in terms of the differential increments in these factors. A lengthy procedure for calculating the resolution function for a point detector was

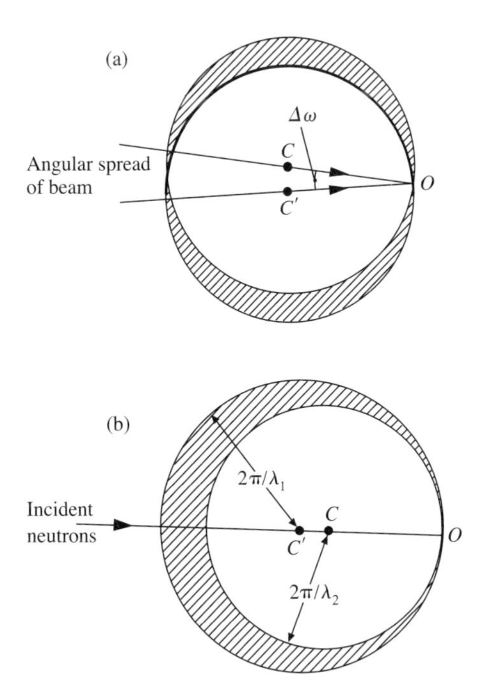

Fig. 7.6 Ewald construction representing the effects of (a) finite beam divergence $\Delta\omega$ and (b) finite wavelength spread $\Delta\lambda = \lambda_2 - \lambda_1$. The Ewald circles occupy the shaded regions: C and C' are the centres of the circles at the extremes of these regions.

first introduced by Cooper and Nathans (1968), while Schoenborn (1983) extended this analysis to cover the resolution function for an extended position-sensitive detector.

A disadvantage of the four-circle diffractometer is the difficulty of mounting ancillary apparatus (cryostats, furnaces, magnets, pressure cells, *etc.*) inside the χ circle. To carry out experiments under non-ambient conditions, it is often preferable to adopt an alternative diffraction geometry. In the *normal-beam geometry* the crystal rotates about a single vertical axis, and the detector rotates about the 2θ-axis but can also be tilted out of the horizontal plane. There are no ω and ϕ circles and so bulky equipment may be placed on the central table of the instrument without obstructing the neutron beam.

7.2.2 Area detector instruments

We have described the traditional experimental procedure, in which a narrow wavelength band is selected from the wide spectrum of neutrons emerging from a nuclear reactor and the Bragg reflections are measured one at a time with a single detector positioned at the appropriate scattering angle. A more rapid survey of reciprocal space can be achieved by using one-dimensional or two-dimensional position-sensitive detectors. A resistive-wire proportional counter (Section 4.5.2) is a one-dimensional PSD, and an image plate or a two-dimensional multiwire proportional counter is a two-dimensional PSD. Fine mesh two-dimensional ^{3}He gas detectors have also been developed for these instruments, but not yet with the spatial resolution of image plates. Image plates are made from flexible plastic sheets coated with a phosphor which is made neutron-sensitive by adding gadolinium. The Gd nuclei act as neutron scintillators by producing a cascade of keV gamma rays, which create colour centres at the detector and are subsequently read out by laser scanning. Area detectors map small regions in reciprocal space, rather than single points, and are particularly effective in studying crystals with large unit cells (McIntyre, 1992). We anticipate that in the future there will be considerable improvements in instrument design, using clusters of PSDs surrounding the sample.

Image plates are employed in the quasi-Laue LADI diffractometer at the Institut Laue Langevin, as illustrated in Fig. 7.7. It operates within a relatively narrow wavelength range of around 1 Å; hence, the term 'quasi-Laue'. The sample is mounted on the cylindrical axis and can be rotated about this axis. The neutron beam enters and leaves by opposing holes in the cylinder, producing Bragg reflections which pass through the aluminium wall and are recorded on image plates, which are curved over a cylinder around the sample and can be read out electronically. The plates use the same storage-phosphor (*i.e.* BaFBr doped with Eu^{2+} ions) as that used in X-ray image plates, but with gadolinium oxide added: the gadolinium nuclei act as neutron scintillators by creating a cascade of gamma rays and conversion electrons. The detector with its reading head tracks slowly in the horizontal direction while the cylinder rotates

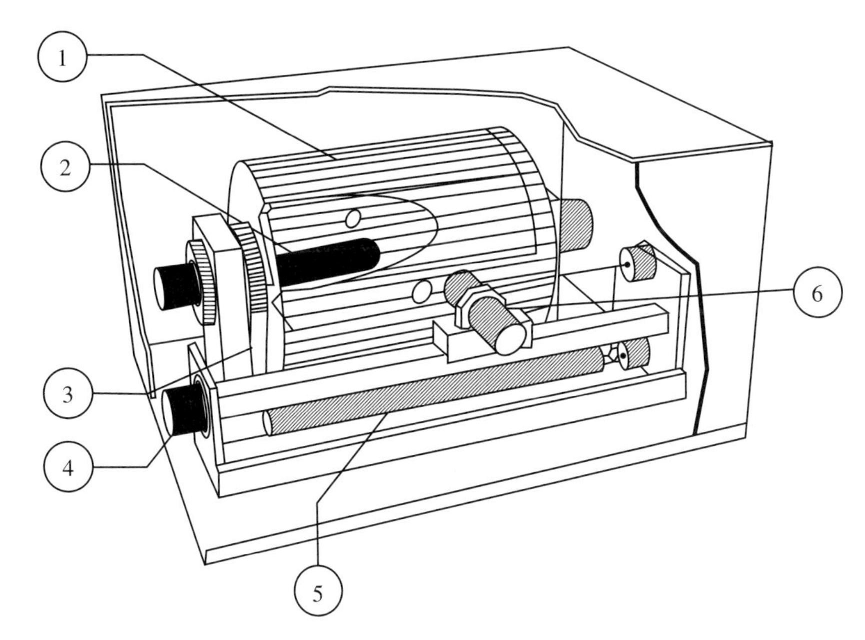

Fig. 7.7 Schematic diagram of the LADI diffractometer. 1: Image plates wrapped round drum. 2: Sample holder. 3: Transmission belt to drive drum. 4: Carrier for reader head with photomultiplier. 5: He–Ne laser. 6: Reader head.

at high speed. The readout time is 3 minutes for an image comprising around 10 million pixels.

A principal drawback of large area detectors is their susceptibility to background from different directions, leading to a relatively poor signal-to-noise ratio. Weak effects in the diffraction pattern are easier to detect with a four-circle instrument, and, by inserting an analyser in the diffracted beam, the four-circle instrument can observe weak peaks which are only 10^{-6} of the intensity of the strongest nuclear Bragg peaks.

7.3 Time-of-flight diffraction

The origins of pulsed neutron diffraction go back to the work of Lowde (1956) and to its subsequent development by Buras *et al.* (1964, 1965). In this method the sample is irradiated by a pulsed beam containing a wide band of wavelengths, and because the neutron beam is pulsed, different wavelengths can be sorted by their time of arrival at the detector.

The linear relationship between the wavelength λ and the time-of-flight t from source to detector is given by the de Broglie expression:

$$\frac{h}{\lambda} = m_{\mathrm{n}} v = m_{\mathrm{n}} \left(\frac{L}{t} \right) \tag{7.7}$$

where L is the total flight path from the source to the sample and on to the detector. If t is measured in μsec, L in metres (m) and λ in Å, we have

$$t(\mu\mathrm{sec}) = 252.78 L(\mathrm{m}) \lambda(\text{Å}) \tag{7.8}$$

Combining eqn. (7.8) with the Bragg eqn. (7.3) gives the times of flight for which *hkl* diffraction peaks are recorded in a detector positioned at

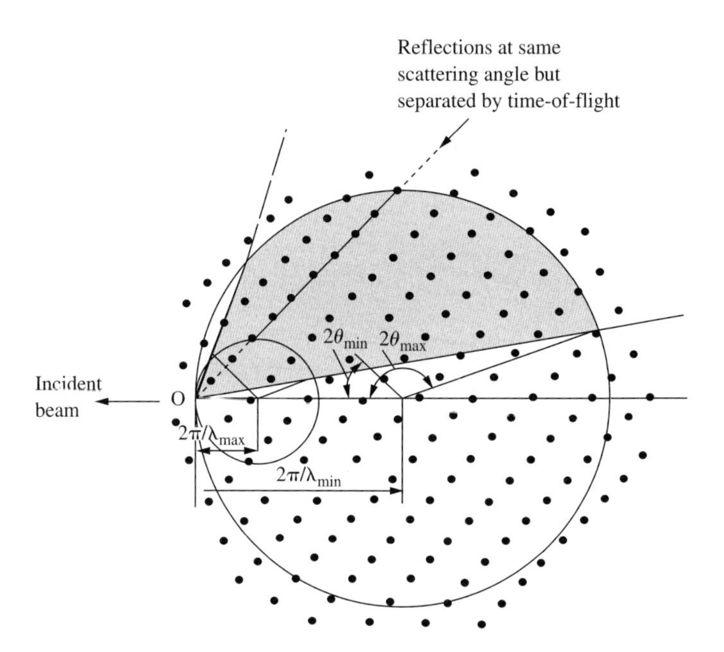

Fig. 7.8 Construction in reciprocal space to illustrate pulsed white-beam diffraction from a stationary single crystal. The Ewald circles are drawn for the maximum and minimum wavelengths, λ_{max} and λ_{min}, and the area detector spans the scattering angles from $2\theta_{min}$ to $2\theta_{max}$. All reciprocal lattice points within the shaded area are observed in a single measurement with a stationary crystal and detector. The diagram shows reflections in one section only of reciprocal space; in three dimensions, further reflections occur above and below this section.

the scattering angle 2θ. With long flight paths, frame overlap between one pulse and the next becomes a problem, but this can be overcome by using disc choppers to reduce the range of wavelengths in the neutron pulse.

The efficiency of data collection on a time-of-flight instrument is further enhanced by using an extended area detector. Figure 7.8 illustrates white-beam diffraction from a single crystal using a detector covering a wide range of scattering angles. If the incident beam contains a band of wavelengths from λ_{min} to λ_{max}, there is a corresponding nest of Ewald spheres with radii from $2\pi/\lambda_{max}$ to $2\pi/\lambda_{min}$. All *hkl* points lying within the volume of reciprocal space encompassed by these spheres will give Bragg reflections in the range of scattering angles from $2\theta_{min}$ to $2\theta_{max}$, and they can be measured with an area detector spanning this angular range. To extend the observations to all *hkl* points within the *limiting sphere*, whose radius is $4\pi/\lambda_{min}$ and whose centre is at the origin of reciprocal space, it is necessary to repeat the intensity measurements at several different settings of the crystal.

The time-of-flight single-crystal diffractometer SXD at the ISIS Laboratory uses the Laue technique to access extended regions of reciprocal space at one crystal setting. A schematic diagram of SXD is shown in Fig. 7.9. It receives neutrons with wavelengths in the range 0.1–6 Å, and is surrounded by position-sensitive detectors which are placed at a variety of scattering angles on either side of the instrument. The combination of a polychromatic incident beam and coverage with a large area detector results in the simultaneous measurement of a large portion of three-dimensional reciprocal space, and permits the study of features

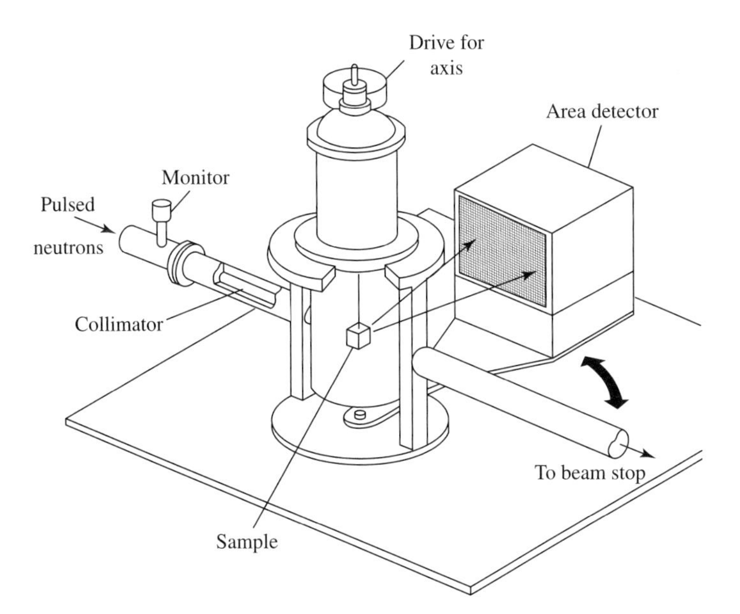

Fig. 7.9 Schematic diagram of the time-of-flight diffractometer SXD at the ISIS pulsed source. The area detector can be positioned at different locations around the sample.

which do not necessarily occur at the Bragg reflections. The instrument has proved to be particularly effective for diffuse scattering studies, as described in the book of Nield and Keen (2001). However, as with the LADI instrument, a relatively large integrated background is generated by the use of a 'white' incident beam and a large area detector. For making precise measurements of Bragg intensities, the four-circle diffractometer, with a monochromatic incident beam and a tightly defined diffracted beam (employing slits or analyser), is often preferred, despite being so time-consuming.

7.4 Reduction of intensities to structure amplitudes

The conversion of the intensity measured on a single-crystal diffractometer to the amplitude of the structure factor $|F_{hkl}|$ depends critically on the type of crystal undcr investigation. At one extreme the sample is a 'perfect crystal', such as dislocation-free silicon (as used in the semiconductor industry) in which phase coherence of the scattered neutron waves is maintained over sheets of atoms spaced many centimetres apart. In the scattering from perfect crystals there is a dynamic exchange of energy between incident and diffracted beams, and the angular width over which diffraction occurs is extremely narrow, perhaps less than 1 min of arc. The theoretical treatment of diffraction by such perfect crystals, known as the *dynamical diffraction theory*, is considered further in Section 11.3.

The vast majority of crystals contain lattice defects or imperfections which destroy phase coherence over very short distances, of the order of

10^{-4} cm. According to Darwin (1922) the crystal can then be treated as an assembly of mosaic blocks, and is known as a *mosaic crystal.* Each block is considered to be a perfect crystal, with dimensions of a few microns, but faults and cracks between the blocks ensure that there is no phase coherence of the waves scattered by different blocks. The angular separation between adjacent blocks is as much as one degree of arc, giving rise to a corresponding width of the Bragg reflections. The theory of scattering by a mosaic crystal is known as the *kinematic diffraction theory* and this theory is employed in this chapter.

In a conventional four-circle instrument, operating with a monochromatic beam and with a single detector, the crystal is rotated at a uniform angular velocity about an axis normal to the diffraction vector **H**. The *integrated intensity* I_{hkl} is the total intensity recorded in the detector as the crystal sweeps through the *hkl* reflecting position. For an *ideal mosaic crystal* (or *ideally imperfect crystal*) this integrated intensity is related to $|F_{hkl}|$ by the expression

$$I_{hkl} = \frac{\lambda^3 \left|F_{hkl}\right|^2}{V_c^2 \sin 2\theta} \tag{7.9}$$

where V_c is the volume of the unit cell. The geometrical factor $1/\sin 2\theta$, the so-called *Lorentz factor*, accounts for the different rates at which the reflections cut across the surface of the Ewald sphere during the rotation of the crystal. A more general Lorentz expression, which is applicable to a four-circle diffractometer equipped with a position-sensitive detector, has been derived by McIntyre and Stansfield (1988).

In the time-of-flight method of single-crystal diffractometry, the crystal is bathed in a beam of white radiation and is stationary during the intensity measurements. It is now necessary to measure the wavelength-dependent quantity $i_0(\lambda)d\lambda$, which is the number of neutrons striking unit area of the sample and with wavelengths lying between λ and $\lambda + d\lambda$. Equation (7.9) relating the integrated intensity I_{hkl} and the quantity $|F_{hkl}|^2$ is then replaced by

$$I_{hkl} = \frac{i_0(\lambda)\,\lambda^4 V \left|F_{hkl}\right|^2}{2V_c^2 \sin^2\theta} \tag{7.10}$$

Equations (7.9) and (7.10) refer to a very small crystal. Before applying the equations to the measured intensities, a correction is made for the reduction in intensity by thermal vibrations (*Debye-Waller factor*). Corrections are also necessary for systematic size effects such as absorption, extinction and multiple diffraction.

7.4.1 Correction for absorption

The effect of absorption is expressed by the equation

$$I = I_0 \exp(-\mu t) \tag{7.11}$$

where I is the intensity after the neutrons have traversed a distance t in the crystal and μ is the macroscopic linear absorption coefficient.

To a good approximation, absorption processes are independent of the physical state of the absorber and so, for a sample containing n different elements, μ can be written as

$$\mu = N \sum_{i=1}^{n} c_i \sigma_{ai} \tag{7.12}$$

where N is the number of atoms per unit volume, c_i is the atomic fraction of the ith element and σ_{ai} is its absorption cross-section. This cross-section is the sum of true absorption due to nuclear capture, and absorption due to incoherent scattering. For hydrogenous materials the dominant contribution is from incoherent scattering. The true absorption and incoherent cross-sections for the elements and their isotopes are listed in Appendix B. Apart from a few exceptions, such as boron, cadmium and gadolinium, the absorption coefficients of the elements are much smaller for neutrons than for X-rays. Thus for most elements μ is less than 1 cm^{-1}, and for samples smaller than 1 mm the attenuation from absorption can usually be ignored.

7.4.2 Correction for extinction

Equations (7.9) and (7.10) are based on the *kinematic* treatment of diffraction, in which it is assumed that the primary beam is attenuated by absorption alone as it passes through the crystal. However, the lattice planes first encountered by the primary beam will reflect some of the incident radiation, so that the inner parts of the crystal have less chance of contributing to the measured intensity I_{hkl} than the surface region. This effect, known as *extinction*, causes a weakening of the diffracted intensity which is akin to an extra absorption process, but is more prominent the stronger is the Bragg reflection.

It is customary to consider extinction to be either *primary* or *secondary*. *Primary extinction* occurs in perfect single crystals containing no imperfections and is accounted for by the *dynamical* treatment of diffraction. Within a perfect crystal the primary beam is attenuated by scattering processes in which there is a strict phase relationship between primary and scattered beams. Most crystals are imperfect and can be considered to consist of very small mosaic blocks, perhaps 1 μm in size. The individual blocks behave as perfect crystals but are too small to attenuate the beam significantly by primary extinction. However, the radiation which reaches a block lying at a depth below the surface has passed through other blocks with the same orientation and is scattered again at the same angular setting. This is known as *secondary extinction*. It is equivalent to an effective increase in the linear absorption coefficient and can be calculated more readily than primary extinction.

The need to correct for extinction is revealed when the measured intensities of strong reflections are found to be systematically less than their calculated values. A popular treatment for the correction of primary and secondary extinctions using monochromatic radiation is that due to Becker and Coppens (1974). In the time-of-flight diffraction method, there is an additional complexity due to the wavelength dependence of extinction (Jauch *et al.*, 1988). Clearly, extinction is less the smaller the crystal, and it may be possible to reduce the extinction by dipping the crystal in liquid nitrogen, producing the desired effect of enhancing the mosaic structure. Extinction may be negligible for small organic crystals or for crystals with large unit cells, but it can be very serious indeed for inorganic crystals with small unit cells.

7.4.3 Avoidance of multiple diffraction

Multiple diffraction occurs when, for a particular orientation of the crystal, the Bragg reflecting condition is satisfied for more than one family of reflecting planes. Referring to Fig. 7.3 this means that two or more reciprocal lattice points such as P lie simultaneously on the surface of the Ewald sphere. If we imagine the crystal to be rotated about the axis OP, where O is the origin of reciprocal space, the Bragg condition for P is still maintained and a second reflection is excited each time the corresponding reciprocal lattice point crosses the surface of the Ewald sphere. Renninger (1937) was the first to show that the measured hkl intensity can be either reduced ('*Aufhellung*') or enhanced ('*Umweganregung*') by the presence of a second reflection. The effects of multiple diffraction can be removed or minimized by rotating the crystal about the axis OP so as to avoid the excitation of another reflection as shown in Fig. 7.10. Rotation of the crystal through a wide range of azimuthal angle around the scattering vector $\mathbf{Q}$ can also be used to check the corrections for absorption and extinction: after correction the observed intensity should be independent of azimuthal angle.

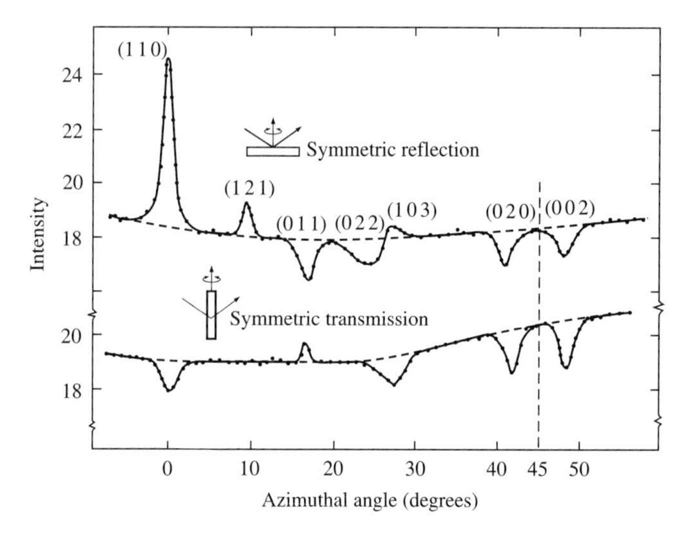

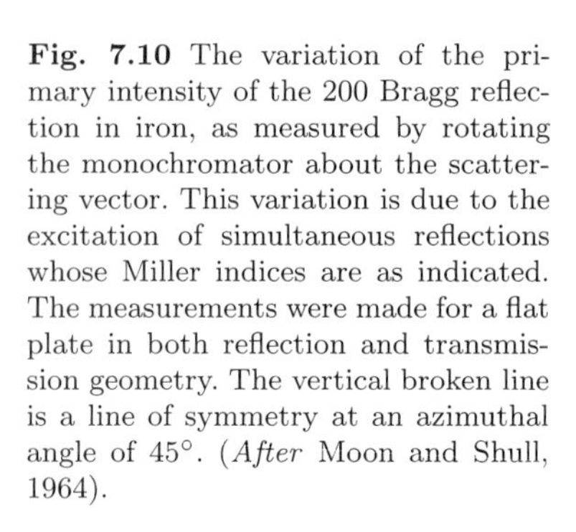
Fig. 7.10 The variation of the primary intensity of the 200 Bragg reflection in iron, as measured by rotating the monochromator about the scattering vector. This variation is due to the excitation of simultaneous reflections whose Miller indices are as indicated. The measurements were made for a flat plate in both reflection and transmission geometry. The vertical broken line is a line of symmetry at an azimuthal angle of 45°. (*After* Moon and Shull, 1964).

7.5 Some examples of studies by single crystal neutron diffraction

There are numerous examples of studies using single-crystal neutron diffraction. In physics they include the investigation of magnetic materials; in chemistry, the study of molecular structures, the nature of the hydrogen bond and the structures of metal hydrides; in biology, the crystallography of proteins and the diffraction from fibres and membranes, usually with some of the hydrogen replaced by deuterium. Magnetic studies are reviewed in the book by Bacon (1975), and chemical studies in the book by Wilson (2000). The application of diffuse neutron diffraction to the study of disordered systems is described in the book by Nield and Keen (2001). We must emphasize that there is a strong symbiosis between neutron and X-ray work; for example, the nature of the chemical bond can be examined by combined X-N work, and in structural studies it is invariably essential to derive the maximum information from X-ray diffraction before embarking on any neutron investigation.

In Chapter 9 we refer to magnetic studies. Here we shall refer to just a few of the investigations in other fields. Creatine phosphate acts as an energy reservoir in muscle. When the phosphate is hydrolysed, energy is released and creatine monohydrate $C_4N_3O_2H_9 \cdot H_2O$ is formed. Frampton *et al.* (1997) have determined the molecular structure and bonding in a single crystal of the monohydrate, using low-temperature pulsed neutron diffraction. Figure 7.11 shows the structure at 123 K with the thermal ellipsoids drawn at 50% probability. Neutrons can examine hydrogen atoms much more readily than X-rays and can yield the *anisotropic* displacement parameters of the hydrogen atoms.

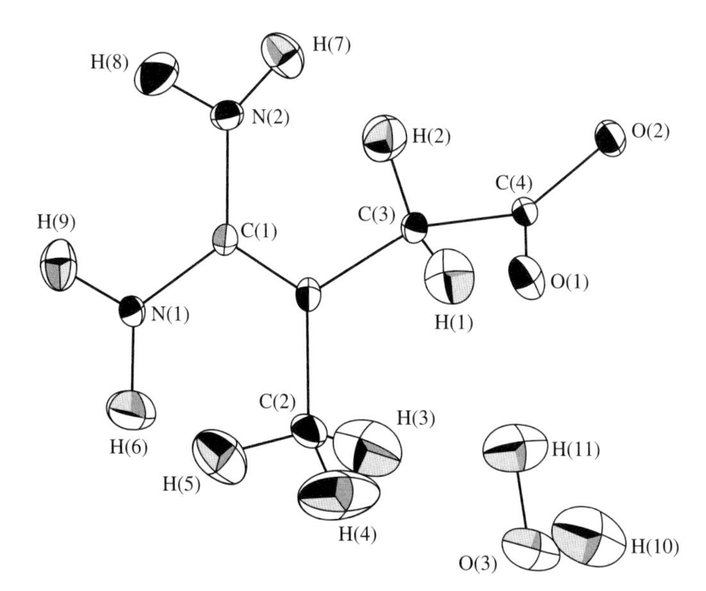

Fig. 7.11 Molecular structure of creatine monohydrate. The anisotropic thermal ellipsoids have been derived for all atoms including H-atoms. (*After* Frampton *et al.*, 1997.)

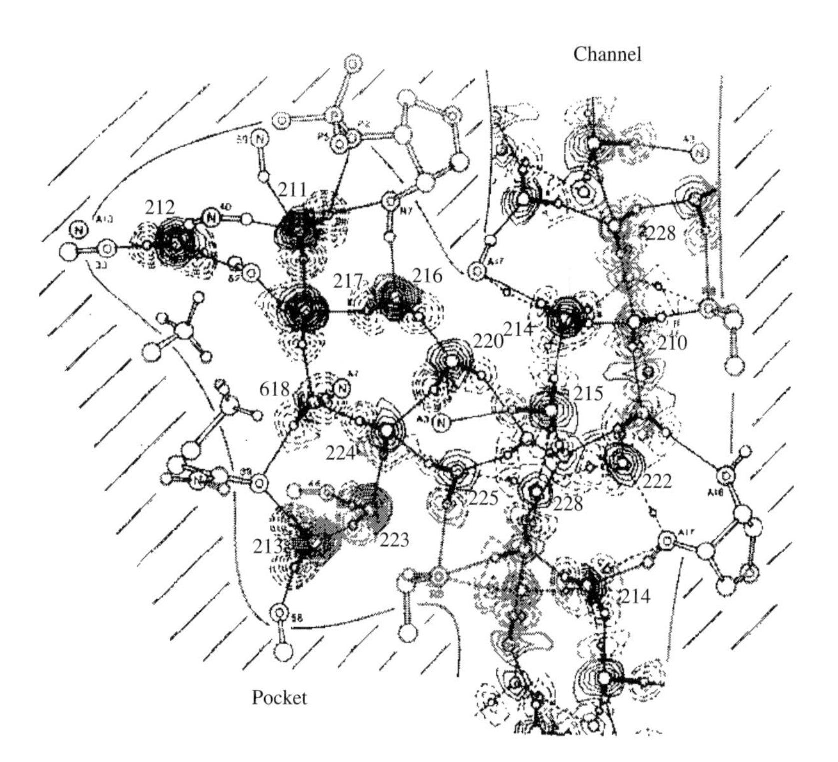

Fig. 7.12 Water networks in vitamin B_{12} coenzyme. Large open circles are oxygen and small broken circles are hydrogen. (*After* Bouquière *et al.*, 1993.)

A high-resolution study of vitamin B_{12} coenzyme, with the molecular formula $C_{72}H_{100}N_{18}O_{17}PCo \cdot 16H_2O$, was undertaken at 15 K by Bouquière *et al.* (1993) using the D19 diffractometer at the ILL in Grenoble. This provides a glimpse of what can be achieved at 1 Å resolution with a neutron crystallographic study of a hydrated biomolecule. The importance of the role of the solvent in the various functions associated with proteins is well known. The aim of the study was to determine the positions of the water molecules and to identify the nature of the water networks around the coenzyme molecules. Over 15,000 reflections were collected of which one third were symmetry-independent. From difference Fourier maps it was found that the solvent structure consists of ordered regions of water and partly disordered regions exhibiting a complex network of hydrogen bonds. Figure 7.12 shows a portion of the water networks, where the 'pocket' is an ordered region with clearly defined solvent sites, and the 'channel' is a disordered region.

The LADI instrument was used by Niimura *et al.* (1997) in a pioneering Laue diffraction study of lysozyme at 2 Å resolution. They employed a broad wavelength bandpass covering the range from 3 Å to 4 Å. Within this bandpass the Laue patterns consisted predominantly of single reflections, thus avoiding the well-known difficulty of a white Laue beam giving overlapping, higher-order *hkl* reflection. A total of 38,000 reflections were collected in 10 days using the ILL neutron source. This achievement heralded a new phase in the neutron study of highly complex crystal structures by dramatically reducing data collection times.

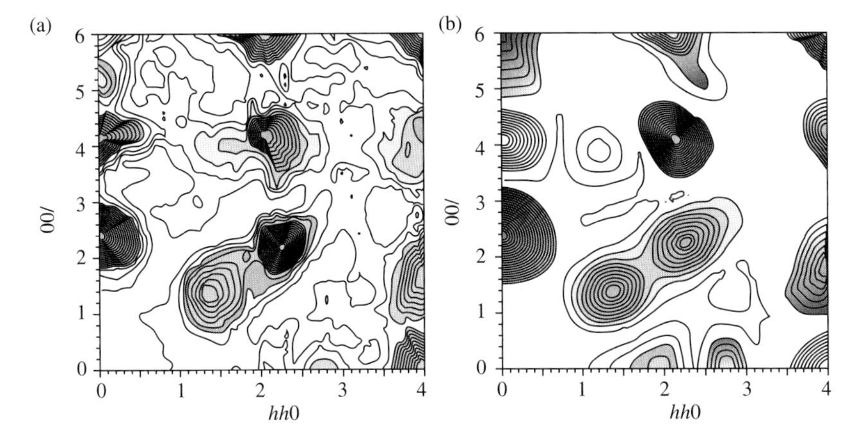

Fig. 7.13 The (110) section of the diffraction pattern from CaF_2 doped with Y_2O_3, showing strong non-Bragg diffuse scattering near the Bragg peaks. The Bragg reflections occur at points in reciprocal space with $h+l=2n$, where n is an integer. (a) Observed pattern. (b) Calculated pattern. (*After* Hull and Wilson, 1992).

Bragg diffraction only gives information about the average structure (*i.e.* average contents of all the unit cells) of a crystal. Local deviations from the average structure are often of particular importance in technological materials. This disorder gives rise to diffuse scattering in general regions of reciprocal space away from the Bragg peaks. The time-of-flight instrument SXD (Fig. 7.9) scans simultaneously large areas of reciprocal space and Fig. 7.12 shows the diffuse scattering pattern obtained with SXD on a single crystal of CaF_2 doped with Y^{3+}. The effect of doping is to replace 2+ cation sites of Ca with 3+ cation sites of Y: in order to maintain charge neutrality this is accompanied by an increase in the concentration of anions, some of which occupy interstitial sites in the fluorite framework. The measurements were made at 20 K in order to reduce the thermal diffuse scattering from lattice vibrations.

In this type of experiment an enormous amount of data is collected in three-dimensional reciprocal space. It would be desirable to compare observed and calculated data in a rigorous least-squares procedure, as is routinely done in analysing Bragg scattering. However, this would require excessive computational power, and any comparison must currently be restricted to sections of data. Figure 7.13(a) refers to the (110) section and Fig. 7.13(b) is the calculated scattering for the same section, based on a model in which the interstitial anions assemble into clusters randomly distributed throughout the lattice. The best fit was obtained for clusters containing up to 12 atoms arranged at the corners of a cuboctahedron.

References

Arndt, U. W. and Willis, B. T. M. (1966) "*Single crystal diffractometry.*" University Press, Cambridge.

Bacon, G. E. (1975) "*Neutron diffraction.*" 3rd edition. University Press, Oxford.

Becker, P. J. and Coppens, P. (1974) *Acta Crystallogr. A* **30** 129–147. "*Extinction within the limit of validity of the Darwin transfer equations.*"

Bouquière, J. P., Finney, J. L., Lehmann, M. S., Lindley, P. F. and Savage, H. F. J. (1993) *Acta Crystallogr. B* **49** 79–89. "*High-resolution study of vitamin B_{12} coenzyme at 15K: structure analysis and comparison with the structure at 279K.*"

Buras, B. and Leciejewicz, J. (1964) *Phys. Status Solidi* **4** 349–355. "*A new method for neutron diffraction crystal structure investigations.*"

Buras, B., Mikke, K., Lebech, B. and Leciejewicz, J. (1965) *Phys. Status Solidi* **11** 567–573. "*The time of flight method for investigations of single-crystal structures.*"

Cooper, M. J. and Nathans, R. (1968) *Acta Crystallogr. A* **24** 481–484, 619–627. "*The resolution function in neutron diffractometry.*"

Darwin, C. G. (1922) *Phil. Mag.* **43** 800–829. "*The reflection of X-rays from imperfect crystals.*"

Ewald, P. P. (1921) *Zeits. für Krist.* **14** 129–156. "*Das reziproke Gitter in der Strukturtheorie.*"

Frampton, C. S., Wilson, C. C., Shankland, N. and Florence, A. J. (1997) *J. Chem. Soc., Faraday Trans.* **93** 1875–1879. "*Single-crystal neutron refinement of creatine monohydrate at 20K and 123K.*"

Giacovazzo, C. (2002) "*Fundamentals of crystallography.*" 2nd edition. University Press, Oxford.

Hull, S. and Wilson, C. C. (1992) *Physica B* **180–181** 585–587. "*The defect structure of anion excess $Ca_{1-x}Y_xF_{2+x}$.*"

Jauch, W., Schultz, A. J. and Schneider, J. R. (1988) *J. Appl. Crystallogr.* **21** 975–979. "*Accuracy of single-crystal time-of-flight neutron diffraction.*"

Lehmann, M. S. and Larsen, F. K. (1974) *Acta Crystallogr. A* **30** 580–584. "*A method for location of the peaks in step-scan-measured Bragg reflexions.*"

Lowde, R. D. (1956) *Acta Crystallogr.* **9** 151–155. "*A new rationale of structure-factor measurement in neutron-diffraction analysis.*"

McIntyre, G. J. (1992) *Neutron News* **3** 15–18. "*Position-sensitive detectors in single-crystal diffractometry.*"

McIntyre, G. J. and Stansfield, R. F. D. (1988) *Acta Crystallogr. A* **44** 257–262. "*A general Lorentz correction for single-crystal diffractometers.*"

Moon, R. M. and Shull, C. G. (1964) *Acta Crystallogr.* **17** 805–812. "*The effects of simultaneous reflections on single-crystal neutron diffraction intensities.*"

Nield, V. M. and Keen, D. A. (2001) "*Diffuse neutron scattering from crystalline materials.*" University Press, Oxford.

Niimura, N., Minezaki, Y., Nonaka, T., Castagna, J. C., Cipriani, F., Hoghoj, P., Lehmann, M. S. and Wilkinson, C. (1997) *Nat. Struct. Biol.* **4** 909–914. "*Neutron Laue diffractometry with an imaging plate provides an effective data collection regime for neutron protein crystallography.*"

Renninger, M. (1937) *Z. Phys.* **106** 141–176. "*Umweganregung: eine Bisher unbeachtete Wechselwirkungserscheinung bei Raumgitterinterferenzen.*"

Schoenborn, B. P. (1983) *Acta Crystallogr. A* **39** 315–321. "*Peak-shape analysis for protein crystallography with position-sensitive detectors.*"

Wilson, C. C. (2000) "*Single crystal neutron diffraction from molecular materials.*" World Scientific, Singapore.

Woolfson, M. M. and Fan, H.-F. (1995) "*Physical and non-physical methods of solving crystal structures.*" University Press, Cambridge.

The powder diffraction method

8

Shortly after the discovery of the diffraction of X-rays by a single crystal, Debye and Scherrer (1916) and Hull (1917) independently showed that a characteristic diffraction pattern is also given by a sample consisting of a polycrystalline powder. This method of examining materials is known as the Powder Diffraction Method. Much crystallographic information is lost by condensing the three dimensions of reciprocal space, which are explored in single-crystal diffraction, into the single dimension of the powder pattern. However, for some materials it may not be possible to prepare a suitable single crystal. For example, zeolites, fast-ion conductors, high-T_c superconductors, fullerenes, *etc.* may be available only in powder form. Other materials may undergo a series of phase transitions on cooling, making it impossible to retain the integrity of a single crystal. In spite of its inherent limitations, powder diffraction can give not only the detailed atomic arrangement, but also the texture and composition of the sample. It can be used for *in situ* studies of chemical reactions and processes. With the increasing power of computers, advances in instrumentation, and the trend towards the study of more complex materials, powder diffraction has undergone a transformation since the 1980s.

In this chapter we shall cover the principles and practice of neutron powder diffraction. The neutron method has evolved, to a large extent, from the earlier X-ray method, but at the outset we must emphasize again an important difference between neutron and X-ray practice. In a typical X-ray experiment, the number of X-ray quanta striking unit area of the sample in a given time is many orders of magnitude larger than the corresponding number of neutrons in a typical neutron experiment. In order to maintain a reasonable counting rate, the samples for neutron studies tend to be considerably larger than those used in X-ray work, although modern instruments using multidetectors, such as GEM at ISIS at the Rutherford Appleton Laboratory or D20 at the Institut Laue Langevin, dramatically reduce the need for 'grams of sample'. The weak interaction of the neutron with matter ensures that there is adequate illumination at all depths of penetration of the sample.

8.1 Principles of the powder diffraction method

In Chapter 7 we noted that there are two experimental procedures for conducting a Bragg scattering experiment with thermal neutrons. In the conventional method of neutron powder diffraction, neutrons of a fixed wavelength are selected by a crystal monochromator; these neutrons are then scattered by the sample, and the intensity $I(\theta)$ of the scattered beam is measured as a function of the scattering angle 2θ. The experimental plot of $I(\theta)$ *versus* θ shows a series of diffraction peaks whose positions along the θ axis are determined by the Bragg equation, $\lambda = 2d \sin \theta$, where d is the spacing of the (hkl) set of planes and λ is the neutron wavelength [see Fig. 8.1(a)].

By reversing the roles of λ and θ, it is possible to measure instead the intensity $I(\lambda)$ as a function of λ at a fixed value of 2θ. As in the case of single-crystal diffraction, these two alternatives give rise to two types of instruments: the fixed-wavelength (angle-dispersive) powder diffractometer and the fixed-angle (energy-dispersive) time-of-flight (TOF) instrument. The first type is usually designed to operate on a reactor source, where a continuous monochromatic beam bathes the sample, and the second type on a pulsed source, where a pulsed white beam strikes the sample. For a pulsed source the time-of-flight t of the neutrons issuing from a single pulse is directly proportional to their wavelength, and so the diffraction pattern is obtained by recording the intensity $I(t)$ as a function of t [see Fig. 8.1(b)].

8.1.1 Angle-dispersive procedure

Figure 8.2 is a schematic illustration of an angle-dispersive instrument. The 'white' neutron beam from the reactor passes through a collimator in the reactor shield, and a particular wavelength is then selected

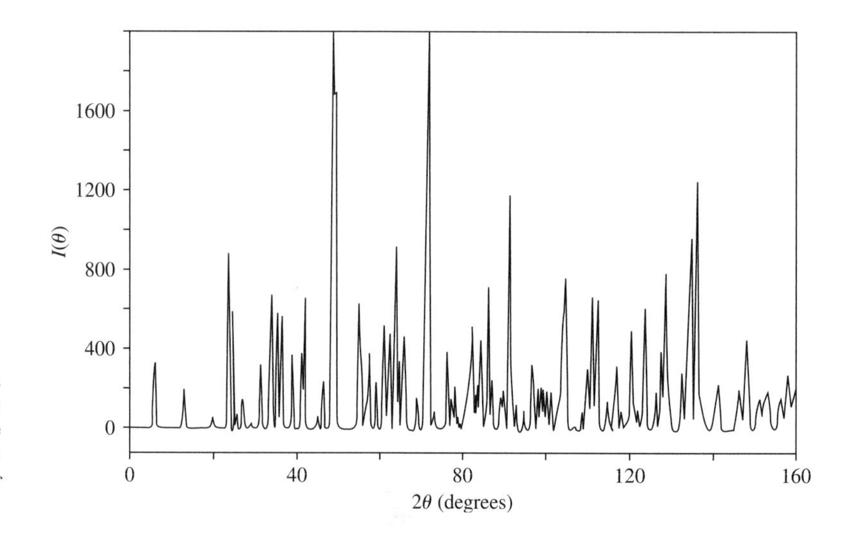

Fig. 8.1 (a) Neutron powder pattern of the superconductor $YBa_2Cu_4O_8$ at 35 K. The ticks along the base of the pattern indicate the calculated positions of the diffraction peaks. (*After* Kaldis *et al.*, 1989.)

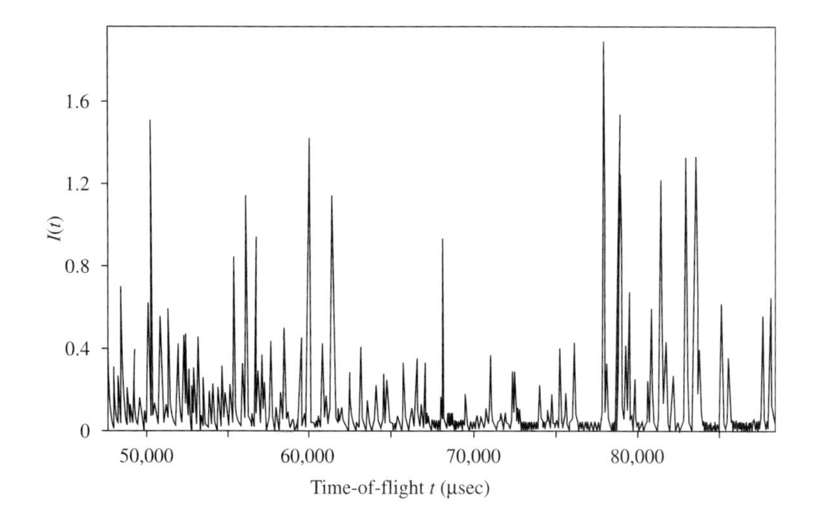

Fig. 8.1 (b) Diffraction pattern of deuterated benzene at 5 K, obtained on the pulsed neutron instrument HRPD at the Rutherford Appleton Laboratory. (*After* David, 1992.)

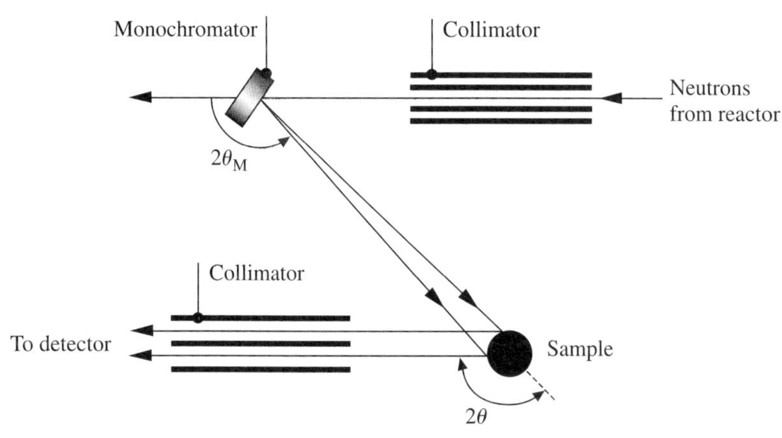

Fig. 8.2 Schematic diagram of an angle-dispersive powder diffractometer.

from this beam by Bragg reflection from a crystal monochromator. The monochromator is embedded in shielding of a heavy metal, such as lead or tungsten, and beyond this there is an outer shielding of a hydrogenous material, such as paraffin surrounded by cadmium or boron. The metal shielding serves to absorb the intense gamma beam from the reactor core, and the outer hydrogenous shielding slows down the fast neutrons in the incident beam before they are absorbed in cadmium or boron. Thus the unwanted radiation is largely removed, and the sample is illuminated by a beam of monochromatic neutrons, which are perhaps a few cm in height and a few mm in width with a sample as small as a few mm^3 in volume. The polycrystalline sample is contained (if necessary) in a thin-walled cylindrical can or holder, usually of vanadium or a null-matrix alloy such as TiZr, which gives a featureless background underneath the diffraction pattern of the sample. The angular divergence of the beam entering the detector is restricted to about 10 *minutes of arc* by means of a Soller slit collimator, and the diffraction pattern is recorded in discrete steps of 2θ where the magnitude of the steps matches the beam divergence. Variations in the number of neutrons striking the

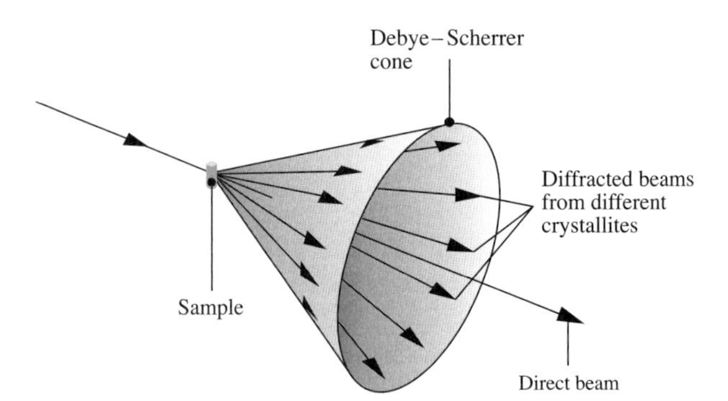

Fig. 8.3 The cone of diffracted beams for a given set of *hkl* planes of Bragg angle θ_{hkl} in a polycrystalline sample illuminated by a monochromatic beam. The semi-angle of the cone is $2\theta_{hkl}$.

sample, due to slight fluctuations in the reactor power, are allowed for by recording the neutron intensity at each step for the same number of counts in the incident-beam monitor. Modern instruments (*e.g.* the D20 diffractometer at the ILL) use multidetectors to increase the rate of collecting data.

Ideally, the sample will consist of a very large number of small crystallites that are in random orientations. A small proportion of these crystallites will be in the correct orientation for Bragg scattering by the (*hkl*) planes, and these diffracted neutrons lie along the surface of a cone, the *Debye–Scherrer cone*, whose apex is at the sample and whose semi-angle is twice the Bragg angle (Fig. 8.3). There is a separate cone for each family of (*hkl*) lattice planes, and the detector records a diffraction peak each time it intercepts an *hkl* cone.

8.1.2 Energy-dispersive procedure

In the fixed-angle technique, a pulsed source delivers a white neutron beam directly on to the sample, replacing the continuous monochromatic beam from a steady-state reactor. As the neutron pulse travels towards the sample, the faster shorter-wavelength neutrons disperse from the slower longer-wavelength neutrons in the incident beam. The wavelength of the scattered radiation recorded by the detector is obtained by measuring the total time of flight from source to detector. The relation between this time of flight t and the wavelength λ is given by

$$t(\mu\text{sec}) = 252.78L(\text{m})\lambda(\text{Å}) \tag{8.1}$$

where L is the total length of the flight path (*i.e.* source–sample distance + sample–detector distance) and the symbols in square brackets indicate units. For example, the TOF is about 5 ms for neutrons of 1 Å wavelength traversing a path length of 20 m.

To achieve good resolution of the Bragg peaks the moderator is thin, only a few cm in extent, and the flight path is long. However, the longer the flight path, the smaller is the wavelength band which is free from frame overlap, and so to observe the complete diffraction pattern several instrument settings are required. Frame overlap is the condition whereby faster neutrons from a later pulse overtake slower neutrons from an earlier pulse, causing contamination of the data. The range of wavelengths $\Delta\lambda$ which is free from overlap is obtained by putting

$$t_{\max} - t_{\min} = \Delta t = \frac{1}{\nu} \tag{8.2}$$

where ν is the pulse frequency of the source. Using eqn. (8.1) $\Delta\lambda$ is given by

$$\Delta\lambda(\text{Å}) = \frac{3,956}{L(\text{m})\nu(Hz)} \tag{8.3}$$

and so, for an instrument with a 20-m flight path and with neutron pulses occurring at intervals of 20 msec, the wavelength range free from overlap is 4 Å. Waveband choppers (Section 5.3.1) inserted into the incident beam will remove this frame overlap. An example of a very long flight-path instrument is the High Resolution Powder Diffractometer, HRPD, at ISIS, which has a path length of 96 m from source to sample. The pulse-repetition frequency is 50 Hz, and so $\Delta\lambda$ in eqn. (8.3) is approximately 0.8 Å: a complete diffraction pattern must either be covered in a number of wavelength sections, each of which is no wider than 0.8 Å, or the choppers in the incident beam must be operated at a lower frequency, say 10 Hz instead of 50 Hz, in order to remove four pulses out of five and to allow the complete pattern to be measured in one setting of the instrument.

Figure 8.4 is a schematic illustration of HRPD. The nickel guide tube in the long incident flight path is slightly curved so that the unwanted gamma rays and fast neutrons are lost in the walls of the guide. The main detector bank is sited as close to back-scattering geometry as possible in order to achieve the highest possible resolution (see Section 8.2). The detector assembly has circular symmetry around the incident beam and is composed of eight sectors. Each sector contains 20 scintillation detector elements covering a range of scattering angles from 170° to 174°,

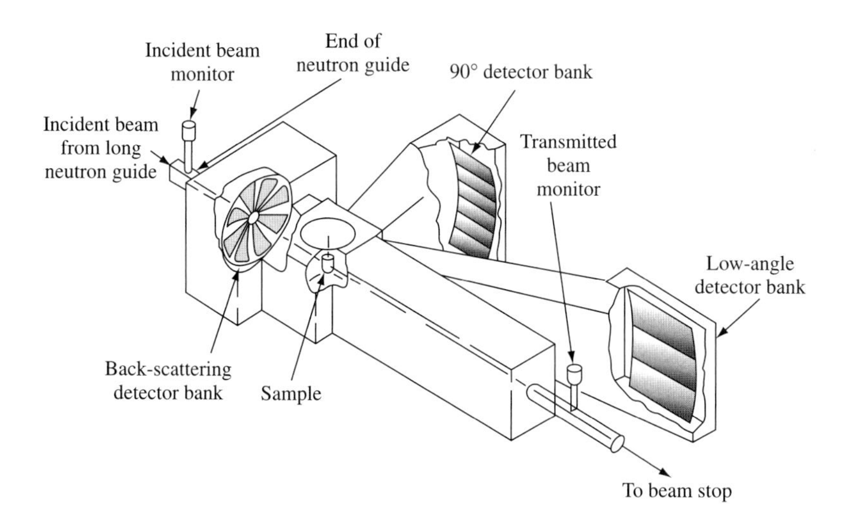

Fig. 8.4 The high-resolution powder diffractometer (HRPD) at the ISIS pulsed neutron source.

and the detectors record slightly different diffraction patterns which are combined by software to form a single intensity pattern $I(\lambda)$. Further banks of detectors are positioned at smaller angles of scattering: these record diffraction peaks at poorer resolution, but because of the lower 2θ-scattering angle they cover longer ranges of d-spacings than the back-scattering bank.

The General Materials Diffractometer (GEM) at the ISIS pulsed neutron source is a powder diffraction instrument, in which 2,500 detectors surround the sample (Williams *et al.*, 1998). The detector geometry is fixed during the measurements, thus simplifying the provision of a special environment, such as a cryostat or pressure cell, around the sample. The detector arrays cover a large area of 12 m^2, ensuring that a high proportion of the neutrons scattered by the sample are counted. It is possible to work with very small samples, down to a few mg in weight, and counting times can be as short as 10 sec, allowing the study of *in situ* chemical changes in the sample. Such an instrument combines, in effect, the angle-dispersive and energy-dispersive methods, thereby maximizing the amount of data collected. Monochromatic beam instruments on reactors can also be furnished with large area detectors; thus, the D2B instrument at the ILL employs position-sensitive detectors in both horizontal and vertical planes without compromising the high resolution.

The High Resolution Fourier Diffractometer HRFD, built at the pulsed source IBR-2 in Dubna (Russia), employs a mechanical Fourier chopper to achieve high resolution with the long moderator pulse ($\approx$350 μsec). The basic idea of this method is to measure the Fourier components of the TOF pattern at discrete, evenly spaced frequencies, and then to reconstitute the pattern off-line from these components. This approach suffers from practical difficulties and has prevented its widespread use, although impressive demonstration measurements on high-T_c superconductors have been made (Hiismäki, 1997).

Instruments such as GEM and D2B can be used to measure the total 4π scattering or *Total Differential Scattering Cross-Section* from a sample. This quantity is given by

$$\frac{d\sigma}{d\Omega} = I^S(Q) + I^D(Q) \tag{8.4}$$

where $I^S(Q)$ is the term from self-scattering and $I^D(Q)$ is the distinct scattering between pairs of different atoms. The self-scattering can be calculated approximately from the nuclear scattering lengths and subtracted from the data to give the distinct scattering. Structural information may then be obtained by a Fourier transformation of $I^D(Q)$, yielding the *differential correlation function*

$$D(r) = \frac{2}{\pi}\int_0^\infty QI^D(Q)\sin(rQ)dQ \tag{8.5}$$

$D(r)$ relates to the average scattering density at a distance r from the origin and gives information about the correlation between pairs of atoms. A peak in $D(r)$ at $r = r_0$ indicates that there is an interatomic distance r_0 which occurs frequently in the sample, and the area under this peak leads to the coordination number for r_0. Although the correlation function interpretation of diffraction data was developed for the study of non-crystalline materials, such as glasses and liquids (Chapter 13), it is also used for the study of disordered crystals. In measuring the whole diffraction pattern containing both Bragg and diffuse scattering, we obtain the 'long-range structure', as determined from the Bragg peaks, and the 'local structure' from the background diffuse scattering. We note that the function $D(r)$ is similar to the Patterson function in crystallography, in that information is derived directly from the diffraction analysis without assuming a model of the structure.

8.2 Resolution of powder diffraction peaks

The Bragg peaks in single crystal diffraction are usually well separated in three-dimensional reciprocal space. Powder diffraction data are collected in one dimension only, that is, as a function of $Q = 4\pi \sin\theta/\lambda$, and good resolution is essential in order to minimize the overlapping of adjacent diffraction peaks. In this section we shall consider various features affecting the resolution of the diffraction pattern in both fixed-wavelength and fixed-angle instruments. The word 'resolution' for a powder experiment usually means, as here, the width of a given diffraction peak and hence the degree of separation or overlap of adjacent peaks. The same word is used in a different sense for a single-crystal experiment, where it refers to the minimum d-spacing observed in the diffraction pattern and hence to the definition of individual features in the Fourier map of atomic density.

8.2.1 Effect of size of mosaic block

There is an ultimate limit to the resolution or sharpness of powder diffraction peaks which is caused by *particle-size broadening*. When all other sources of broadening have been removed, this effect remains. Each mosaic block of a crystallite in the powder sample consists of parallel planes of atoms repeating at regular intervals d (Fig. 8.5). Particle-size broadening arises because of the finite number of reflecting planes within each block: the smaller the number, the greater is the width of the diffraction peak. The effect is analogous to the broadening of the peaks of a diffraction grating which increases as the number of lines in the grating is reduced.

A simple derivation of the extent of this 'intrinsic broadening' is obtained by considering a parallel beam of monochromatic radiation, which strikes the planes in one mosaic block at a glancing angle θ. If θ is equal to the Bragg angle θ_B, the path difference between successive planes is exactly one wavelength, given by $\lambda = 2d \sin\theta_B$. The total path difference between the rays reflected from the top and the bottom of the

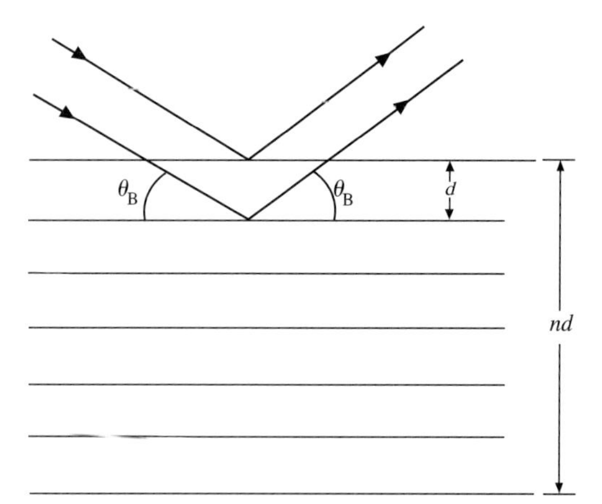

Fig. 8.5 Bragg scattering from parallel planes within one mosaic block of the crystal. nd is the total thickness of the block.

mosaic block is

$$n\lambda = 2d\sin\theta_B \tag{8.6}$$

where n is the total number of planes and $nd(=\Delta)$ is the thickness of the block. At the Bragg angle the planes in the entire block are scattering in phase.

Suppose now that θ is slightly increased to give an extra path difference between the top and bottom of the block of one wavelength. The top half of the block would then be 180° out of phase with the bottom half and the contributions of the two halves to the diffracted intensity will cancel one another out. Thus this condition occurs when θ_B is augmented by $\delta\theta_B$ in accordance with the expression

$$(n+1)\mathrm{p}\lambda = 2nd\sin(\theta_B + \delta\theta_B) \tag{8.7}$$

and the range of θ over which Bragg intensity occurs is $\pm\delta\theta_B$ with $\delta\theta_B$ given by eqn. (8.7). Combining this equation with eqn. (8.6), and assuming that $\delta\theta_B \ll 1$, gives

$$\lambda = 2nd\cos\theta_B\delta\theta_B \quad \text{or} \quad \delta\theta_B = \frac{\lambda}{2\Delta\cos\theta_B} \tag{8.8}$$

A rotation of the sample by $\delta\theta_B$ results in a rotation of the reflected beam by twice this angle, and so the angular width of the diffraction peak is

$$\delta\,(2\theta_B) = \frac{\lambda}{\Delta\cos\theta_B} \tag{8.9}$$

More exact treatments lead to the replacement of this equation by

$$\delta = K\frac{\lambda}{\Delta\cos\theta_B} \tag{8.10}$$

where K is a numerical constant, fairly close to unity, and whose precise value depends on the shape and the orientation of the mosaic blocks.

For a large perfect crystal such as dislocation-free silicon, the width of the Bragg reflection is of the order of a few seconds of arc. Most crystals reflect over a range much greater than this, indicating that the sizes Δ of the mosaic blocks lie roughly within the range

$$100\ \text{Å} < \Delta < 10{,}000\ \text{Å}$$

Thus for a powder with a block size of 1,000 Å, which is examined with a fixed-wavelength instrument operating at a wavelength of 1 Å and a Bragg angle of 60°, the particle-size broadening is about 0.1°. There will be little point in reducing the beam divergence below 0.1°, as this will cause a drop in intensity with no compensating gain in resolution.

8.2.2 The focusing effect: fixed-wavelength instruments

An important property of a fixed-wavelength instrument employing a crystal monochromator is the so-called *focusing effect*, which arises from the finite angular spread of the incident beam. The width of the Bragg reflections increases with this angular spread but passes through a minimum when the incident beam at the monochromator and the reflected beam at the detector are parallel.

A simplified treatment of the focusing effect is as follows. A more detailed account is given by Arndt and Willis (1966).

Suppose that the beam striking the monochromator has a divergence $\pm\alpha$ relative to its central path, and that the monochromator (which we assume to be a perfect crystal) is set at a Bragg angle θ_M to reflect the neutrons in the central path (Fig. 8.6) at a wavelength λ_M. Then the Bragg angles at the monochromator for the neutrons following the extreme paths in the scattering plane are $\theta_M + \alpha$ and $\theta_M - \alpha$ Differentiating the Bragg equation, an incremental change in Bragg angle $\Delta\theta$ is accompanied by a change in the reflected wavelength $\Delta\lambda$ given by

$$\Delta\lambda = \lambda_M \cot\theta \cdot \Delta\theta \tag{8.11}$$

Thus $\Delta\lambda$ is the range of wavelengths selected from the 'white beam' of neutrons issuing from the reactor source. The reflection of the beam by the monochromator at an angle $\theta_M + \alpha$ occurs at a wavelength of $\lambda_M(1+\alpha\cot\theta_M)$ and reflection at $\theta_M - \alpha$ occurs at $\lambda_M(1 - \alpha\cot\theta_M)$. Thus the monochromator introduces a correlation between wavelength and glancing angle, such that the shorter wavelengths in the primary beam are reflected at lower angles and the longer wavelengths at higher angles.

Let us now place the powder sample in the beam reflected by the monochromator, so that the sample diffracts the beam at a Bragg angle θ_B. In the 'parallel configuration' [Fig. 8.6(a)], the sample reflects the radiation in the same way as the monochromator: longer wavelengths strike the reflecting planes of the sample at higher glancing angles and the shorter wavelengths at lower angles. The entire beam from the

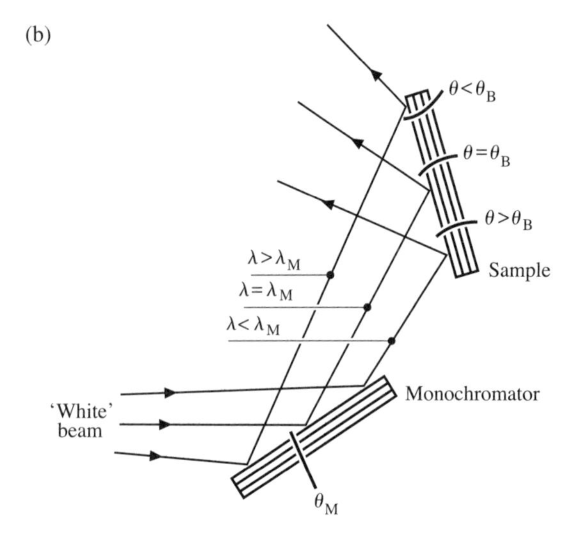

Fig. 8.6 (a) 'Parallel' and (b) 'antiparallel' configurations for Bragg scattering at a fixed wavelength.

monochromator tends to be reflected by a stationary sample. On the other hand, in the 'antiparallel configuration' [Fig. 8.6(b)] the longer wavelengths are incident at lower angles, and so the sample must be turned through a finite angle of rotation in order to reflect the whole beam. The width of the *rocking curve*, obtained by scanning the reflection through the angle 2θ, is considerably greater for the anti-parallel position. A powder diffractometer will normally operate in the parallel position, for which the width of the rocking curve passes through a minimum when the incident and twice reflected beams are parallel.

A detailed treatment of the resolution of powder diffraction peaks was derived 50 years ago by Caglioti *et al.* (1958). They calculated the dependence of the full-width at half-maximum (FWHM) of the powder peaks on the beam divergences between the source and monochromator (α_1), between the monochromator and sample (α_2), and between the sample and detector (α_3). They also allowed for the influence of the mosaic spread of the monochromator β. Writing H as the full width at half maximum (FWHM), Caglioti *et al.* derive the expression

$$H^2 = U \tan^2 \theta + V \tan \theta + W \tag{8.12}$$

where

$$U = \frac{4\left(\alpha_1^2\alpha_2^2 + \alpha_1^2\beta^2 + \alpha_2^2\beta^2\right)}{\tan^2\theta_{\mathrm{M}}\left(\alpha_1^2 + \alpha_2^2 + 4\beta^2\right)} \tag{8.13}$$

$$V = -\frac{4\alpha_2^2\left(\alpha_1^2 + 2\beta^2\right)}{\tan\theta_{\mathrm{M}}\left(\alpha_1^2 + \alpha_2^2 + 4\beta^2\right)} \tag{8.14}$$

and

$$W = \frac{\alpha_1^2\alpha_2^2 + \alpha_1^2\alpha_3^2 + \alpha_2^2\alpha_3^2 + 4\beta^2\left(\alpha_2^2 + \alpha_3^2\right)}{\alpha_1^2 + \alpha_2^2 + 4\beta^2} \tag{8.15}$$

From eqn. (8.12) the peak width H passes through a minimum value when

$$\tan\theta - \frac{V}{2U} \tag{8.16}$$

or

$$\tan\theta = \tan\theta_{\mathrm{M}} \frac{\alpha_2^2\left(\alpha_1^2 + 2\beta^2\right)}{\alpha_1^2\alpha_2^2 + \alpha_1^2\beta^2 + \alpha_2^2\beta^2} \tag{8.17}$$

If the initial beam divergence is small ($\alpha_1 \ll 1$), then strict focusing occurs at the parallel position (Fig 8.6(a)) when $\theta = \theta_{\mathrm{M}}$, in accordance with the simplified treatment given earlier.

One might imagine that the best choice of take-off angle $2\theta_{\mathrm{M}}$ for the monochromator would be about 90°, so that focusing would occur near the middle of the 2θ range of the diffraction pattern. However, it is preferable to choose a rather larger value of $2\theta_{\mathrm{M}}$, because the peaks in the pattern are more closely spaced at higher values of the scattering angle 2θ, and also because the peak width increases more rapidly for $2\theta > 2\theta_{\mathrm{M}}$ than for $2\theta < 2\theta_{\mathrm{M}}$.

Figure 8.7 shows the variation of peak width with scattering angle for powder samples of terbium iron garnet and Y_2BaNiO_5. The measurements were taken on the instrument D1A at the ILL, using a Bragg angle at the monochromator of 66°. The minimum of these curves occurs at a scattering angle 2θ close to $2\theta_{\mathrm{M}}$, and the resolution rapidly degrades when 2θ exceeds $2\theta_{\mathrm{M}}$.

8.2.3 Time-of-flight instruments

In a TOF diffractometer the resolution is determined primarily by the uncertainty in the distance traversed by the neutrons between the moderator and the detector (ΔL), the uncertainty in the corresponding TOF (Δt), and the uncertainty in the scattering angle ($\Delta 2\theta$). The factor ΔL arises mainly from the finite thickness of the moderator, but the

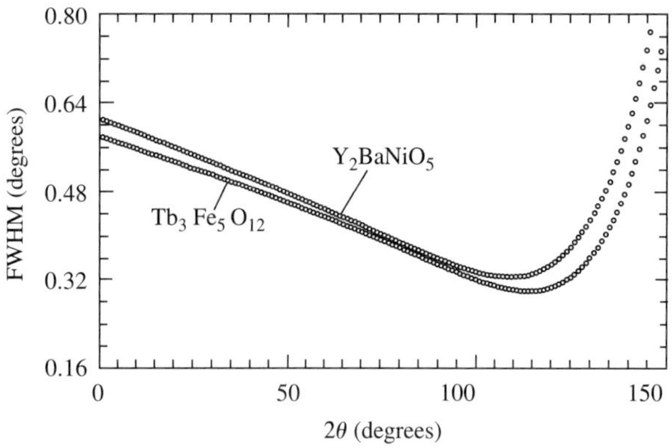

Fig. 8.7 Experimental resolution curves of the D1A instrument at the ILL, Grenoble. The FWHM passes through a minimum when the scattering angle 2θ is slightly less than $2\theta_{\mathrm{M}}$.

finite thicknesses of the sample and detector also contribute to ΔL. Uncertainty in the TOF occurs on account of the many collisions in the moderator which are required to reduce the neutron energy to the thermal region. Finally, $\Delta 2\theta$ is caused by variations in the angular divergence of the neutrons as they pass from moderator to sample and from sample to detector.

In theory, the Bragg peaks occur at sharply defined values of Q, but because of the uncertainties in lengths, times and angles, a particular time channel will record intensity over a finite range ΔQ, giving rise to a broadening of the diffraction peaks. Defining the instrumental resolution as $\Delta Q/Q$, and differentiating Bragg's equation in the form

$$Q = 4\pi\left(\frac{\sin\theta_{\mathrm{B}}}{\lambda}\right) = \frac{4\pi \mathrm{m}_n}{\mathrm{h}}\left(\frac{\mathrm{L}\sin\theta_{\mathrm{B}}}{\mathrm{t}}\right) \tag{8.18}$$

gives

$$\frac{\Delta Q}{Q} = -\frac{\Delta t}{t} + \frac{\Delta L}{L} + \cot\theta \cdot \Delta\theta \tag{8.19}$$

Assuming that the three terms on the right-hand side give independent contributions to the resolution, we obtain (since $\Delta Q/Q = \Delta d/d$)

$$\left(\frac{\Delta d}{d}\right)^2 = \left(\frac{\Delta t}{t}\right)^2 + \left(\frac{\Delta L}{L}\right)^2 + (\cot\theta_{\mathrm{B}} \cdot \Delta\theta_{\mathrm{B}})^2 \tag{8.20}$$

Equation (8.20) shows that the highest resolution is achieved in TOF diffraction by using long incident flight paths and scattering angles close to 180° (*i.e.* back scattering). Moreover, since Δt, the pulse width of neutrons emitted by the moderator, is approximately proportional to wavelength and t is precisely proportional to λ, then $\Delta t/t$ (and thus $\Delta d/d$) is broadly constant across the whole diffraction pattern recorded on a TOF diffractometer (Fig. 8.8). This contrasts with fixed-wavelength instruments for which the optimum resolution occurs near the focusing position and then rises rapidly at higher scattering angles.

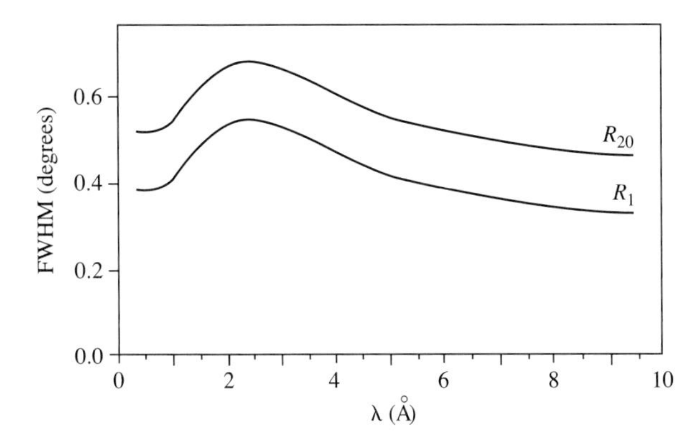

Fig. 8.8 Resolution of the time-of-flight instrument HRPD at ISIS. R_1 refers to the innermost detector at $2\theta = 178°$ and R_{20} to the outermost detector at $2\theta = 170°$.

8.3 Structure refinement from powder data

The solution of a crystal structure from a diffraction pattern normally requires the indexing of the pattern, the determination of the dimensions of the unit cell, the identification of the space group, and the solution of the phase problem (see Section 7.1.4). The approximate positions of the atoms in the unit cell are then derived by Fourier analysis. The final step is the refinement of this approximate model of the structure. The earlier steps are often accomplished by the analysis of X-ray diffraction data from a single crystal, but the final refinement step is particularly well suited to the analysis of neutron powder data.

The refinement is undertaken by the method invented by Hugo Rietveld (1969) and subsequently applied to the refinement of many thousands of crystal structures. Rietveld introduced the method originally for the refinement of powder patterns recorded with neutrons of fixed wavelength, but the method was later extended to the analysis of data obtained with either neutrons or X-rays and with the scattered radiation measured at a fixed wavelength or at a fixed scattering angle. A summary of the method is given by Albinati and Willis (2006) and a detailed review in the book edited by Young (1993). A useful set of practical guidelines for newcomers to the field of Rietveld refinement has been published by the *International Union of Crystallography* (McCusker *et al.*, 1999). If the data are of sufficiently high quality, up to 200 structural parameters can be successfully refined.

In the Rietveld method, the intensity profile of the entire diffraction pattern is calculated and compared with the observed profile. It is not necessary to derive the integrated intensities of individual diffraction peaks, one at a time, and so the method can be applied to any diffraction pattern no matter how severe is the overlap between adjacent peaks. The Rietveld algorithm minimizes by a least-squares process the function

$$\sum_i w_i \left| y_{io} - y_{ic} \right|^2 \tag{8.21}$$

where y_{io} is the observed intensity at the ith incremental step of the pattern, y_{ic} is the calculated intensity and w_i is the weight assigned to an individual observation. If the background intensity is small, and if the only source of error in measuring the intensities is that from counting statistics, a suitable weight for the observed intensity is given by

$$w_i = (y_{io})^{-1} \tag{8.22}$$

The summation in eqn. (8.21) is over all the steps at which the intensity is measured. The calculated intensity y_{ic} at the ith step is given by the sum of the contributions at that step in the pattern from all the neighbouring Bragg reflections k and from the background y_{ib}:

$$y_{ic} = s \sum_{hkl} \left(\left[m_k L_k \left| F_k \right|^2 f(2\theta_i^k) + y_{ib} \right] \right) \tag{8.23}$$

where s is a scale factor, m_k is the multiplicity (*i.e.* number of symmetry-equivalent planes of the kth reflection), L_k is the geometrical *Lorentz factor*, $2\theta_i$ is the scattering angle for the ith step, $2\theta_k$ is twice the Bragg angle of the kth reflection and $f(2\theta_i^k)$ is a function describing the shape of this reflection.

A fundamental problem of the Rietveld method is the formulation of a suitable function for the peak shape. For a neutron diffractometer operating at a fixed wavelength the peak-shape function is usually assumed to be a Gaussian of the form

$$f(2\theta_i^k) = \frac{2}{H_k} \sqrt{\frac{\ln 2}{\pi}} \exp\left(-\frac{4 \ln 2 \, (2\theta_i - 2\theta_k)}{H_k^2} \right) \tag{8.24}$$

where H_k is the FWHM of the peak. The angular dependence of the FWHM is given by Caglioti's eqn. (8.12):

$$H_k^2 = U \tan^2\theta + V \tan\theta + W$$

where U, V, W are treated as adjustable parameters allowing for intrinsic line broadening.

Many Bragg reflections can contribute to the intensity at the ith step of the pattern. The tails of a Gaussian function fall off rapidly beyond its half-width, and at a distance from the maximum of 1.5 times the FWHM the function has a value of only 0.2% of its peak value. Hence little error is introduced by assuming that the Gaussian in eqn. (8.24) extends over a range of $\pm 1.5\ H_{hkl}$ and that it is cut off outside this range. By combining this criterion

$$|2\theta_i - 2\theta_k| < 1.5 H_k \tag{8.25}$$

with eqn. (8.23), we can calculate the total intensity at a point $2\theta_i$ of the pattern and the calculation can be successively repeated at intervals of, say, 0.02° in θ_i. A least-squares refinement is then carried out to minimize the difference between the observed and calculated intensities and to optimize the parameters defining the peak-shape function and the parameters defining the crystal structure. The structural parameters include the dimensions of the unit cell, the positions of the atoms in the unit cell and their thermal displacement factors (or Debye–Waller factors).

The Rietveld method has also been applied widely to the refinement of TOF data. In this case the diffraction profiles are appreciably asymmetric on account of the time structure of the incident neutron pulse. The peak-shape function in eqn. (8.24) is replaced by a time function $f(t)$ which is asymmetrical and cannot be described by a single Gaussian. In an early approach (Von Dreele *et al.*, 1982) several Gaussians and an exponential function are convoluted together to describe $f(t)$. In spite of the extra number of variable parameters compared with a fixed-wavelength analysis, the Rietveld method has proved to be very successful for the refinement of TOF powder data.

The progress of the refinement can be followed by calculating the weighted profile R_{wp} and the goodness-of-fit GoF. These indices are defined by the expressions

$$R_{wp} = \left[\frac{\sum_i w_i \left|\mathbf{y}_{io} - y_{ic}\right|^2}{\sum_i w_i y_{ic}^2}\right]^{1/2} \tag{8.26}$$

and

$$\text{GoF} = \sum_i \frac{w_i \left(y_{io} - y_{ic}\right)^2}{(N - P)} \tag{8.27}$$

where N is the number of data points in the diffraction pattern and P is the number of refined parameters. The numerator in both these indices is the quantity being minimized. The GoF should approach the ideal value of unity, and R_{wp} should be close to the statistically expected R factor, which is given by

$$R_E = \left[\frac{(N - P)}{\sum_i w_i y_{io}^2}\right]^{1/2} \tag{8.28}$$

The power of Rietveld analysis is illustrated in Fig. 8.9, which shows the TOF diffraction pattern of a mixture of corundum, fluorite and zincite. The enlarged pattern for d-spacings in the range 0.5 Å $d <$ 1.0 Å is also shown. The entire pattern was recorded on HRPD at ISIS in less than 1 hour, and Rietveld analysis led to the determination of phase fractions which were accurate to $\pm 0.3\%$.

There are various problems that frequently arise in applying the Rietveld method and these are discussed by McCusker *et al.* (1999). These include making a proper allowance for the background [y_{ib} in eqn. (8.23)] and taking into account any preferred orientation of the

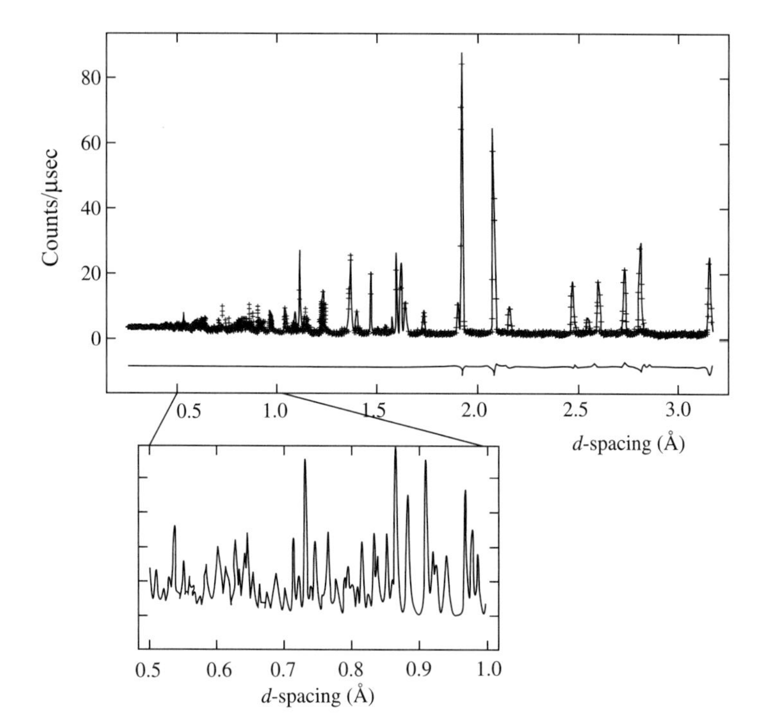

Fig. 8.9 Time-of-flight powder pattern of α-Al_2O_3, CaF_2 and ZnO. The lower plot in the upper pattern is the difference between the observed and calculated counts.

crystallites in the sample. (Ideally the crystallites should be distributed throughout the sample with random orientations.) The background may be determined by measuring regions of the pattern which are free from Bragg reflections. This procedure assumes that the background varies smoothly with $\sin\theta/\lambda$, whereas this is not strictly true: thermal diffuse scattering (TDS), for example, occurs at all values of Q but rises to a maximum at the Bragg positions. An alternative approach is to include a separate background function in the refinement (Richardson, 1993). If the background is not accounted for satisfactorily, the temperature factors may be incorrect or even negative (and thus meaningless).

Preferred orientation is a formidable problem which can drastically affect the measured intensities. A simple correction formula for plate-like morphology was given by Rietveld (1969) in his original paper. Ahtee *et al.* (1989) have shown how the effects of preferred orientation can be included in the refinement by expanding the expression describing the distribution of orientations in spherical harmonics. A general model for describing the orientation texture has been proposed by Popa (1992). These models may give only a crude approximation to the correction, and steps should be taken to minimize the preferred orientation by rotating the sample during data collection or by mixing the sample with a suitable diluent. Because of the large sample volume (up to 20 mm in diameter) and the high penetration of neutrons, the problem tends to be less severe than for X-ray diffraction.

8.4 Neutron strain scanning

Neutrons can be used to solve various problems in engineering. For example, small-angle neutron scattering (Chapter 10) gives information about microstructure, voids and sizes of precipitates in materials. The most widespread application of powder neutron diffraction in engineering is in the measurement of internal strain.

The technique for strain scanning was first developed at the Harwell Laboratory, UK, in the early 1970s, and since then it has been used worldwide by the materials science and engineering communities. X-rays are also used, but neutrons possess the ability to penetrate much more deeply inside the sample. Thus the $1/e$ depth for steel is about 1 cm for 1 Å neutrons but is less than 10 μm for X-rays of similar wavelength: neutrons can be employed as a probe for measuring bulk properties whereas X-rays are restricted to the examination of surface properties. The neutron technique is described in early review articles by Webster (1991) and Hutchings (1992).

Neutron Strain Scanning uses the crystal lattice as an atomic gauge to measure strain distributions in volumes of less than 1 mm^3. The principle of the method is illustrated in Fig. 8.10. The white beam from a reactor is monochromated by a large single crystal, and this beam passes through a Soller slit assembly before striking the sample. The neutrons scattered through an angle 2θ then pass through another Soller slit assembly which is positioned close to the detector. The incident and scattered beams are defined in area by neutron-absorbing masks, and the 'gauge volume' sampled by the diffractometer consists of the region defined by the intersection of these two tightly collimated beams. The direction in which the strain is measured is normal to the reflecting planes, that is, along the direction of the scattering vector **Q**. By moving a sample through the gauge volume the distribution of strain throughout the sample is measured. Strain is a second rank tensor and so at

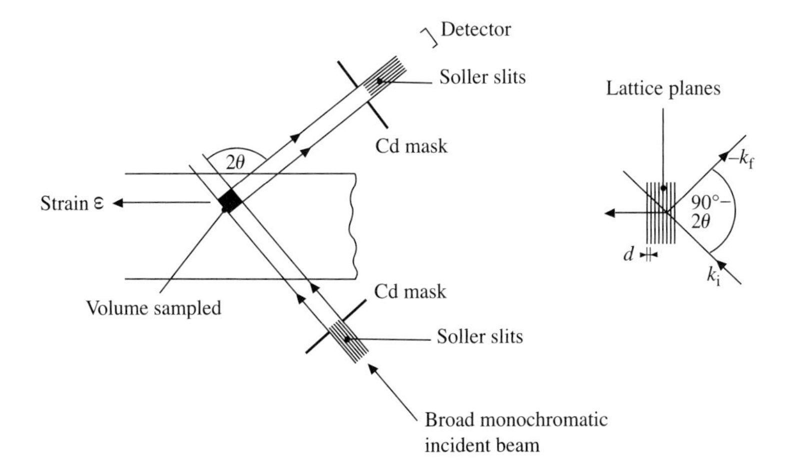

Fig. 8.10 The principles of strain measurement at a fixed neutron wavelength, showing the definition of gauge volume and the direction in which strain is measured.

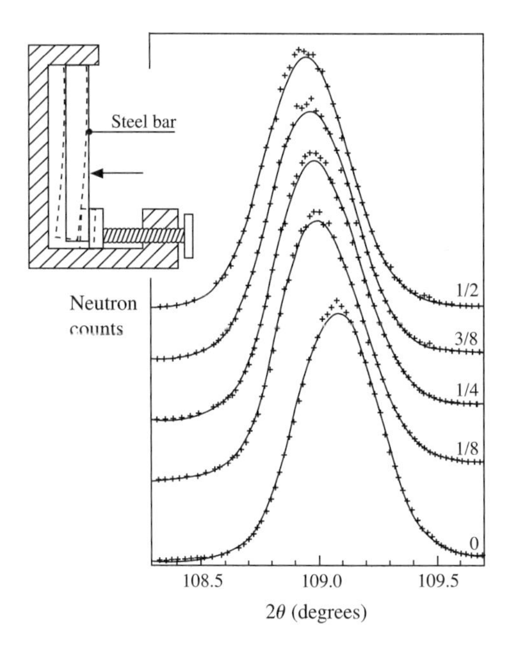

Fig. 8.11 Data showing the (211) diffraction peak of a mild steel bar for different strains using the D1A diffractometer at the ILL. The fractional labels on each curve refer to the number of turns of the vice screw shown (inset). The solid lines are the best fit of a Gaussian function to the data. (*After* A. J. Allen *et al.*, 1985.)

least six directionally independent measurements are required to determine the components of the strain tensor. Once the principal strains are known, the associated three-dimensional stress can be calculated using the appropriate compliance tensor.

If the measured spacing for a particular (hkl) reflection is d, and d_0 is the strain-free spacing, then the strain ε averaged over the gauge volume is $\varepsilon = (d - d_0)/d_0$. For a monochromatic beam of neutrons, a small change Δd in the lattice spacing results in a change $\Delta\theta$ in the angular position of the Bragg reflection, so that the strain in the direction of $\mathbf{Q}$ is given by

$$\varepsilon = \Delta d/d = -\cot\theta\Delta\theta \tag{8.29}$$

Peak shifts are at most a few tenths of a degree, and so good instrumental resolution is essential: for steel a peak shift of 0.01° at a scattering angle of 90° indicates a residual stress of about 20 Mpa. Figure 8.11 shows some measurements made by Allen *et al.* (1985) of the elastic strain in a mild steel bar which was clamped at one end and then subjected to a macroscopic strain by a vice screw exerted at the other end. The strains in a welded joint are shown in Fig. 8.12.

The TOF technique using pulsed neutrons is particularly well suited to the measurement of strain, because many (hkl) reflections of a polycrystalline material can be measured simultaneously at the same setting of the diffractometer. The strain is given by

$$\varepsilon = \frac{\Delta t}{t} \tag{8.30}$$

where Δt is the difference in total flight times for the strained and unstrained sample. ENGIN-X is a pulsed neutron instrument at the ISIS

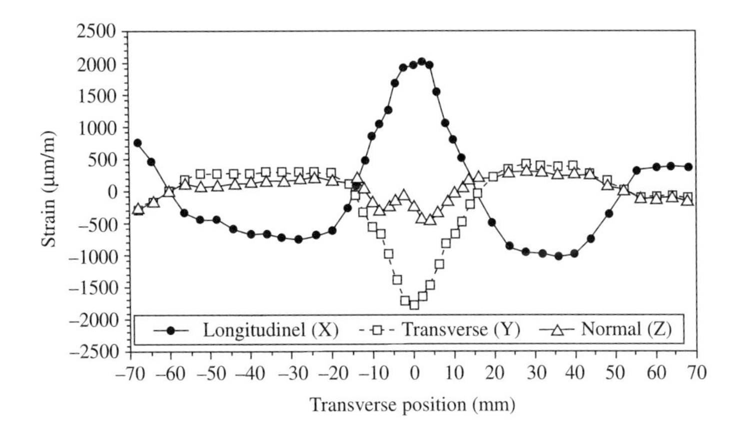

Fig. 8.12 Variation of the longitudinal, transverse and normal strain across a welded steel joint. (*After* Webster, 2003.)

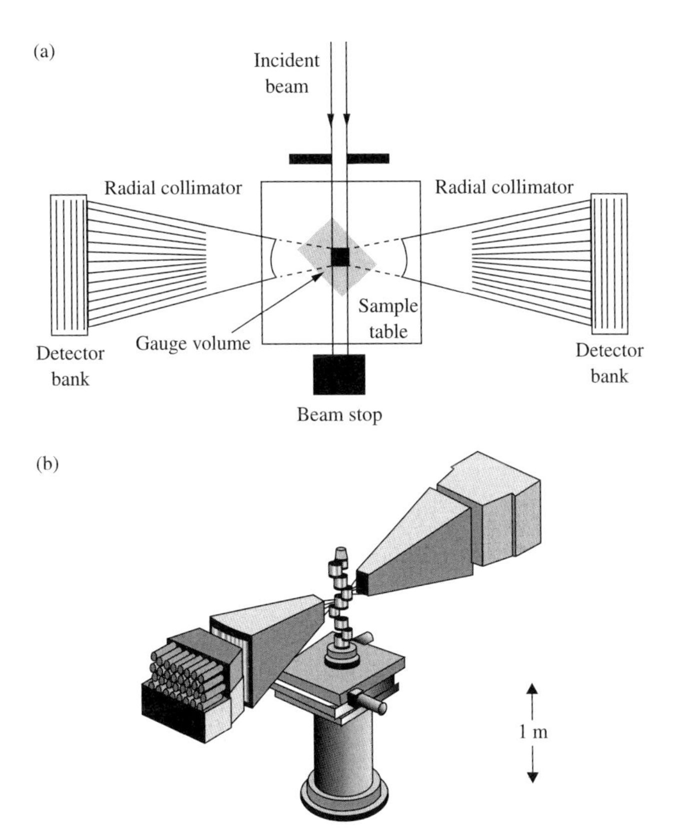

Fig. 8.13 (a) Measurement of strain by time-of-flight neutron diffraction. (b) The ENGIN-X instrument.

source of the Rutherford Appleton Laboratory, which has been designed solely with the aim of making strain measurements in large engineering samples. The design of the instrument is illustrated in Fig. 8.13(a) and a photograph in Fig. 8.13(b). Cold neutrons from a liquid methane moderator pass along a curved guide of 50 m length to the sample. The outgoing beam is defined by two sets of radial collimators, beyond which are two detector banks centred on horizontal scattering angles of ±90°. Testing can be performed at temperatures up to 1,200°C, and samples weighing up to 1.5 tonnes can be scanned along all three $x\,y\,z$ axes.

The SALSA strain imaging instrument at ILL is a modern purpose-built diffractometer operating on a reactor source and employing a novel hexapod sample orientation table.

The use of neutron diffraction to determine residual strains is now an established technique employed by the engineering community. A large number of measurements have been made on materials such as steels, brasses, aluminium and ceramics, and these results have provided information on the resistance to fracture of engineering components. The fact that large materials must be transported to a central neutron facility for examination is a constraint limiting the attraction of the technique.

8.5 Some chemical applications of powder diffraction

Neutron powder diffraction is a technique used widely by the neutron community. With an instrument such as GEM, equipped with numerous detectors surrounding the sample, data can be acquired rapidly and with high resolution. Figure 8.14 shows a TOF diffraction pattern from this instrument which was recorded in 10 sec. The fixed-wavelength ultra-high count-rate diffractometer D20 at the ILL has a comparable performance and has been used for rapid thermodiffractometry.

A wide range of chemical problems has been tackled by powder diffraction, including the exploration of phase diagrams, the kinetics of phase transitions, *in situ* studies of reaction kinetics, the study of non-stoichiometric compounds, high-pressure studies, determination of the structures of high-temperature oxide superconductors and the determination of magnetic structures. Here we shall select a few of these types of problem.

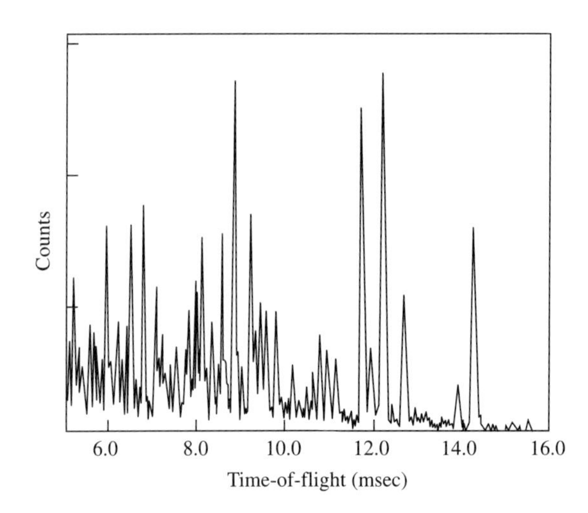

Fig. 8.14 Diffraction pattern of yttrium aluminium garnet (YAG) in a vanadium can of 8 mm diameter.

The structures have been determined of high-temperature oxide superconductors for which single crystals are not available (Hewat, 1990, 1992; Kamiyama *et al.*, 1994). In these superconductors, X-rays are scattered principally by the heavy metal atoms (such as barium or bismuth or the rare earths) but the oxygen atoms are of particular importance in understanding the superconducting properties. Thermal neutrons are scattered strongly by oxygen atoms, and there have been many neutron powder studies of $YBa_2Cu_3O_{7-\delta}$ and of related compounds with yttrium replaced by rare-earth ions. $YBa_2Cu_3O_{7-\delta}$ has a superconducting transition temperature, T_c, which is strongly dependent on δ: for δ close to zero T_c is 90 K, but T_c decreases rapidly for $\delta > 0.2$. The neutron powder work (Capponi *et al.*, 1987; Beyers and Shaw, 1989; Jorgensen *et al.*, 1990) has revealed the total oxygen content and the distribution of oxygen in the unit cell and shown that the reduction in T_c is accompanied by the formation of ordered oxygen vacancies. Earlier X-ray work on single crystals led to difficulties of interpretation on account of twinning: powder studies yield *d*-spacings of the sample and are unaffected by twinning.

A second type of problem is the study materials at high pressure. Condensed matter in the solar system includes more than 90% by mass of hydrogenous molecules such as hydrogen and water under pressures in excess of 10 GPa (100 kbar). Even on the earth a large amount of H_2O exists in minerals under high pressure in the crust and upper mantle. Neutron diffraction is a sensitive probe for determining the variation with pressure of bond lengths involving hydrogen. A high-pressure cell, known as the Paris–Edinburgh press, has been adapted to the TOF technique with fixed scattering angle and no moving components. When ordinary ice I*h* is compressed to 1.5 GPa at 77 K, it transforms to high-density amorphous ice (HDA) which remains metastable to structural transitions at ambient pressure. *In situ* neutron studies of the pressure dependence of the structure and recrystallization of HDA indicate that the isothermal compression is achieved by a 20% contraction of the second nearest neighbour oxygen–oxygen coordination shell (Klotz *et al.*, 2005). The structure of methane hydrate has been examined by Loveday *et al.* (2001) at pressures up to 10 GPa. Methane hydrate is believed to be the major CH_4-containing phase in Titan, the largest moon of Saturn. Figure 8.15 shows neutron diffraction patterns at increasing pressure from the deuterated hydrate. There are structural transitions at about 1 and 2 GPa to new hydrate phases which remain stable to at least 10 GPa. This work throws light on the reason for the continuing replenishment of Titan's atmospheric methane.

8.6 Future prospects

Powder samples are usually more readily available than single crystals, and data collection with powders is more rapid, making it cheaper

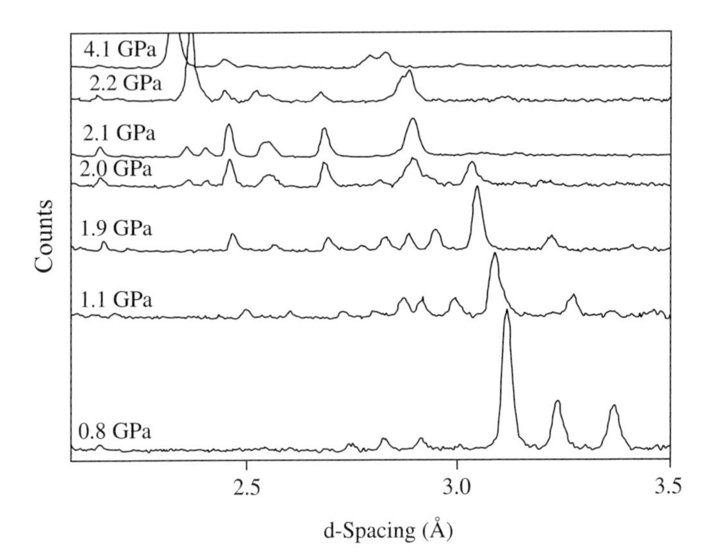

Fig. 8.15 Diffraction patterns showing the three phases of methane hydrate. (*After* Loveday *et al.*, 2001.)

than the collection of single-crystal data. There is increasing interest in measuring the whole diffraction pattern to obtain not only the 'long-range structure', as determined from the Bragg peaks, but also the 'local structure' determined from the diffuse scattering in the pattern.

New analytical techniques include the maximum entropy method (MEM), Monte Carlo techniques, and 'direct space' or reverse Monte Carlo (RMC) methods. The RMC technique was originally designed for the study of liquids and amorphous materials, and its extension to neutron powder diffraction data is described by Mellegard and McGreevy (1999). The RMC approach uses a plausible model of the scattering system, which may contain several thousands of atoms, and after Fourier transformation the calculated intensity is compared with the observed diffraction pattern. The atoms are then moved randomly and these moves are accepted in the model if the fit with the calculated cross-section is improved.

Progress in powder diffraction goes hand in hand with progress in developing new algorithms and computer programs for processing the data. More complex types of experiment are being increasingly tackled, such as the study of *in situ* chemical reactions and studies at high pressure and at high magnetic fields. These are exciting times in both neutron and X-ray powder diffraction, where new ideas can be rapidly tested using cheap desktop computing power, and where the Internet is available for distributing codes and programs.

References

Ahtee, M., Nurmela, M. and Suortti, P. (1989). *J. Appl. Crystallogr.* **22** 261–268. "*Correction for preferred orientation in Rietveld refinement.*"

Albinati, A. and Willis, B. T. M. (2006) *Int. Tables Crystallogr.* Volume C 710–712. "*The Rietveld method.*"

Allen, A. J., Hutchings, M. T. and Windsor, C. G. (1985) *Adv. Phys.* **34** 445–473. "*Neutron diffraction methods for the study of residual stress fields.*"

Arndt, U. W. and Willis, B. T. M. (1966) "*Single crystal diffractometry.*" University Press, Cambridge.

Caglioti, G., Paoletti, A. and Ricci, F. P. (1958) *Nucl. Instrum. Methods* **3** 223–228. "*Choice of collimators for a crystal spectrometer for neutron diffraction.*"

Capponi, J. J., Chaillout, C., Hewat, A. W., Lejay, P., Marezio, M., Nguyen, N., Raveau, B., Soubeyroux, J. L., Tholence, J. L. and Tournier, R. (1987) *Europhys. Lett.* **3** 1301–1307. "*Structure of the 100K superconductor* $YBa_2Cu_3O_7$ *between 5 and 300K by neutron powder diffraction.*"

David, W. I. F. (1992) *Physica B* **180–181** 567–574. "*Transformations in neutron powder diffraction.*"

Debye, P. and Scherrer, P. (1916) *Physik. Zeitschr.* **47** 277–283. "*Interferenzen an regellos orientierten Teilchenim Röntgenlicht.*"

Hewat, A. W. (1990) *Neutron News* **1** 28–34. "*Neutron powder diffraction and oxide superconductors.*"

Hewat, A. W. (1992) *Nuclear Sci. Eng.* **110** 408–416. "*Neutron diffraction, structural inorganic chemistry, and high- temperature superconductors.*"

Hiismäki, P. (1997) "*Modulation spectrometry of neutrons with diffractometry applications.*" World Scientific, Singapore.

Hull, A. W. (1917) *Phys. Rev.* **10** 661–696. "*A new method of X-ray analysis.*"

Hutchings, M. T. (1992) *Neutron News* **3** 14–19. "*Neutron diffraction measurement of residual stress fields-the engineer's dream come true?*"

Jorgensen, J. D., Veal, B. W., Paulikas, A. P., Nowicki, L. J., Crabtree, G. W., Claus, H. and Kwok, W. K. (1990) *Phys. Rev. B* **41** 1863–1877. "*Structural properties of oxygen-deficient* $YBa_2Cu_3O_{7-\delta}$."

Kaldis, E., Hewat, A. W., Hewat, E. A., Karpinski, J. and S. Rusiecki, S. (1989) *Physica* C **159** 668–680. "*Low temperature anomalies and pressure effects on the structure and* T_c *of the superconductor* $YBa_2Cu_3O_7$."

Kamiyama, T., Izumi, F., Takahashi, H., Jorgensen, J. D., Dabrowski, B., Hitterman, R. L., Hinks, D. G., Shaked, H., Mason, T. O. and Seabaugh, M. (1994) *Physica C* **229** 377–388. "*Pressure induced structural changes in* $Nd_{2-x}Ce_xCuO_4$ *(*$x = 0$ *and 0.165).*"

Klotz, S., Strässle, Th., Saita, A. M., Rousse, G., Hamel, G., Nelmes, R. J., Loveday, J. S. and Guthrie, M. (2005) *J. Phys. Condens. Matter* **17** S967–S974. "*In situ neutron diffraction studies of high density amorphous ice under pressure.*"

Loveday, J. S., Nelmes, R. J., Guthrie, M., Belmonte, S. A., Allan, D. R., Klug, D. D., Tse, J. S. and Handa, Y. P. (2001) *Nature* **410** 661–663. "*Stable methane hydrate above 2GPa and the source of Titan's atmospheric methane.*"

McCusker, L. B., Von Dreele, R. B., Cox, D. E., Louer, D. and Scardi, P. (1999) *J. Appl. Crystallogr.* **32** 36–50. "*Rietveld refinement guidelines.*"

Mellergard, A. and McGreevy, R. L. (1999) *Acta Crystallogr.* **A55** 783–789. "*Reverse Monte Carlo modelling of neutron powder diffraction data.*"

Popa, N. C. (1992) *J. Appl. Crystallogr.* **25** 611–616. "*Texture in Rietveld refinement.*"

Richardson, J. W. (1993) *Background modelling in Rietveld analysis* in "*The Rietveld Method.*" pp. 102–110, edited by Ray Young. (*IUCR Monographs on Crystallography.*) Oxford University Press, Oxford.

Rietveld, H. M. (1969) *J. Appl. Crystallogr.* **2** 65–71. "*A profile refinement method for nuclear and magnetic structures.*"

Von Dreele, R. B., Jorgensen, J. D. and C. G. Windsor, C. G. (1982) *Appl. Crystallogr.* **15** 581–589. "*Rietveld refinement with spallation neutron powder diffraction data.*"

Webster, P. J. (1991) *Neutron News* **2** 19–22. "*Neutron strain scanning.*"

Webster, P. J. (2003) *J. Phys. IV France* **103** 305–319. "*Analyse des constraintes résiduelles et de la texture dans les matériaux: applications industrielles.*"

Williams, W. G., Ibberson, R. M., Day, P. and Enderby, J. E. (1998) *Physica B* **241–243** 234–236. "*GEM: General materials diffractometer at ISIS.*"

Young, R. A. (editor) (1993) "*The Rietveld method.*" Oxford University Press, Oxford.

Polarized neutrons and magnetic neutron diffraction

9

9.1 Introduction

Atoms such as iron, cobalt, manganese and nickel, in the first transition series of elements, have incomplete $3d$ electronic shells with unpaired electrons. These unpaired electrons give rise to a resultant magnetic moment, and the interaction of this atomic moment with the magnetic moment of the neutron gives rise to *magnetic neutron scattering*. In this chapter, we shall be concerned with the magnetic scattering of polarized neutron beams by magnetic materials. Magnetic neutron scattering is always accompanied by nuclear neutron scattering, as described in Chapter 7.

The neutron has a spin quantum number of $1/2$, with an associated angular momentum and magnetic moment. Any component of the angular momentum along an external magnetic field, or quantization axis, can take only one of two values, *viz.* $\pm\hbar/2$. There are only two possible orientations of the neutrons, either parallel or antiparallel to the field (Fig. 9.1), and these are known as the *up* and *down* states of the neutron. If a neutron beam contains n_+ neutrons in the up-state and n_- in the down-state, the polarization P of the beam is defined as

$$P = \frac{n_+ - n_-}{n_+ + n_-} \tag{9.1}$$

Hence $P = \pm 1$ describes a completely polarized beam and $P = 0$ an unpolarized beam. This definition of polarization is satisfactory for experiments in which there is a single quantization axis. However, in the technique of three-dimensional neutron polarimetry (see Section 9.5.4) the polarization is a vector quantity.

9.2 Some fundamental concepts of magnetic scattering

Before describing the methods of polarizing the beam, we shall consider a few basic concepts of neutron scattering from magnetically ordered

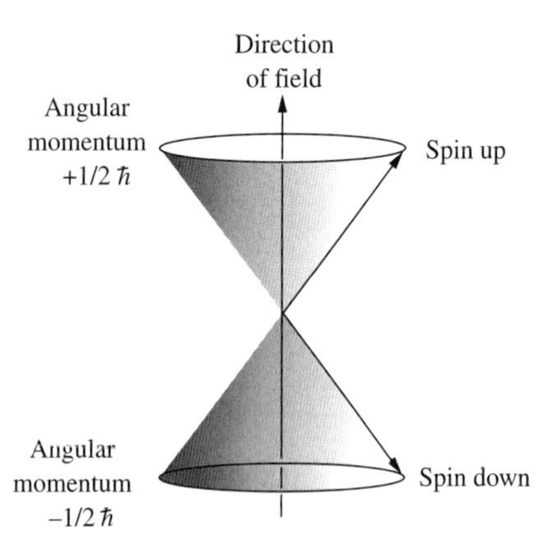

Fig. 9.1 Two possible spin orientations of a neutron in a magnetic field.

single crystals. Most of the theory required for understanding the principles of magnetic scattering is to be found in a remarkable set of papers by Halpern and Johnson (1937a, 1937b, 1939). The first two are short papers giving the principal results of a quantum mechanical treatment of magnetic neutron scattering, and the third paper gives a full derivation of these results. The theoretical results given in this section are also derived in Section 9.4 of the book by Squires (1996).

The magnetic moment of the neutron is $-\gamma\mu_N$, where the *nuclear magneton* μ_N is

$$\mu_N = \frac{e\hbar}{2m_n} \tag{9.2}$$

where m_n is the mass of the neutron, and $\gamma\,(=1.913)$ is the gyromagnetic ratio which is the proportionality constant between the magnetic moment and the angular momentum. μ_N is over 1,000 times smaller than the magnetic moment of an electron and yet the amplitude of magnetic neutron scattering from an atom is of the same order of magnitude as the amplitude of nuclear neutron scattering. The magnetic scattering length, p, of an atom is related to the atomic spin S by

$$p = \left(\frac{\mu_0}{4\pi}\right)\left(\frac{e^2}{m_e}\right)\gamma S f(\mathbf{Q}) \tag{9.3}$$

where $f(\mathbf{Q})$ is the magnetic form factor which is unity for $\mathbf{Q} = 0$ (see Fig. 2.10). It is assumed that the orbital component of the atomic moment is completely quenched. μ_0 is the permeability of a vacuum and the factor $(\mu_0/4\pi)(e^2/m_e)$ is the classical radius of an electron, which is 2.8×10^{-15} m. The magnitude of p approaches that of the nuclear scattering amplitude b, which is in the range 0.5–1.0×10^{-14} m for most atoms.

The structure factor $F_M(\mathbf{Q})$ for elastic magnetic scattering from a single crystal is given by

$$F_{\mathrm{M}}(\mathbf{Q}) = \sum_j f_j\,(\mathbf{Q}) p_j \mathbf{q}_j \exp\{i\mathbf{Q}\cdot\mathbf{r}_j\}\exp(-W_j) \qquad (9.4)$$

where the summation is over the magnetic atoms j in the unit cell, p_j is the magnetic scattering length of atom j whose position in the cell is given by the vector $\mathbf{r}_j$, and W_j is the exponent of the Debye–Waller temperature factor. $\mathbf{q}_j$ is the *magnetic interaction vector*, or *Halpern vector*, defined by

$$\mathbf{q}_j = \hat{\mathbf{Q}}\left(\hat{\mathbf{Q}}\cdot\hat{\boldsymbol{\mu}}_j\right) - \hat{\boldsymbol{\mu}}_j \qquad (9.5)$$

$\hat{\mathbf{Q}}\cdot\hat{\boldsymbol{\mu}}_j$ in eqn. (9.5) is the scalar product of the unit vectors $\hat{\mathbf{Q}}$ and $\hat{\boldsymbol{\mu}}_j : \mu_j$ is along the direction of the magnetic moment of atom j and for Bragg scattering $\mathbf{Q}$ is perpendicular to the hkl reflecting planes (Fig. 9.2).

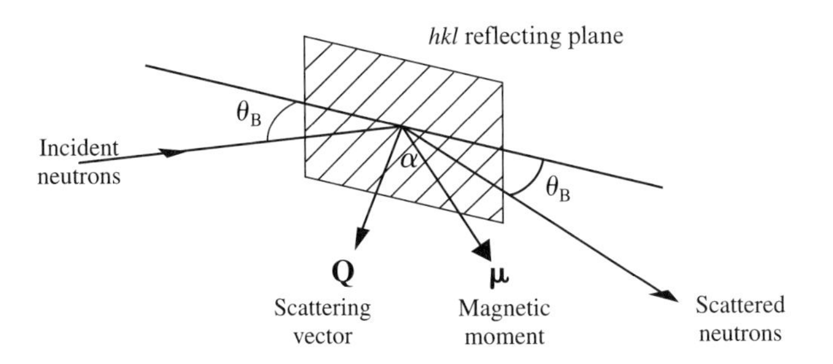

Fig. 9.2 The magnetic interaction vector lies in the plane of $\mathbf{Q}$ and $\boldsymbol{\mu}$ and is perpendicular to $\mathbf{Q}$. Its magnitude is the sine of the angle (α) between $\mathbf{Q}$ and $\boldsymbol{\mu}$.

Equation (9.5) shows that, in order for an atom to contribute to the scattered intensity, it must possess a component of its magnetic moment perpendicular to the scattering vector $\mathbf{Q}$. Hence for a ferromagnetic or antiferromagnetic crystal with moments aligned along a single direction, there is no magnetic scattering when the moments are along the direction of the scattering vector and maximum scattering when the moments are perpendicular to $\mathbf{Q}$. This is illustrated in the cubic crystal of Fig. 9.3, where there is maximum magnetic scattering from the (100) planes and zero scattering from the (010) planes.

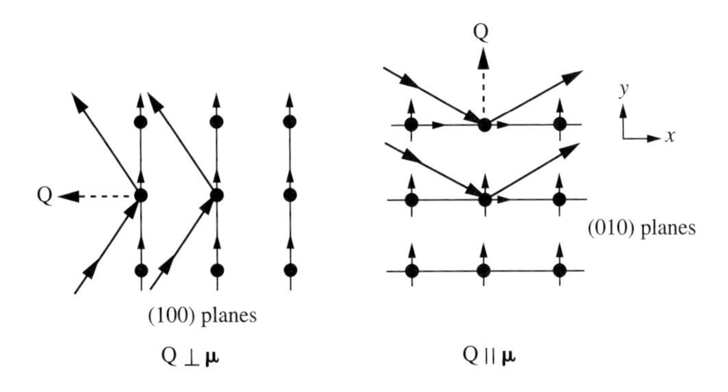

Fig. 9.3 Scattering from cubic crystal with atomic moments $\boldsymbol{\mu}$ aligned vertically. $\mathbf{Q}$ is the scattering vector.

9.3 Production of a polarized neutron beam

There are several ways of producing a beam of neutrons which is partially or completely polarized. Different methods are appropriate in the cold, thermal and epithermal regions of the energy spectrum, and the choice for a particular instrument is determined by the intended application. Here we shall refer briefly to the three principal methods: Bragg reflection from a magnetized single crystal, total reflection from a magnetized mirror or supermirror, and transmission through a polarizing filter.

9.3.1 Polarizing single crystals

Ferromagnetic single crystals can simultaneously polarize and monochromatize a neutron beam. The principle is illustrated in Fig. 9.4. A magnetic field $\mathbf{B}$ is applied perpendicular to the scattering vector $\mathbf{Q}$ of the (hkl) reflecting planes. The field saturates the atomic moments of the crystal along the field direction, and there is maximum magnetic scattering because the atomic moments are at right angles to the scattering vector.

The differential scattering cross-section for Bragg reflection from the (hkl) planes is

$$\frac{\mathrm{d}\sigma}{\mathrm{d}\Omega} = F_{\mathrm{N}}^2(\mathbf{Q}) + 2\left(\hat{\mathbf{P}}\cdot\hat{\boldsymbol{\mu}}\right)|F_{\mathrm{N}}(Q)|\,|F_{\mathrm{M}}(Q)| + F_{\mathrm{M}}^2(\mathbf{Q}) \qquad (9.6)$$

where F_{N} and F_{M} are the nuclear and magnetic structure factors. The unit vector $\hat{\mathbf{P}}$ describes the polarization of the incident beam with respect to $\mathbf{B}$ and $\hat{\boldsymbol{\mu}}$ is a unit vector along the direction of the atomic moments. The magnitude of P is $+1$ for up states of the neutron and -1 for down states. Hence $\hat{\mathbf{P}}\cdot\hat{\boldsymbol{\mu}} = +1$ for neutrons polarized parallel to the field and $(\hat{\mathbf{P}}\cdot\hat{\boldsymbol{\mu}}) = -1$ for neutrons polarized antiparallel to the

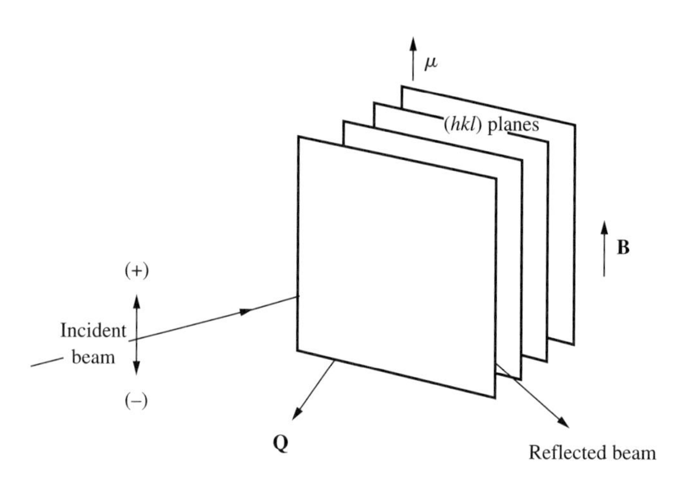

Fig. 9.4 Geometry of a polarizing monochromator.

Table 9.1 Properties of some polarizing crystal monochromators.

	$Co_{0.92}Fe_{0.08}$	Cu_2MnAl	Fe: ^{57}Fe
Matching (hkl) with $F_N = F_M$	(200)	(111)	(110)
d-Spacing	1.76 (Å)	3.43 (Å)	2.03 (Å)
Take-off angle $2\theta_M$ at 1 Å	33.1°	16.7°	28.6°
Maximum wavelength	3.5 Å	6.9 Å	4.1 Å

field. Equation (9.6) then becomes

$$\frac{d\sigma}{d\Omega} = |F_N(Q) + F_M(Q)|^2 \tag{9.7}$$

for neutrons in the up-state, and

$$\frac{d\sigma}{d\Omega} = |F_N(Q) - F_M(Q)|^2 \tag{9.8}$$

for neutrons in the down-state. Clearly, the reflected beam is completely polarized if the nuclear and magnetic structure factors are equal.

Single crystals of iron and nickel are unsuitable as polarizers because their nuclear structure factors are too large to be balanced by their magnetic structure factors. An exception is the isotope ^{57}Fe which has a lower nuclear scattering length than natural Fe and the matching condition $F_N(\boldsymbol{Q}) = F_M(\boldsymbol{Q})$ can be achieved by mixing Fe and ^{57}Fe in appropriate proportions. This condition is also satisfied for a few alloy crystals: for example, the (200) reflection of the alloy $Co_{0.92}Fe_{0.08}$, and the (111) reflection of the Heusler alloy Cu_2MnAl (see Table 9.1). The main drawback of $Co_{0.92}Fe_{0.08}$ is the relatively high absorption of cobalt: it has been largely superseded by the Heusler alloy but care must be taken with this alloy to avoid contamination of the (111) reflection with the second-order ($\lambda/2$) nuclear scattering from the strong (222) reflection. The maximum possible wavelength occurs at a take-off angle $2\theta_M$ of 180°: at a wavelength of 1 Å the take-off angle of Cu_2MnAl is small and so the diffraction data from a reactor source with this monochromator are obtained at relatively poor resolution (see Section 8.2.2).

9.3.2 Polarizing mirrors

The production of polarized neutrons by total reflection from ferromagnetic mirrors was proposed originally by Halpern (1949) and subsequently demonstrated by Hughes and Burgy (1951) with a cobalt mirror.

For a ferromagnetic material the neutron refractive index n (see Chapter 11) is given by

$$n = 1 - \frac{\lambda^2 N}{2\pi} \left(\bar{b} \pm p\right) \tag{9.9}$$

where $\bar{b}$ is the mean coherent scattering length, N is the number density of scattering nuclei and p is the magnetic scattering length. The $+$ and $-$ signs in eqn. (9.9) refer to neutrons with moments parallel and antiparallel to the applied field. The critical glancing angle θ_c for total reflection is given by

$$n = \cos\theta_c \approx 1 - \frac{1}{2}\theta_c^2 \tag{9.10}$$

and combining this with eqn. (9.9) gives

$$\theta_c^{\pm} = \lambda\left[\left(\frac{N}{\pi}\right)(\bar{b} \pm p)\right]^2 \tag{9.11}$$

There are two critical angles, θ_c^+ and θ_c^-, and the beam is fully polarized if the glancing angle of incidence lies between these angles. A major limitation of such a mirror is that the critical angles are very small, typically 10 minutes of arc. Consequently, the grazing angle of incidence and the angular range of polarization are small and a long and cumbersome device is required to polarize a wide beam. For example, to polarize a beam 2 cm wide the mirror must be several metres in length.

A more practical polarizer which partially overcomes the above limitation uses thin films made by evaporating on to substrates of low surface roughness, such as float glass or polished silicon. A two-dimensional multi-structure is made from alternate layers of a magnetic material A and a non-magnetic material B. The reflectivity of the multi-structure is proportional to

$$R = [N_A(b_A \pm p_A) - N_B b_B]^2 \tag{9.12}$$

where N is the number of atoms in each layer. With a judicious choice of N, b_A and p_A it is possible to have $R = 0$ by matching the refractive indices of the magnetic and non-magnetic layers for one of the polarization states of the neutrons, so that this state is not reflected. Modern deposition techniques allow a ready adjustment of the refractive index and matching is easily achieved. These non-reflected neutrons can be absorbed at very low angles in a layer of Gd. If the spacing of the bilayers is d, there is a Bragg peak of polarized neutrons at a scattering angle 2θ given by $\sin\theta = \lambda/2d$. The thin films are deposited on to a substrate such as float glass or polished silicon.

A broadband polarized beam is obtained by using polarizing *supermirrors* (see Section 5.1), in which there is a gradient of lattice spacing of the bilayers, resulting in an extension of the range of θ/λ values beyond those expected for normal mirror reflections. Hoghøj *et al.* (1999) describe the neutron polarizing Fe/Si supermirrors at the Institut Laue-Langevin (ILL). In Fig. 9.5, we show some results obtained by Stunault *et al.* (2006) with a Fe/Si supermirror deposited on a single crystal silicon wafer. To ensure that the neutrons are reflected at least once, the mirror can be gently bent forming a 'neutron beam bender'. Polarizing multilayers can reach a polarization efficiency of 97%, but their use is restricted to wavelengths in excess of 2 Å on account of the low angles of incidence.

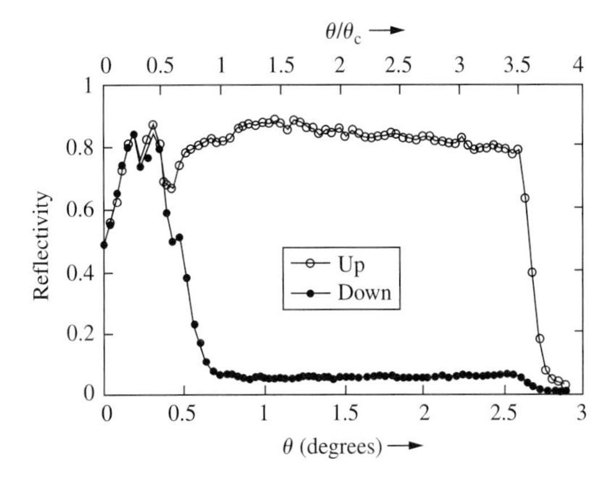

Fig. 9.5 The reflectivity of a Fe/Si supermirror for neutrons in the up-state and down-state. The apparently poor performance at low-scattering angles is an artefact caused by under-illumination. (*After* Stunault *et al.*, 2006.)

9.3.3 Polarizing filters

An extensive description of polarizing filters is given in the article by Anderson and Schärpf (1999). A polarizing filter removes one of the spin states from the incident beam, allowing the other spin state to be transmitted, albeit with some attenuation. This can be carried out either by preferential absorption or preferential scattering. If T^+ and T^- are the transmittance of the *up* and *down* neutron states, the polarizing efficiency P is

$$P = \frac{T^+ - T^-}{T^+ + T^-} \tag{9.13}$$

and the total transmittance T is

$$T = \frac{T^+ + T^-}{2} \tag{9.14}$$

The polarizing efficiency increases with the thickness of the filter but the transmittance decreases. A compromise must be made therefore between polarization P and transmittance T. The 'quality factor' $P\sqrt{T}$ is often used to optimize the filter thickness (see Fig. 9.6), with the optimum thickness chosen to correspond with the maximum value of $P\sqrt{T}$.

The most widely used polarization filters exploit the spin-dependence of the absorption cross-section of polarized nuclei at their nuclear resonance energy, and provide polarization over a wide range of energy and scattering angles. The total cross-section for the two neutron spin states is given by

$$\sigma_\pm = \sigma_0 \pm \sigma_p \tag{9.15}$$

where σ_0 is the spin-independent cross-section for absorption and scattering and σ_p is the polarization cross-section. It can be shown that to achieve a polarization in excess of 95% the ratio σ_o/σ_p must be greater than 0.65.

Alvarez and Bloch (1940) produced the first polarizing filter by passing a neutron beam through magnetized iron plates. The cross-section from spin-dependent Bragg scattering approaches 10 barns near the Bragg

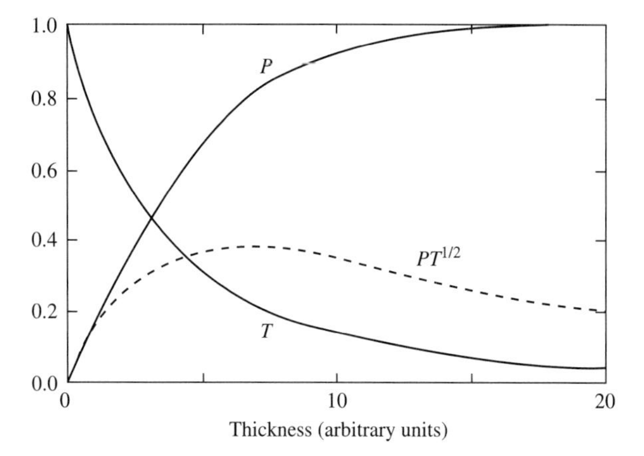

Fig. 9.6 The neutron polarization P and transmission T for a ^{3}He filter which has 55% nuclear polarization. The peak of the factor $PT^{1/2}$ indicates that the optimum polarization of the neutrons is about 80% for a transmission of about 30%. (*After* Heil *et al.*, 1999.)

cut-off at 4 Å, and for wavelengths just below 4 Å the transmittance is about 30% with polarizing efficiencies up to 50%.

Nuclei with strong resonance energies in the range 0.1–1 eV include ^{149}Sm, ^{151}Eu, ^{163}Dy and ^{165}Ho and nuclear polarization is achieved by cooling to very low temperatures (≈10 mK) in a high magnetic field. A ^{149}Sm filter has a polarizing efficiency close to unity in a narrow wavelength range (0.85–1.10 Å) but the transmittance is only 15%. ^{3}He gas at high pressure is an ideal material for a neutron polarizer (Andersen *et al.*, 2005). In such a filter the very high spin-dependent neutron capture cross-section of ^{3}He gas is dominated by the resonance capture of neutrons with down spin. The ^{3}He nuclei can be polarized by optical pumping using high power lasers, followed by compression of the gas. The Tyrex filling station at the ILL for nuclear polarization of ^{3}He gas can provide about 1.3 bar/h of polarized ^{3}He gas. This device uses metastable-exchange optical pumping to polarize the ^{3}He gas at a low pressure of ≈1 mbar. The gas is then compressed up to high pressure (≈4 bar) using a hydraulic titanium-alloy piston compressor. The polarized gas is transported to the instrument in a magnetic shield and decanted into the polarizer or analyser. Relaxation of the polarization arises from collisions with the walls of the container and from stray magnetic fields, and so the filter must be recharged from time to time. Under optimum conditions recharging is made at daily intervals. Figure 9.6 shows the polarization and transmission of a neutron beam passing through a filter with 55% nuclear polarization. Optimum polarization of the neutrons is about 80% even at this relatively low polarization of the nuclei (Heil *et al.*, 1999).

9.4 Neutron spin flippers and devices for guiding the polarization

A polarized beam will tend to become depolarized during its passage through a region of zero field. In order to keep the neutron spins aligned along a definite quantization axis, it is necessary to use a guide field of the order of $0.01\,T$ to produce a well-defined field over the whole flight path of the beam.

A device which reverses the polarization direction with respect to the guide field is called a *π-flipper.* A *$\pi/2$-flipper* produces a 90° rotation, turning the polarization at right angles to the guide field. Instruments operating with polarized neutron beams require one or more such devices for changing the direction of polarization. To understand their operation we consider first the action of a magnetic field on the magnetic moment of the neutron. The neutron carries a magnetic moment $\boldsymbol{\mu} = -1.913\mu_{\mathrm{N}}$, where μ_{N} is the nuclear magneton. In a magnetic field $\mathbf{B}$ exerts a torque $\mathbf{T}$ on the magnetic moment, and is the vector product

$$\mathbf{T} = \boldsymbol{\mu} \times \mathbf{B} \tag{9.16}$$

This torque would tend to turn the magnetic moment towards the field direction. However, the magnetic moment is associated with angular momentum, and because torque is proportional to the rate of change of angular momentum, we can write

$$\frac{\mathrm{d}\boldsymbol{\mu}}{\mathrm{d}t} \propto \mu \times \mathbf{B} \tag{9.17}$$

The equation of motion (9.17) governs the change in neutron spin state, and the effect of the field is to cause μ to *precess* about $\mathbf{B}$ rather than to turn towards $\mathbf{B}$. If μ is parallel to the local field $\mathbf{B}$, the vector product $\boldsymbol{\mu} \times \mathbf{B}$ is zero and there is no change in neutron spin state. The angular frequency of precession is known as the Larmor angular frequency ω_{L} and is given by

$$\omega_{\mathrm{L}}\ (\mathrm{kHz}) = 2.916 \times 10^{-4}\ B\ (\mathrm{Tesla}) \tag{9.18}$$

Thus for a field of 0.1 mT the Larmor angular frequency is 1.832×10^{4} rad/sec (or 2,916 kHz). If the magnetic field rotates at a frequency ω_{B} which is much less than the Larmor frequency, the component of the polarization parallel to the field maintains its direction along the field while the transverse component precesses around $\mathbf{B}$. This is an adiabatic change of polarization with the system always very close to equilibrium, and to guarantee adiabaticity it is necessary that the ratio $\omega_{\mathrm{L}}/\omega_{\mathrm{B}} > 10$, where ω_{B} is the frequency of rotation of the guide field. If the field $\mathbf{B}$ changes non-adiabatically, and at a rate much faster than the Larmor frequency, to a field $\mathbf{B}'$ lying along another direction, the polarization vector begins to precess about this new direction.

These processes are illustrated in Fig. 9.7. In slowly varying fields the neutron follows the field direction adiabatically, as shown in Fig. 9.7(a) where the spin turns through 90°. On the other hand, when the rotational frequency is much more than the Larmor frequency ($\omega_{\mathrm{B}}/\omega_{\mathrm{L}} > 10$), the spin direction cannot follow the field and the polarization vector remains fixed in space when the field direction is changed. Such a field flip allows us to change the relative orientation of the field and the neutron

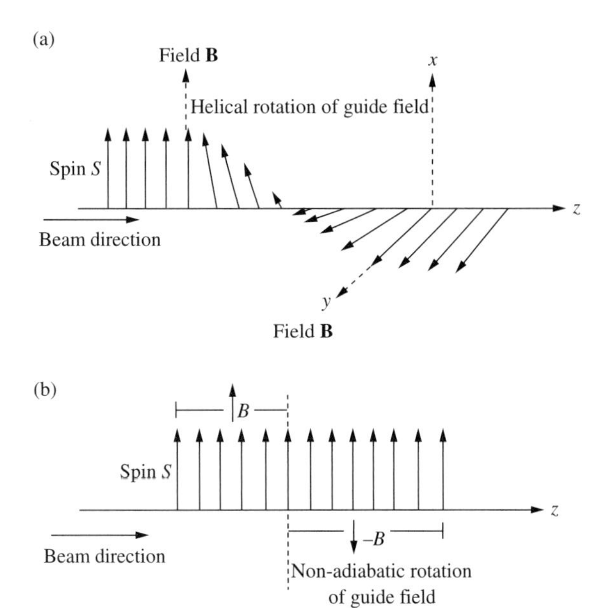

Fig. 9.7 (a) Adiabatic rotation of the magnetic guide field **B** from x direction to y direction. The neutron spins follow the slowly varying guide field. (b) Non-adiabatic rotation of the guide field from **B** direction to –**B** direction.

spin. This non-adiabatic rotation, as illustrated in Fig. 9.7(b), effectively flips the neutron spin by 180° with respect to the guide field.

Jones and Williams (1978) have described the construction and performance of the Drabkin two-coil and the Dabbs-foil π-flippers. Both flippers act non-adiabatically with a measured flipping efficiency close to 100%. The Drabkin flipper is very simple, consisting of two d.c. coils arranged coaxially about the neutron beam so as to produce magnetic fields of opposite polarity. The fields cancel midway between the coils and it is here that 'field flipping' occurs. Flipping takes place independent of the neutron velocity and so the device can be used to flip the neutrons in a white beam. It has the advantage that there is no material in the beam but it has the drawback of being restricted to thin beams only. The Dabbs aluminium foil flipper does not have this advantage but it can operate with a wider neutron beam.

The Mezei spin flipper (Hayter, 1978) uses both non-adiabatic and adiabatic processes to project the polarization of the beam on to any arbitrary direction. The Mezei coil is a thin rectangular solenoid, arranged at an angle to the guide field (Fig. 9.8). Before entering the coil the spins are parallel to a homogeneous external field. Immediately on entry they experience a resultant internal field pointing in a new direction and consisting of the external field plus the coil field. Larmor precession starts around this resultant field and the neutrons leave the coil with a new spin direction determined by the resultant field, the thickness of the coil and the neutron velocity. For example, a 1 Å neutron will be turned through π radians in a distance inside the coil of 1 cm if the resultant internal field is perpendicular to the guide field and of magnitude only 7 mT.

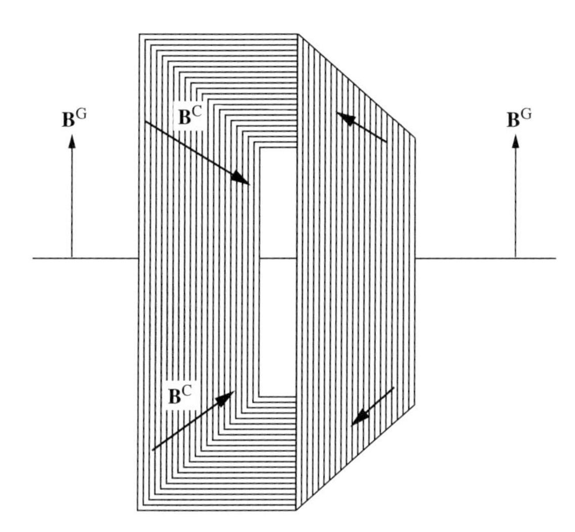

Fig. 9.8 Mezei coil flipper. $\mathbf{B}^{\mathrm{G}}$ is the guide field and $\mathbf{B}^{\mathrm{C}}$ the coil field. Inside the thin rectangular coil the field is the resultant of $\mathbf{B}^{\mathrm{G}}$ and $\mathbf{B}^{\mathrm{C}}$.

There are many other designs of spin flippers and spin turners, and these are well described in the book of Williams (1988). A more recent development is the system used in the CRYOPAD (see Section 9.5.4). Here the *Meissner effect*, in which a magnetic field is expelled from a superconductor, is used to separate the fields of the incident and scattered neutron beams, so that the two beams can be polarized independently of each other and in any direction in space.

9.5 Experimental methods

The study of magnetic materials by neutron scattering is described in the book by Izyumov *et al.* (1990) and in the more recent book edited by Chatterji (2006). This study often begins with a conventional powder diffractometer. The particular advantages of this instrument are high flux and good coverage at low-Q. However, to unravel the details of the magnetic structure it is usually necessary to examine a single-crystal sample and to use an experimental technique employing polarized neutrons. The simplest technique is the measurement of *flipping ratios* using a single spin flipper. In *unidirectional polarization analysis*, there are two spin flippers and the scattered intensity is measured for spin flip and non-spin flip along a single axis. Finally, in *three-dimensional neutron polarimetry* the final polarization of the scattered beam is measured independently of the spin state of the incident beam.

Table 9.2 summarizes, in the order of increasing sophistication, the experimental methods using polarized neutrons.

9.5.1 Unpolarized neutrons

Relatively simple magnetic structures can be investigated by powder diffraction using unpolarized neutrons. Powder diffractometers such as D1B (Chapter 8) have an effective cut-off in Q arising from the magnetic

Table 9.2 Different techniques used for studying the properties of magnetic materials.

Technique	Sample type	Experimental measurement	Applications
Unpolarized neutrons	Powder or single crystal	Total cross-section (nuclear and magnetic)	Magnetic properties of simple collinear ferromagnets and antiferromagnets
Polarized neutrons with one spin flipper	Ferromagnetic single crystal	Flipping ratios	Magnetic form factors; spin density distributions
Uniaxial polarization analysis	Antiferromagnets and paramagnets	Four spin-state cross-sections (↑↑,↑↓, ↓↓, ↓↑)	Separation of nuclear and magnetic cross-sections
Spherical neutron polarimetry	Single crystals	Cross-section for neutron spin along any direction	Complex non-collinear structures

form factor $f_j(\mathbf{Q})$, and this maximum value of $Q = 4\pi \sin(\theta)/\lambda$ is about 6 Å^{-1}. Good resolution is necessary at low-Q as many systems, such as helimagnetic structures, have very long periodicities.

The first study of a magnetic structure with neutrons was carried out by Shull and Smart (1949) on powdered manganese oxide, MnO. This is a simple antiferromagnet with spins arrayed on two interpenetrating lattices, A and B. The spins on A are all parallel and are antiparallel to the spins on B. Another simple antiferromagnet is $KMnF_3$ whose cubic crystal structure is shown in Fig. 9.9. The spins are associated with the manganese atoms, which lie at the corners of a simple cubic lattice (Scatturin *et al.*, 1961). In both MnO and $KMnF_3$ adjacent Mn atoms along the cubic-axis directions have opposite spins; the magnetic unit cell has twice the cell edge of the chemical cell and so many of the nuclear and magnetic reflections are separated (Fig. 9.10).

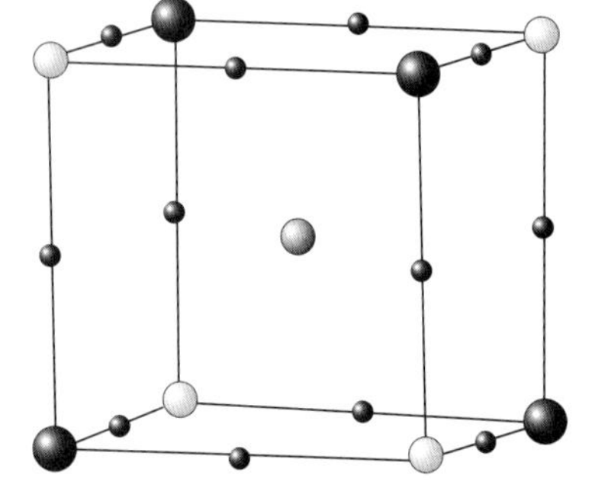

Fig. 9.9 The unit cell of $KMnF_3$ possessing the cubic perovskite structure. The spins of the manganese atoms in sublattice A are antiparallel to the spins of the manganese atoms in sublattice B.

The rare-earth metals have magnetic structures in which the magnetic moments are arrayed in a 'helifan', which is intermediate between a helix and a fan (Cowley and Bates, 1988). Different helifan structures are obtained by changing the temperature or by varying the applied magnetic field, and these structures may have commensurate regions of atomic spin separated by discommensurations or 'spin slips'. The discovery of the *giant magneto-resistance effect* has stimulated further studies of rare-earth superlattice systems by both neutron and X-ray scattering techniques (Cowley, 2004).

An example of a powder study with a high-resolution time-of-flight diffractometer is the determination of the antiferromagnetic structure

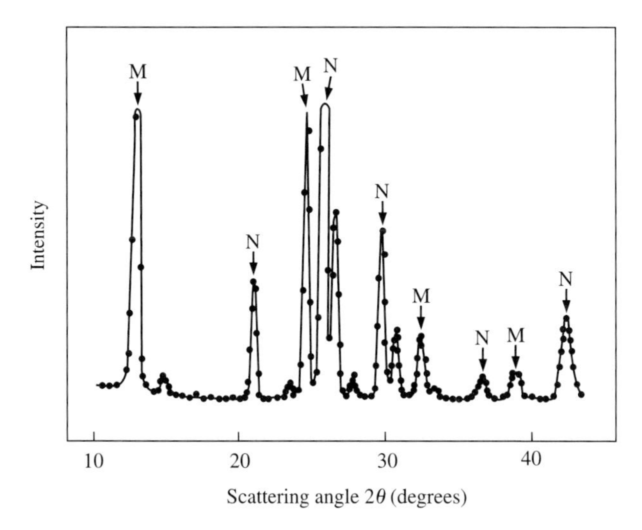

Fig. 9.10 Neutron diffraction powder pattern of $KMnF_3$ at liquid helium temperature. The peaks labelled M are magnetic reflections with half-integer indices, and the peaks labelled N are nuclear reflections with integral indices. The remaining peaks contain contributions from both magnetic and nuclear scattering. (*After* Scatturin *et al.*, 1961.)

of ferric arsenate at 4.2 K (Forsyth *et al.*, 1999). The magnetic structure is incommensurate with the chemical structure, and the dominant antiferromagnetic coupling is between Fe^{3+} ions, linked through tetrahedral arsenate groups. Many of the magnetic interactions in this system are highly *frustrated* with the spins having relative orientations which are nearly orthogonal to one another.

9.5.2 Polarized neutron scattering: measurement of flipping ratios

The measurement of flipping ratios has been used to determine the magnetic form factor $f(\mathbf{Q})$ and the spin density distribution in ferromagnetic materials. Figure 9.11 shows the basic design of the instrument for this type of experiment. The polarizing monochromator is magnetized in a field of several thousand *Oe* and the reflected beam passes along a collimator in which there is a uniform magnetic guide field (of a few hundred *Oe*) along the beam direction. A spin flipper is placed in the incident beam between the monochromator and the sample.

The total elastic scattering cross-section is measured for the two cases for which the neutron spins incident on the sample are aligned parallel and antiparallel to the direction of guide field. For neutrons polarized parallel to the field, the cross-section is

$$\frac{\mathrm{d}\sigma}{\mathrm{d}\Omega} = [F_{\mathrm{N}}(\mathbf{H}) - \sin\alpha F_{\mathrm{M}}(\mathbf{H})]^2 \tag{9.19}$$

and for neutrons polarized antiparallel to the field it is

$$\frac{\mathrm{d}\sigma}{\mathrm{d}\Omega} = [F_{\mathrm{N}}(\mathbf{H}) + \sin\alpha F_{\mathrm{M}}(\mathbf{H})]^2 \tag{9.20}$$

where α is the angle between the applied field and the direction of the scattering vector $\mathbf{Q}$. The measured *flipping ratio* r is the ratio of these

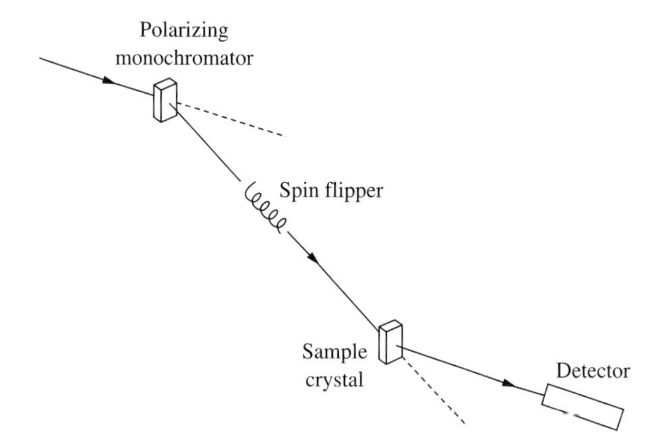

Fig. 9.11 Schematic diagram of a polarized neutron diffractometer.

two cross-sections:

$$\begin{aligned} r &= [F_{\mathrm{N}}(\mathbf{H}) - \sin\alpha F_{\mathrm{M}}(\mathbf{H})]^2 \div [F_{\mathrm{N}}(\mathbf{H}) + \sin\alpha F_{\mathrm{M}}(\mathbf{H})]^2 \\ &= [(1 - \sin\alpha \cdot \gamma)/(1 + \sin\alpha \cdot \gamma)]^2 \end{aligned} \quad (9.21)$$

where γ is the ratio $F_{\mathrm{M}}(\mathbf{H})/F_{\mathrm{N}}(\mathbf{H})$. [It has been assumed that the structure factors in eqns (9.19)–(9.21) are real, that is, that the structures are centro-symmetric.] The flipping ratio is obtained by measuring the intensities with and without a neutron flipper in the beam. It is customary to arrange for the magnetization of the sample to be normal to the scattering vector, so that $r = [(1-\gamma)/(1+\gamma)]^2$. Equation (9.21) is a quadratic equation giving two solutions for γ. $F_{\mathrm{N}}(\mathbf{H})$ is obtained from the known crystal structure and it is then an easy matter to derive the corresponding value of $F_{\mathrm{M}}(\mathbf{H})$.

The power of this polarized neutron technique in measuring magnetic structure factors can be illustrated in the case of nickel. In an experiment using unpolarized neutrons the measured cross-section is the average of (9.19) and (9.20):

$$\frac{\mathrm{d}\sigma}{\mathrm{d}\Omega} = F_{\mathrm{N}}^2(\mathbf{H}) + \sin^2\alpha \cdot F_{\mathrm{M}}^2(\mathbf{H}) \quad (9.22)$$

The maximum possible change in this cross-section will occur when the applied field is changed from being perpendicular ($\alpha = \pi/2$ in eqn. (9.21)) to being parallel ($\alpha = 0$) to the scattering vector $\mathbf{Q}$. For the (111) reflection of nickel this intensity change is about 2%; for the (400) reflection it has fallen to 0.1% on account of the reduction from the magnetic form factor. These are very small effects, which are affected by systematic errors such as absorption. By contrast, in the polarized beam experiment the flipping ratio is 1.7 for the (111) reflection, and so the two intensity measurements with polarized neutrons differ by 70%. Moreover, systematic errors from absorption and the Debye–Waller factor cancel out in measuring the flipping ratio of the stationary crystal.

Flipping ratio measurements for ferromagnets give values of the magnetic structure factors $F_{\mathrm{M}}(\mathbf{H})$, which can then be Fourier transformed

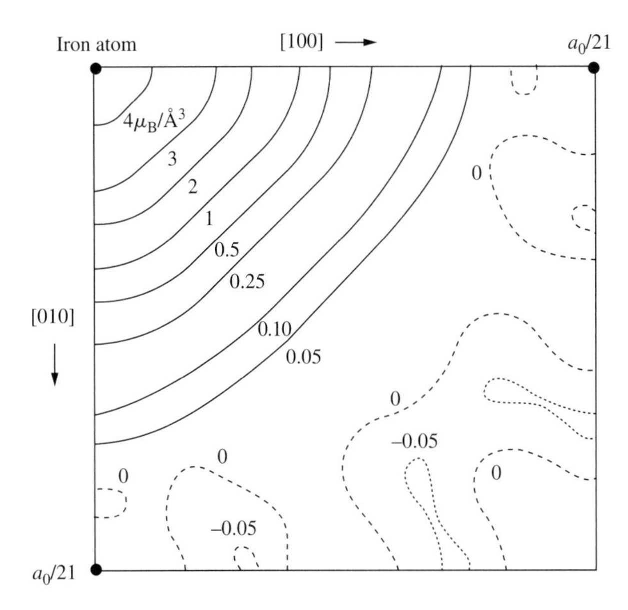

Fig. 9.12 Spin density for iron within the [001] plane. Broken lines are negative contours. (*After* Shull and Yamada, 1962.)

to give three-dimensional spin density distributions. The technique was pioneered by Shull at the MIT reactor and quickly taken up at neutron scattering centres around the world. Figure 9.12 shows some results obtained by Shull and Yamada (1962) from accurate data collected on the first 26 magnetic structure factors of iron. The contours of spin density in the [001] plane are not circular, that is, the spin density around the iron atom is aspherical. The data indicate that the occupancy ratio of the t_{2g}/e_g orbitals is close to unity, in contrast to the ratio of 3/2 demanded by spherical symmetry.

A second type of measurement is the determination of the magnetic form factor $f(\mathbf{H})$. One example of this is the experiment of Lander *et al.* (1984) on plutonium antimonide. PuSb has the cubic rock-salt structure which is ordered ferromagnetically at low temperatures. A single crystal was placed in a cryomagnet and mounted on the polarized-beam diffractometer D3 at the ILL. Flipping ratios obtained at 10 K as a function of the scattering vector Q gave the results illustrated in Fig. 9.13. Analysis using relativistic Dirac–Fock wave functions to calculate the radial probability distributions of the plutonium ions showed that there are two major components (A and B in Fig. 9.13) to the magnetic form factor. Most form factors fall off monotonically with increasing Q, much like curve A, but the presence of the B component leads to a maximum in the form factor at $\sin\theta/\lambda \approx 0.2$ Å^{-1}. This component is due to the very large orbital moment associated with the trivalent ground state of the plutonium ion.

The flipping-ratio technique has been used not only on systems with aligned spins, but also on *paramagnets* in high external magnetic fields. Combined spin and charge density measurements by neutron and X-ray techniques have been carried out on molecular-based compounds that

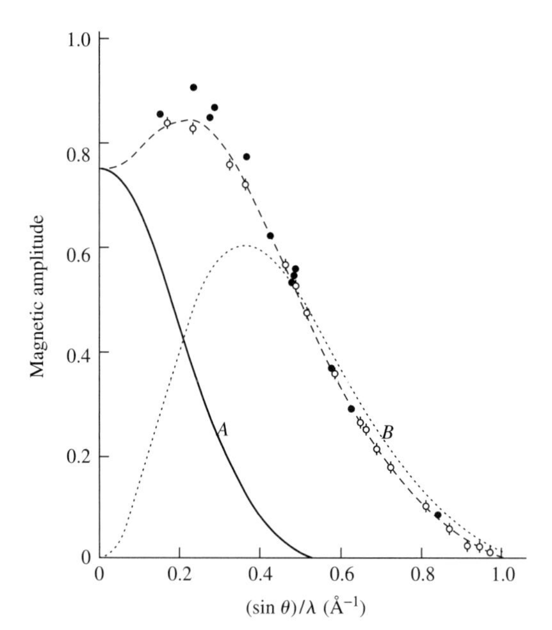

Fig. 9.13 The magnetic amplitude of ferromagnetic PuSb in which the moment direction is along [001]. Open points correspond to reflections with **Q** lying in the (001) plane and closed points to reflections out of this plane. The broken curve is the best fit for all reflections. (*After* Lander *et al.*, 1984.)

are paramagnets (Pontillon *et al.*, 1999; Gillon *et al.*, 2002). Polarized neutron diffraction has provided unique information on the magnetic properties of systems as complex as yttrium (III) semiquinonato complex Y $(HBPz_3)_2$ (DTBSQ) (Claiser *et al.*, 2005).

9.5.3 Uniaxial polarization analysis

The technique of polarization analysis was pioneered by Moon *et al.* (1969). They used a triple-axis spectrometer (see Chapter 14) with Co–Fe polarizing crystals in both the primary beam and the final beam before the detector. Spin flippers were mounted before and after the sample. With this configuration it was possible to measure the four cross-sections $\uparrow\uparrow, \uparrow\downarrow, \downarrow\downarrow, \downarrow\uparrow$ where $\uparrow$ is the spin state with the flipper off and $\downarrow$ is the state with the flipper on. The guide field was constant throughout the experiment and the scattered polarization was analysed in this single direction. Moon *et al.* show that polarization analysis can be used (a) to separate nuclear and magnetic Bragg scattering in antiferromagnets, (b) to separate magnon and phonon scattering in ferromagnets and antiferromagnets, and (c) separate coherent and spin-incoherent nuclear scattering.

Hicks (1996) has given a comprehensive review of experiments with neutron polarization analysis, including the derivations of the scattering cross-sections and the final polarizations of the scattered beam. In evaluating these experiments, it is essential to bear in mind the following simple *rules of thumb*:

(i) The nuclear scattering is always non-spin-flip ($\uparrow\uparrow$ or $\downarrow\downarrow$).

(ii) The magnetic scattering is both spin flip ($\uparrow\downarrow$ or $\downarrow\uparrow$) and non-spin flip. The non-spin flip scattering is given by those components of

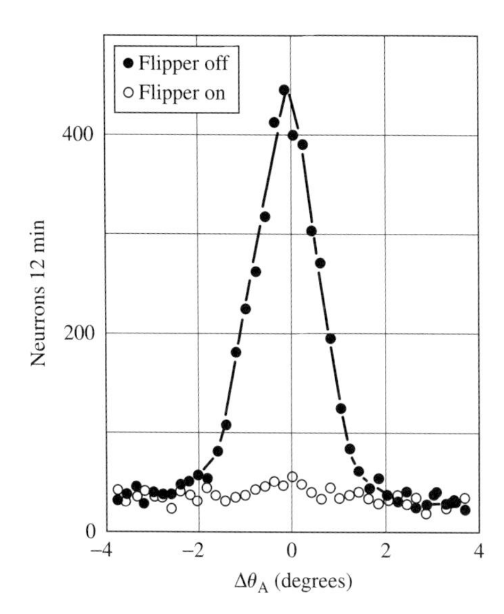

Fig. 9.14 The nuclear isotopic incoherent scattering from nickel which is suppressed when the flipper in the scattered beam is switched on. (*After* Moon *et al.*, 1969.)

the magnetization parallel to the neutron spin, and the spin flip scattering by those components of the magnetization perpendicular to the neutron spin.

(iii) If the polarization is parallel to the scattering vector, the magnetization in the direction of the polarization will not be observed since the Halpern vector [eqn. (9.5)] is zero. Hence all magnetic scattering is spin-flip.

In Fig. 9.14 we give an example of the first rule of thumb. Nickel has a large nuclear isotopic incoherent scattering because of the large number of isotopes with different scattering lengths and zero spin. Figure 9.14 shows that all scattering preserves the initial polarization and that there is no scattering with the flipper on.

The uniaxial polarization method has been widely used in the study of magnetic correlations in spin glasses, antiferromagnets and frustrated systems. Using a ^{3}He spin filter (see Section 9.3.3), Heil *et al.* (2002) extracted both the nuclear and magnetic cross-sections of an amorphous sample of ErY_6Ni_3. The technique has also been used to study the low-temperature magnetic phase transitions in $Gd_2Ti_2O_7$ (Stewart *et al.*, 2004) and spin correlations in the paramagnetic phase of La_2CuO_4 (Toader *et al.*, 2005).

9.5.4 Spherical neutron polarimetry

The experimental technique in Section 9.5.3 was *uniaxial* as the neutron quantization axis was fixed. To measure all three Cartesian components of the polarization vector, the incident and scattered guide fields must be kept independent of each other by placing the sample in a zero-field environment. The CRYOPAD (Cryogenic Polarization Device) at the ILL is capable of achieving this vectorial control of the neutron

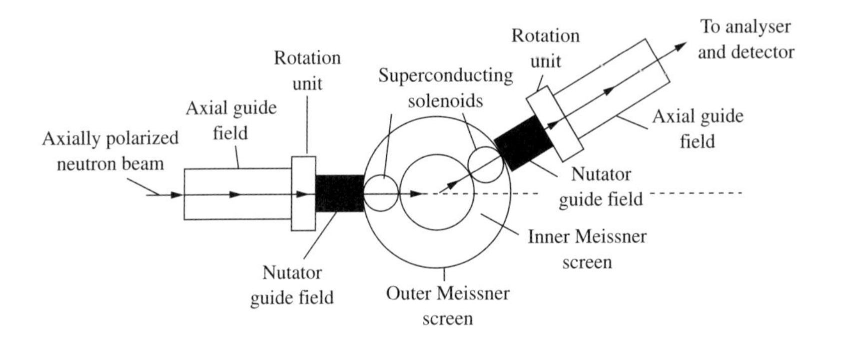

Fig. 9.15 Horizontal cross-section of the CRYOPAD.

polarization, so that all three components of both incident and scattered neutron beams can be determined.

The initial CRYOPAD was designed by Tasset (1989) and a later version is described by Tasset *et al.* (1999). The incident beam polarization can be set in any desired direction, and the magnitude and direction of the polarization of the scattered beam derived under computer control. The sample chamber is maintained in a zero-field state: this prevents any precession of the polarization from stray magnetic fields around the sample and ensures that the only change in polarization is that due to scattering from the sample. This zero field is achieved by surrounding the sample with thin superconducting screens which isolate (on account of the *Meissner effect*) the magnetic segments for the incoming and scattered neutron beams. The polarization within these segments is controlled by *nutator guide fields* and mobile precession solenoids, which can align the polarization along any direction of space. A horizontal cross-section of CRYOPAD is shown in the diagram of Fig. 9.15. The device operates in conjunction with a spectrometer containing the final analyser and detector.

The theory of measurements using the spherical neutron polarimeter is given in the papers of Brown *et al.* (1993), Tasset *et al.* (1999) and Brown (2001). The SNP technique has been used to solve a number of magnetic problems which proved to be intractable by other means. These include the measurement of the absolute magnetic moment direction in antiferromagnetic systems (Tasset *et al.*, 1988), the precise determination of antiferromagnetic form factors (Brown *et al.*, 1999) and the solution of the magnetic structure of some intermetallic compounds (Brown, 2001).

Polarized neutron techniques have many advantages over those employing unpolarized neutrons. However, their potential will not be fully realized until sources are built which are appreciably brighter than today's sources.

References

Alvarez, L. W. and Bloch, F. (1940) *Phys. Rev.* **57** 111–122. "*A quantitative determination of the neutron moment in absolute nuclear magnetons.*"

Andersen, K. H., Chung, R., Guillard, V., Humblot, H., Jullien, D., Lelièvre-Berna, E., Petoukhov, A. and Tasset, F. (2005) *Physica B* **356** 103–108. "*First results from the new polarised ^3He filling station at the ILL.*"

Anderson, I. S. and Schärpf, O. (1999) *Neutron techniques: beam-definition devices* in Volume C of "*International Tables for Crystallography.*" 2nd edition. pp. 427–439. Kluwer Academic Publishers, Dordrecht, the Netherlands.

Brown, P. J., Forsyth, J. B. and Tasset, F. (1993) *Proc. Roy. Soc. A* **442** 147–160. "*Neutron polarimetry.*"

Brown, P. J., Forsyth, J. B. and Tasset, F. (1999) *Physica B* **267–268** 215–220. "*Precision determination of antiferromagnetic form factors.*"

Brown, P. J. (2001) *Physica B* **297** 198–203. "*Polarised neutrons and complex antiferromagnets.*"

Chatterji, T. (editor) (2006) "*Neutron scattering from magnetic materials.*" Elsevier Press, Amsterdam.

Claiser, N., Souhassou, M., Lecomte, C., Gillon, B., Carbonera, C., Caneschi, A., Dei, A., Gatteschi, D., Bencini, A., Pontillon, Y. and Lelièvre-Berna, E. (2005) *J. Phys. Chem. B* **109** 2723–2732. "*Combined charge and spin density experimental study of the yttrium (III) semiquinonato complex $Y(HBPz_3)_2(DTBSQ)$ and DFT calculations.*"

Cowley, R. A. and Bates, S. (1988) *J. Phys. C Solid State Phys.* **21** 4113–4124. "*The magnetic structure of holmium.*"

Cowley, R. A. (2004) *Physica B* **350** 1–10 "*Magnetic structures and coherence of rare-earth superlattices.*"

Forsyth, J. B., Wright, J. P., Marcos, M. D., Attfield, J. P. and Wilkinson, C. (1999) *J. Phys. Condensed Matter* **11** 1473–1478. "*Helimagnetic order in ferric arsenate.*"

Gillon, B., Mathonière, C., Ruiz, E., Alvarez, S., Cousson, A., Rajendiran, T. and Kahn, O. (2002) *J. Amer. Chem. Soc.* **124**, 14433–14441. "*Spin densities in a ferromagnetic bimetallic chain compound: polarised neutron diffraction and DFT calculations.*"

Halpern, O. and Johnson, M. H. (1937a) *Phys. Rev.* **51** 992. "*On the theory of neutron scattering by magnetic substances.*"

Halpern, O. and Johnson, M. H. (1937b) *Phys. Rev.* **52** 52–53. "*Magnetic scattering of slow neutron.*"

Halpern, O. and Johnson, M. H. (1939) *Phys. Rev.* **55** 898–923. "*On the magnetic scattering of neutrons.*"

Halpern, O. (1949) *Phys. Rev.* **76** 1130–1133. "*On magneto optics of neutrons and some related phenomena.*"

Hayter, J. B. (1978) *Z. Physik B* **31** 117–125. "*Matrix analysis of neutron spin echo.*"

Heil, W., Dreyer, J., Hofmnn, D., Humblot, H., Lelievre-Berna, E. and Tasset, F. (1999) *Physica B* **267–268** 328–335. "*^3He neutron spin filter.*"

Heil, W., Andersen, K. H., Cywinski, R., Humblot, H., Ritter, C., Roberts, T. W. and Stewart, J. R. (2002) *Nucl. Instrum. Methods A* **485** 551–570. "*Large-angle polarisation analysis at thermal neutron wavelengths using a ^{3}He spin filter.*"

Hicks, T. J. (1996) *Adv. Phys.* **45** 243–298. "*Experiments with neutron polarisation analysis.*"

Hoghøj, P., Anderson, I.S., Siebrecht, R., Graf, W. and Ben-Saidane, K. (1999) *Physica B* **267–268** 355–359. "*Neutron polarising Fe/Si mirrors at ILL.*"

Hughes, D. J. and Burgy, M. T. (1951) *Phys. Rev.* **81** 498–506. "*Reflection of neutrons from magnetised mirrors.*"

Izyumov, Y. A., Naish, V. E. and Ozerov, R. P. (1990) "*Neutron diffraction of magnetic materials.*" Consultants Bureau, New York.

Jones, T. L. J. and Williams, W. G. (1978) *Nucl. Instrum. Methods* **152** 463–469. "*Non-adiabatic spin flippers for thermal neutrons.*"

Lander, G. H., Delapalme, A., Brown, P. J., Spirlet, J. C., Rebizant, J. and Double Vogt, O. (1984) *Phys. Rev. Lett.* **53** 2262–2265. "*Magnetization density of 5f electrons in ferromagnetic PuSb.*"

Moon, R. M., Riste, T. and Koehler, W. C. (1969) *Phys. Rev.* **181** 920–931. "*Polarisation analysis of thermal neutron scattering.*"

Pontillon, Y., Caneschi, A., Gatteschi, D., Sessoli, R., Ressouche, E., Schweizer, J. and Lelièvre-Berna, E. (1999) *J. Amer. Chem. Soc.* **121** 5342–5343. "*Magnetic density in an iron (III) magnetic cluster. A polarised neutron investigation.*"

Scatturin, V., Corliss, L., Elliott, N. and Hastings, J. (1961) *Acta Crystallogr.* **14** 19–26. "*Magnetic structures of 3d transition metal double fluorides, $KMeF_3$.*"

Shull, C. G. and Smart, J. S. (1949) *Phys. Rev.* **76** 1256–1257. "*Detection of antiferromagnetism by neutron diffraction.*"

Shull, C. G. and Yamada, Y. (1962) *J. Phys. Soc., Japan* **17**(Suppl. BIII) 1–6. "*Magnetic electron configuration in iron.*"

Stewart, J. R., Ehlers, G., Wills, A. S., Bramwell, S. T. and Gardner, J. S. (2004) *J. Phys. Condens. Matter* **16** L321–L326. "*Phase transitions, partial order and multi-k structures in $Gd_2Ti_2O_7$.*"

Stunault, A., Andersen, K. H., Roux, S., Bigault, T., Ben-Saidane, K. and Rønnow, H. M. (2006) *Physica B* **385–386** 1152–1154. "*New solid-state polariser bender for cold neutrons.*"

Squires, G. L. (1996) "*Introduction to the theory of thermal neutron scattering.*" Dover Publications, New York, USA.

Tasset, F., Brown, P. J. and Forsyth, J. B. (1988) *J. Appl. Phys.* **63** 3606–3608. "*Determination of the absolute magnetic moment direction in Cr_2O_3 using generalized polarisation analysis.*"

Tasset, F. (1989) *Physica B* **156–157** 627–630. "*Zero field neutron polarimetry.*"

Tasset, F., Brown, P. J., Lelièvre-Berna, E., Roberts, T., Pujol, S., Allibon, J. and Bourgeat-Lami, E. (1999) *Physica B* **267–268** 69–74. "*Spherical neutron polarimetry with Cryopad-II.*"

Toader, A. M., Goff, J. P., Roger, M., Shannon, N., Stewart, J. R. and Enderle, M. (2005) *Phys. Rev. Lett.* **94** 197–202. "*Spin correlations in the paramagnetic phase and ring exchange in* La_2CuO_4."

Williams, W. G. (1988) "*Polarised neutrons.*" Oxford University Press, Oxford.

Small-angle neutron scattering

10

The theory and practice of small-angle X-ray scattering (SAXS) dates back to the 1930s and is well described in the classic book of Guinier and Fournet (1955). Small-angle neutron scattering (SANS) came much later. It was not until the early 1970s, when position-sensitive detectors on cold neutron guides became available, that systematic studies began using small-angle neutron scattering, but after this relatively late start the technique rapidly became one of the most popular and productive of all neutron scattering methods. SANS has been applied to a very wide range of problems, including the study of polymer conformations and morphology, the study of biological structures, the characterization of voids and precipitates in alloys, and the study of flux-line lattices in superconducting materials. The technique of 'contrast variation' is largely responsible for the success of SANS, especially in the study of soft condensed matter and biological systems. There are over 30 SANS instruments in operation worldwide at both reactor and spallation sources. In spite of an apparent wealth of instruments, the demand for beam time considerably outstrips the time available.

Small-angle elastic scattering by a sample provides information about the size, shape and orientation of the components in the sample. Unlike most of the other methods discussed in this book, SANS is not directly concerned with the scattering from individual atoms. The scattering takes place instead from aggregates of atoms, and the structural information one seeks is at a coarser level than the atomic level. The range of Q covered is from about 5×10^{-3} to 0.5 Å^{-1}, representing a range in real space ($=2\pi/Q$) from 10 to 1,000 Å. The intensity of scattering depends on the scattering length densities of the elementary volumes in this size range. In the case of neutron scattering, these densities are readily modified by altering the isotopic composition of the sample to give contrast variation.

There are other important differences between neutrons and X-rays, which are to the advantage of one technique or the other. The energy of an X-ray photon of 1.5 Å wavelength is more than 10^5 times greater than a neutron of the same wavelength, and so radiation damage of biological samples is less of a problem with neutrons than with X-rays. Neutron radiation is very penetrating: it can be used to examine properties of samples several cms thick and can probe samples encased in furnaces, cryostats or pressure cells, as well as other bulky equipment such as extrusion nozzles and chemical reaction vessels. The neutron has a magnetic moment which interacts with the spin and orbital moments of unpaired electrons, allowing magnetic properties to be examined. On the other hand, neutron sources are expensive to operate compared with laboratory X-ray sources. Furthermore, the neutron beam itself is relatively weak and so it cannot be collimated in a SANS experiment to a diameter of less than a 1 mm. In contrast, X-rays can be focused down to a few μm, and it is possible to measure the scattering much closer to $Q = 0$ with X-rays than with neutrons. The small X-ray spot size lends itself more conveniently to rastering of the sample, which is valuable in cases where the sample is inhomogeneous.

10.1 Theory of small-angle scattering

A generalized expression for the small-angle scattering from any sample is

$$\frac{\mathrm{d}\sigma\,(Q)}{\mathrm{d}\Omega} = N_\mathrm{p} V_\mathrm{p}^2 \bar{\rho}_\mathrm{p}^2 P(Q) S(Q) \tag{10.1}$$

where N_p is the number of identical particles, each of volume V_p, in the sample. Q is $4\pi(\sin\theta_\mathrm{p})/\lambda$ where 2θ is the scattering angle. The term $\bar{\rho}_\mathrm{p}^2$ is the particle *contrast*, $P(Q)$ is the *shape factor* or *form factor* and $S(Q)$ is the *interparticle structure factor*. These last three terms contain all the information on the size, shape and interactions of the scattering particles, and we shall consider each of them in turn.

10.1.1 The scattering contrast term

Suppose that the scattering takes place by particles (*e.g.* large protein molecules) dissolved or dispersed in a solvent such as water. The scattering process is governed by the interference between the waves scattered by regions of linear dimensions l such that $l \approx 2\pi/Q_\mathrm{max}$, where Q_max is the maximum observed value of Q in the data. The volume of these regions is called the 'resolution volume' v. The distance l is large compared with interatomic distances but small compared with the linear size of the particles. The local scattering length density $\rho(\mathbf{r})$ is obtained by summing the coherent scattering lengths of all atoms in the resolution volume v, centred on position $\mathbf{r}$, and dividing by that volume. Thus $\rho(\mathbf{r})$ is given by

$$\rho(r) = \frac{1}{v}\sum_i b_i(\mathbf{r}) \tag{10.2}$$

where $b_i(\mathbf{r})$ is the scattering length of the atom i at the position $\mathbf{r}$.

The amplitude $A(Q)$ scattered by each particle *in vacuo* is obtained by multiplying $\rho(\mathbf{r})$ by the phase factor $\mathrm{e}^{i\mathbf{Q}\cdot\mathbf{r}}$ and integrating over the

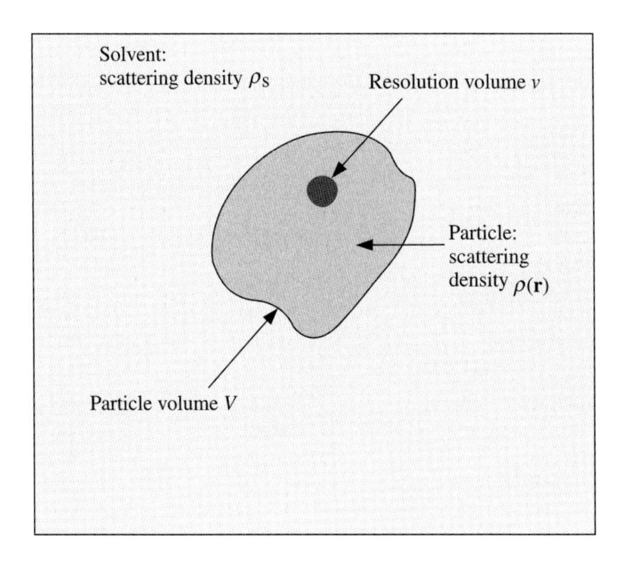

Fig. 10.1 Particle of mean scattering density $\rho(\mathbf{r})$ immersed in a medium of density ρ_S.

volume V_p of the particle:

$$A(Q) = \int_{V_p} \rho(\mathbf{r}) e^{i\mathbf{Q}\cdot\mathbf{r}}\, d\mathbf{r} \tag{10.3}$$

If the particle is in a solvent with mean scattering length density ρ_S, eqn. (10.3) is replaced by

$$A(Q) = \int \left(\rho(\mathbf{r}) - \rho_S\right) e^{i\mathbf{Q}\cdot\mathbf{r}}\, d\mathbf{r} \tag{10.4}$$

Figure 10.1 illustrates the meanings of the various terms in eqns (10.2)–(10.4).

The average scattering density of the particle compared to that of the surroundings is referred to as the *contrast*, and is given by

$$\bar{\rho}_P = \rho_V - \rho_S \tag{10.5}$$

where ρ_V is the average scattering density within the volume V. If $\rho_V = \rho_S$, there is no scattering and we have *contrast matching*. The scattering lengths of hydrogen and deuterium are of opposite sign: by mixing the hydrogenous and deuterated forms of the solvent in the appropriate ratio, the scattering from a particular component in the sample can be removed by matching it with the solvent. The same procedure can be applied to the study of a layer adsorbed on a substrate: by matching the substrate to the dispersion medium, scattering is restricted to the adsorbed layer alone.

Table 10.1 gives the scattering length densities of some common solvents and polymers in fully hydrogenated and perdeuterated forms. The table can be used to calculate the appropriate proportions of these forms in a matching experiment. This calculation requires that the density of the solvent or polymer molecules is accurately known.

Table 10.1 Scattering length densities of solvents and polymers.

	$\bar{\rho}_P$ (protonated) ($\times 10^{10}$ cm^{-2})	$\bar{\rho}_P$ (deuterated) ($\times 10^{10}$ cm^{-2})
Solvent		
Water	−0.56	+6.38
Octane	−0.53	+6.43
Cyclohexane	−0.28	+6.70
Toluene	+0.94	+5.66
Chloroform	+2.38	+3.16
Polymer		
Polybutadiene	−0.47	+6.82
Polyethylene	−0.33	+8.24
Polystyrene	+1.42	+6.42
Polymethylmethacrylate	+1.10	+7.22

10.1.2 The form factor term $P(Q)$

$P(Q)$ describes the influence of the *shape* of the scattering particle on the observed cross-section. At $Q = 0$ it is unity. We assume that the particles are randomly oriented throughout the sample. Analytical expressions of $P(Q)$ exist for many possible shapes (sphere, core of a shell, ellipsoid, rod, plate, *etc.*) and in Table 10.2 we quote the expressions for just three of these shapes. J_1 is a first-order Bessel function and S_i is the Sine integral function.

The calculation of the form factor for the case of a spherical particle of radius R is as follows. From eqn. (10.4) the amplitude of scattering is

$$A(Q) = \left\langle \int \left(\rho(\mathbf{r}) - \rho_S\right) \mathrm{e}^{i\mathbf{Q}\cdot\mathbf{r}}\, \mathrm{d}\mathbf{r} \right\rangle = \bar{\rho}_P \left\langle \int \mathrm{e}^{i\mathbf{Q}\cdot\mathbf{r}}\, \mathrm{d}\mathbf{r} \right\rangle \tag{10.6}$$

where the angular brackets indicate an average value over all orientations of $\mathbf{Q}$. Using spherical polar coordinates $r\theta\phi$, where θ is the angle between $\mathbf{Q}$ and $\mathbf{r}$, the element of volume $\mathrm{d}\mathbf{r}$ is $r^2 \sin\theta \mathrm{d}r\mathrm{d}\theta\mathrm{d}\phi$ and $\mathbf{Q}\cdot\mathbf{r}$ becomes $Qr\cos\theta$. Thus we can write

$$\left\langle \int \mathrm{e}^{i\mathbf{Q}\cdot\mathbf{r}}\, \mathrm{d}\mathbf{r} \right\rangle = \int_0^R \mathrm{d}r \int_0^{2\pi} \mathrm{d}\varphi \int_0^{\pi} \mathrm{d}\theta\ \mathrm{e}^{i\mathrm{Q}\cdot\mathrm{r}\cos\theta} r^2 \sin\theta \tag{10.7}$$

Integrating over the two angular variables θ, ϕ yields

$$\left\langle \int \mathrm{e}^{i\mathbf{Q}\cdot\mathbf{r}} \mathrm{d}\mathbf{r} \right\rangle = \int_0^R \frac{\sin Qr}{Qr} 4\pi r^2 \mathrm{d}r \tag{10.8}$$

and so

$$A(Q) = \bar{\rho}_P \int_0^R \frac{\sin Q\mathrm{r}}{Q\mathrm{r}} 4\pi r^2 \mathrm{d}r \tag{10.9}$$

Table 10.2 Form factors for a few particles of different shape.

Shape of the particle	Form factor $P(Q)$
Sphere of radius R	$\left[\frac{3[\sin(QR) - QR\cos(QR)]}{(QR)^3}\right]^2$
Disc of small thickness and radius R	$\frac{2}{(QR)^2}\left[1 - \frac{J_1(2QR)}{QR}\right]$
Rod of small cross-section and length L	$\frac{2S_i(QL)}{QL} - \frac{\sin^2(QL/2)}{(QL/2)}$

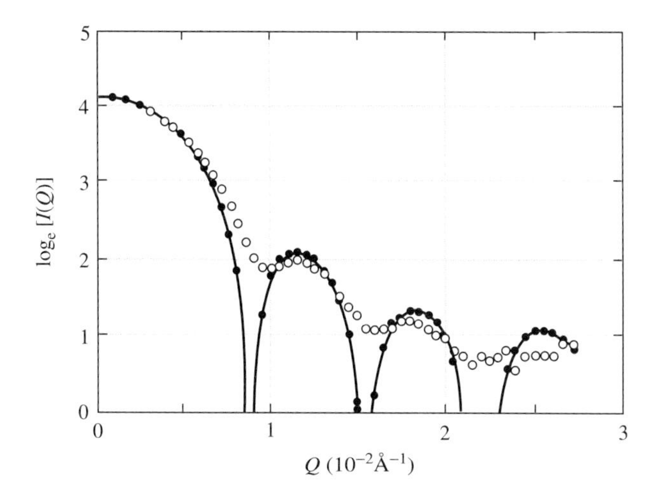

Fig. 10.2 Small-angle scattering from spherical particles of PMMA-H in deuterated water. The open points are the observed counts, and the closed points are the counts corrected for resolution. The full line is the theoretical curve of eqn. (10.12) assuming a radius $R = 492$ Å. (*After* Wignall *et al.*, 1988.)

The integration over r yields:

$$A(Q) = 4\pi\bar{\rho}_{\rm p}R^3\left[\frac{\sin(QR) - QR\cos(QR)}{(QR)^3}\right] \tag{10.10}$$

The scattered intensity $I(\mathbf{Q})$ from a single spherical particle is $|A(\mathbf{Q})|^2$ or $A(\mathbf{Q})\ A^*(\mathbf{Q})$, where $A^*(\mathbf{Q})$ is the complex conjugate of $A(\mathbf{Q})$:

$$I(Q) = (\bar{\rho}_{\rm p})^2\,V_{\rm p}^2\left[\frac{3\left[\sin(QR) - QR\cos(QR)\right]}{(QR)^3}\right]^2 \tag{10.11}$$

The volume $V_{\rm p}$ of the spherical particle is $4\pi R^3/3$ and so the form factor is

$$P(Q) = \left[\frac{3\left[\sin(QR) - QR\cos(QR)\right]}{(QR)^3}\right]^2 \tag{10.12}$$

This form factor is illustrated by the full line of Fig. 10.2. It has discrete maxima separated by cusps, which are at positions close to

$$QR = \pi(n + 1/2) \tag{10.13}$$

where n is an integer. [More exactly, the zeroes occur when $\tan QR = QR$, *i.e.* when $QR = \pi(1.4303,\ 2.4590,\ 3.4709,\ 4.4774,\ 5.4818,\ \ldots)$.] The open points in Fig. 10.2 refer to the small-angle scattering of neutrons from spherical particles of polymethylmethacrylate (PMMA) of uniform size immersed in heavy water. After correcting for resolution, these points can be fitted to the calculated curve to give a mean particle radius R of 492 Å.

10.1.3 The structure factor term *S(Q)*

The interparticle structure factor term $S(Q)$ accounts for the influence on the measured cross-section of local order between the scattering particles. In Chapter 13, we shall encounter $S(Q)$ in the context of scattering from a liquid. Here we can use the expression eqn. (13.8) for a liquid by replacing the atoms by particles:

$$S(Q) = 1 + \frac{4\pi N}{Q} \int_0^\infty r\,[g_{\mathrm{P}}(r) - 1]\sin(Qr)\mathrm{d}r \tag{10.14}$$

where N is the number of particles per unit volume and $g_{\mathrm{P}}(r)$ is the pair-distribution function representing the probability that a pair of particles is a distance r apart. $g_{\mathrm{P}}(r)$ can be obtained from eqn. (10.14) by Fourier inversion, but this is not possible, in general, because the scattering is not known at $Q \to 0$, due to the beam stop, and because it is not known at $Q \to \infty$.

An alternative method is to use the calculated form of $S(Q)$ for particles interacting with a spherically symmetric potential. Expressions for $S(Q)$ have been derived for several types of potential: hard-sphere, sticky hard-sphere, screened Coulomb, *etc.* For very dilute systems with no interaction between the particles $S(Q) = 1$.

10.2 The scattering 'laws'

In this section we shall describe three scattering laws (due to Guinier, Zimm and Porod) which are often used to analyse small-angle neutron and X-ray data. The laws refer to different regions of the scattering vector $\mathbf{Q}$.

10.2.1 The domain of very small Q

For very small values of Q eqn. (10.4) can be expanded to give

$$A(\mathbf{Q}) = \int \left(\rho(\mathbf{r}) - \rho_S\right)\left(1 + i\mathbf{Q}\cdot\mathbf{r} - \frac{1}{2}(\mathbf{Q}\cdot\mathbf{r})^2 + \cdots\right)\mathrm{d}\mathbf{r} \tag{10.15}$$

so that the intensity averaged over all orientations of $\mathbf{Q}$ is

$$I(\mathbf{Q}) = \left\langle \left| \int_V \left(\rho(\mathbf{r}) - \rho_S\right) \left(1 + i\mathbf{Q}\cdot\mathbf{r} - \frac{1}{2}\left(\mathbf{Q}\cdot\mathbf{r}\right)^2 + \cdots\right) \mathrm{d}\mathbf{r} \right|^2 \right\rangle \tag{10.16}$$

The term in $\mathbf{Q}\cdot\mathbf{r}$ vanishes if the origin of $\mathbf{r}$ coincides with the centre of the particle of volume V. As $Q \to 0$, we can ignore terms in (10.16) of higher order than $(\mathbf{Q}\cdot\mathbf{r})^2$ to obtain the expression

$$I(Q) = (\bar{\rho})^2 \left[\int \left(1 - \frac{1}{2}\left\langle(\mathbf{Q}\cdot\mathbf{r})^2\right\rangle\right) \mathrm{d}\mathbf{r}\right]^2 \tag{10.17}$$

But $(\mathbf{Q}\cdot\mathbf{r})^2 = (Q^2r^2)\cos^2\phi$, where ϕ is the angle between $\mathbf{Q}$ and $\mathbf{r}$, and the mean value of $\cos^2\phi$ is 1/3. Hence eqn. (10.17) reduces to

$$I(Q) \approx (\bar{\rho})^2 V^2 \left[1 - \frac{1}{3}\frac{Q^2}{V}\int_V r^2\,\mathrm{d}\mathbf{r}\right] \tag{10.18}$$

In the last step we have assumed that $QR_\mathrm{G} \ll 1$, where R_G is the *radius of gyration* of a particle of volume V. R_G is defined, in analogy to the radius of gyration in classical mechanics, by the expression

$$R_\mathrm{G} = \frac{\int_V r^2 \mathrm{d}\boldsymbol{r}}{\int_V \mathrm{d}\mathbf{r}} = \frac{1}{V}\int_V r^2\,\mathrm{d}\mathbf{r} \tag{10.19}$$

Recalling that $\exp(-\alpha) \approx 1 - \alpha$ if $\alpha \ll 1$, eqn. (10.18) can be rewritten in the form

$$I(Q) = I(0)\exp\left(-\frac{R_\mathrm{G}^2 Q^2}{3}\right) \tag{10.20}$$

with $I(0)$ representing the 'forward scattering' (obstructed by the beam stop) as $Q \to 0$.

Equation (10.20) is known as the *Guinier law*. The law states that the logarithm of the intensity scattered by an assembly of identical particles, when plotted *versus* Q^2, gives a straight line of slope $-R_\mathrm{G}^2/3$. It is valid within a restricted range of Q given by $QR_\mathrm{G} < 1$; this range covers the initial portion of the first maximum in Fig. 10.2. In Table 10.3 we give formulae relating the radius of gyration to the external dimensions of some simple triaxial bodies.

The Guinier plot for the spherical virus, turnip yellow mosaic virus, is shown in Fig. 10.3 (Jacrot, 1976). From the slope of the straight line the radius of gyration R_G is calculated as 123 Å. For a spherical particle the ratio R_G/R is $\sqrt{3/5}$ and so the radius of the virus is 159 Å.

An alternative method of deriving the radius of gyration for dilute solutions of chain molecules has been suggested by Zimm (1948). The scattering from such a polymer system is characterized by the Debye scattering function (Debye and Bueche, 1952), which is given by

$$K\frac{c}{I(Q)} = \frac{1}{M}\left(1 + R_\mathrm{G}^2 Q^2/3\right) + 2A_2\mathrm{c} \tag{10.21}$$

Table 10.3 Radius of gyration of triaxial bodies.

Body	$(R_G)^2$
Sphere of radius, R	$(3/5)R^2$
Hollow sphere of radii, R_1 and R_2	$(3/5)(R_2^5 - R_1^5)/(R_2^3 - R_1^3)$
Ellipsoid of semi-axes, a, b, c	$(a^2 + b^2 + c^2)/5$
Parallelepiped of edges, a, b, c	$(a^2 + b^2 + c^2)/12$
Cylinder of radius a and height b	$(6a^2 + b^2)/12$

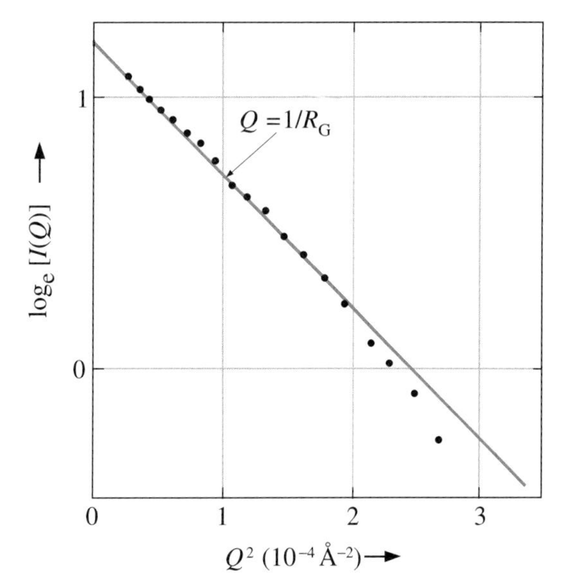

Fig. 10.3 Guinier plot for a spherical virus (turnip yellow mosaic virus). (*After* Jacrot, 1976.)

provided that $Q^2R_G^2 < 1$. c is the concentration of the polymer in the solvent, M is the average molecular weight of solute and solvent, A_2 is the second *virial coefficient* and K is a constant. The extra term $2A_2c$ in (10.21) is the lead term of the interference function between polymer molecules in a relatively dilute solution. A Zimm plot extrapolated to zero c at constant Q gives a line at infinite dilution from which the radius of gyration is derived. Extrapolation at constant c to zero Q gives the value of $2A_2c$, and a third extrapolation to zero Q and zero c gives the molecular weight M. An example of the *Zimm plot* is shown in Fig. 10.4.

10.2.2 The domain of large Q

Experimental measurements are usually made over a much larger range of Q than the limited range covered by the Guinier law. To determine the nature of the scattering at large Q, we introduce the autocorrelation function $P(\mathbf{r})$ defined by

$$P(\mathbf{r}) = \frac{1}{V} \int \rho(\mathbf{r}')\rho(\mathbf{r} + \mathbf{r}')\, d\mathbf{r}' \tag{10.22}$$

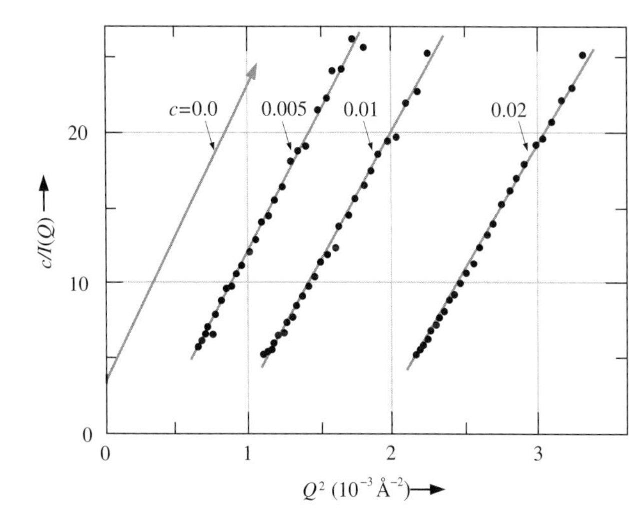

Fig. 10.4 Zimm plot for three concentrations of deuterated polystyrene in solid polystyrene. (*After* Cotton *et al.*, 1974.)

$P(\mathbf{r})$ is obtained by multiplying the scattering densities at any two points separated by a distance $\mathbf{r}$ and averaging this quantity over the entire volume V of the particle. The intensity $I(Q)$ scattered by an individual particle is $|A(\mathbf{Q})|^2$ with $A(Q)$ given by eqn. (10.3):

$$I(\mathbf{Q}) = \iint \bar{\rho}(\mathbf{r})\,\bar{\rho}(\mathbf{r}')\,\mathrm{e}^{i\mathbf{Q}\cdot(\mathbf{r}-\mathbf{r}')}\,\mathrm{d}\mathbf{r}\,\mathrm{d}\mathbf{r}' \tag{10.23}$$

and this can also be written

$$I(\mathbf{Q}) = V\int P(\mathbf{r})\mathrm{e}^{i\mathbf{Q}\cdot\mathbf{r}}\,\mathrm{d}\mathbf{r} \tag{10.24}$$

Thus the measured intensity is the Fourier transform of $P(\mathbf{r})$. [In crystallography $P(\mathbf{r})$ is known as the Patterson function.]

Suppose that ϕ_V is the volume fraction of the particles embedded in the medium and that ϕ_S is the volume fraction of the medium itself. The function $\rho(\mathbf{r}')\rho(\mathbf{r}+\mathbf{r}')$ in eqn. (10.22) is equal to $\phi_V^2\rho_V^2$ when $\mathbf{r}$ and $\mathbf{r}'$ lie entirely within the particle, is $\phi_S^2\rho_S^2$ when they lie entirely within the medium, and is $\phi_V\phi_S\rho_V\rho_S$ when they lie across the boundary between the two phases. For a dilute system $\phi_V \ll \phi_S$ and so the first of these three contributions to the autocorrelation function can be ignored. The second contribution is a constant quantity which extends over nearly the whole system: on taking the Fourier transform in eqn. (10.13) it yields a δ-function at $Q = 0$, and this scattering is intercepted by the beam stop and is of no importance. There remains only the third contribution, which gives information about the characteristics of the interface between the particle and the medium. An exact calculation (Porod, 1951) shows that for particles with a surface area S the intensity $I(Q)$ for $Q \gg 1/R_G$ is proportional to SQ^{-4}. This relation is known as *Porod's law*. By using this law the external area S of the particle can be obtained by plotting $Q^4I(Q)$ *versus* Q and extrapolating to $Q \to \infty$.

Figure 10.5 shows some results of Foret *et al.* (1992) illustrating the application of Porod's law. The samples are silica gels, which consist

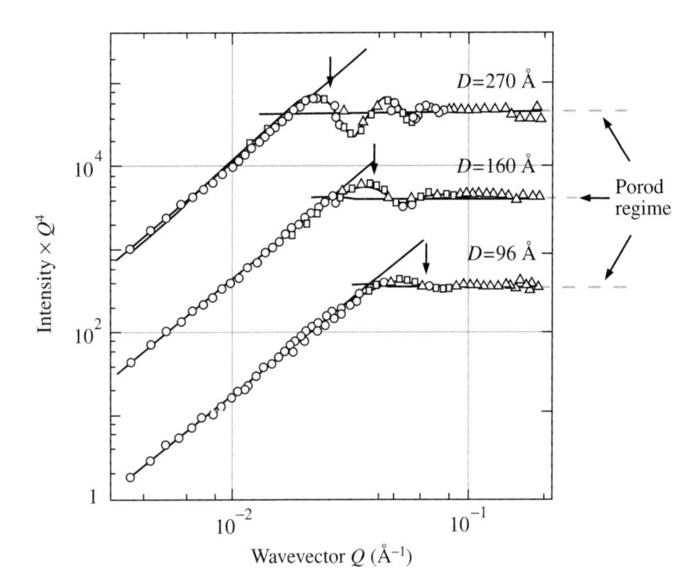

Fig. 10.5 Log–log plot of $Q^4I(Q)$ *versus* Q for three colloidal samples of spherical silica particles. At large Q, $Q^4I(Q)$ tends to a constant value. The crossover from surface to volume scattering occurs near the arrows at the values $Q \approx 2\pi vD$, where D is the particle diameter. (*After* Foret *et al.*, 1992.)

of spherical particles of amorphous silica dispersed in an alkali–water solvent. The figure is a log–log plot for three samples of different particle radius. In the Porod regime at large values of Q, the quantity $Q^4I(Q)$ is constant. The crossover from surface to volume scattering is expected to occur when the neutron wavelength is close to the diameter D of the particle, that is, for $Q \approx 2\pi/D$. These values of Q are shown as arrows in Fig. 10.5 for the three particle sizes, whose diameters D were determined independently by transmission electron microscopy.

10.3 Experimental considerations

10.3.1 Design of SANS instruments

A striking feature of the apparatus used for small-angle scattering is its size. The distance from moderator to sample can be as much as 120 m, and the beam line is buried beneath a massive shield over 1 m thick and weighing many tonnes. The flux of cold neutrons is attenuated by air scattering to the extent of about 10% for each 1 m path length, and so it is essential to have a vacuum in the flight path of the neutrons.

The SANS instrument D11, situated at the end of the curved cold neutron guide H15 of the high flux reactor at the Institut Laue-Langevin (ILL), is an instrument specifically designed for small-angle scattering experiments. A schematic diagram of D11 is given in Fig. 10.6. The instrument has a continuously variable sample-to-detector distance (1.1–38 m) and a variable collimation (12 discrete distances between 40.5 and 1.5 m) allowing access to a wide range of momentum transfer Q. Neutrons from the vertical cold source of ILL pass down a 100 m curved neutron guide, and are then monochromated ($\Delta\lambda/\lambda \approx 9\%$) by a

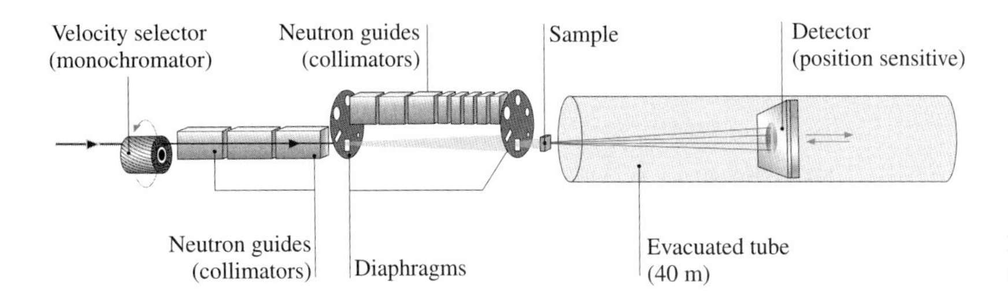

Fig. 10.6 Components of the D11 instrument.

velocity selector consisting of a 30 cm long drum with rotating helical slits. The mean wavelength, which is determined by the speed of rotation of the drum, lies between 4.5 and 40 Å. A series of movable glass guides provides further collimation before the beam strikes the sample in a very low background environment at a distance of 140 m from the cold source. The position-sensitive detector is 96×96 cm^2 in area and consists of 16,384 cells, each 0.75 cm^2 in area, located in an evacuated aluminium tank behind the sample and directly in line with the incident beam. The sample-to-detector distance can take a selection of values from 1.1 to 38 m, and provides, with the given detector size, a dynamic range ($Q_{\mathrm{max}}/Q_{\mathrm{min}}$) of $\approx$15 for one setting. The total available range of momentum transfer (from the longest detector distance and longest possible wavelength to the shortest detector distance and wavelength) is 0.0003 Å$^{-1} < Q < 1$ Å^{-1}. Provided that the scattering from the sample is isotropic, that is, independent of the direction of $\mathbf{Q}$, the intensity is derived by summing over all detector elements at the same radius r from the beam centre. From a knowledge of the sample-to-detector distance and the neutron wavelength this one-dimensional $I(r)$ versus r plot is then converted to $I(Q)$ versus Q.

The D11 instrument has been in operation for over two decades, and there are numerous other SANS instruments at other neutron sources. New types of instrument have been developed, including the ultra-small-angle instrument U-SANS in Vienna (Radulescu *et al.*, 2005) and the spin-echo small-angle instrument SESANS at the Technical University, Delft. Small-angle scattering instruments are listed by Konrad Ibel in the 'World Directory of SANS Instruments' (http://www.ill.fr/lss/ SANS_WD/sansdir.html).

Different considerations apply to the design of an instrument operating on a pulsed source. The Q range of a small-angle scattering instrument can be covered using a single detector operating at a fixed scattering angle. Frame overlap, whereby fast neutrons from a later pulse overtake slow neutrons from an earlier pulse, sets a limit to the usable wavelength range $\Delta\lambda$ and to the path length L from source to detector, although disc choppers can reduce the effective pulse repetition frequency, allowing both $\Delta\lambda$ and L to be increased, as required. The maximum range of wavelength without overlap is given by

$$\Delta\lambda(\text{Å}) = \frac{T\,(\mu\text{sec})}{252.82L\ (\text{m})} \tag{10.25}$$

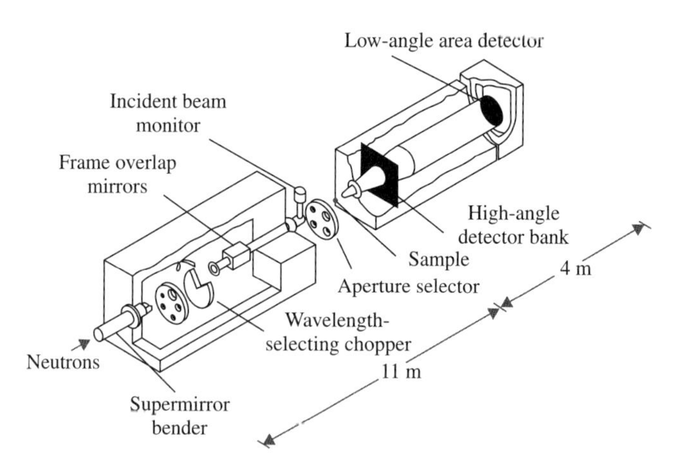

Fig. 10.7 LOQ time-of-flight SANS instrument at ISIS.

where T is the repetition period of the source. Thus for $L = 20$ m, the usable wavelength range is 4 Å with a 50 Hz pulsed source ($T = 2 \times 10^4$ μsec), but if two out of three pulses are suppressed $\Delta\lambda$ increases to 12 Å. The influence of gravity causes the neutron trajectories to be straighter for the faster neutrons than for the slower neutrons, thus giving rise to a blurring of the signal at the detector. This reduction in Q resolution of a small-angle instrument has been considered by Boothroyd (1989), who showed that the gravity contribution to the resolution is unlikely to exceed 5%.

Figure 10.7 shows the layout of the LOQ instrument at the pulsed neutron source ISIS. It is used to investigate the shape and size of large molecules or small particles with dimensions in the range 10–1,000 Å. The 'white' incident beam is combined with time-of-flight detection techniques to give LOQ a wide range in scattering vector $\mathbf{Q}$, and the whole range is accessible in a single experimental measurement without any need to reconfigure the instrument.

10.3.2 Sample size

There are various factors affecting the choice of the area and thickness of the sample. Typically the sample is irradiated by a beam of area 1–2 cm^2, which is defined by an iris diaphragm of cadmium (see Fig. 10.6) located close to the sample. The choice of thickness depends on the total cross-section, which includes absorption and incoherent scattering (especially from hydrogen) as well as coherent small-angle scattering. To reduce the effect of multiple scattering the sample must be as thin as possible, but ultimately the thickness may be limited by counting statistics. The optimum thickness for a hydrogenous sample will be about 1 mm, and 5–10 mm for a sample with hydrogen exchanged with deuterium.

If the sample is a liquid or a solution, it can be contained in a silica cell. Silica is nearly transparent a to neutrons, giving little small-angle scattering, and is unaffected by organic solvents. The following sequence of measurements are necessary for an experiment on a single liquid sample in which the contrast is varied by labelling the sample with a suitable isotope: (a) original unlabelled sample, (b) labelled sample, (c) empty sample cell, (d) water cell and (e) empty water cell. The purpose of the water cell is to normalize the data to an absolute scale, but this cannot be performed on time-of-flight data because of the variation of the cross-section with wavelength (see King, 1999).

10.3.3 Data analysis

A straightforward method of analysis is to postulate a plausible model in real space, to calculate the scattered intensity and to compare it with the observed $I(\mathbf{Q})$. The analysis is simplest for monodispersed systems at infinite dilution; there is then no interference between particles, and the scattering from the whole system reduces to the scattering from a single particle. Theoretical scattering curves have been derived for many common shapes (Section 10.1.2) and one searches for the curve which matches the experimental data. Knowing the shape it is then possible to derive the radius of gyration and the volume of the particle.

An alternative procedure, described by Stuhrmann (1970), is not limited to simple shapes. Stuhrmann analyses the scattering density $\rho(\mathbf{r})$ and the intensity factor $I(\mathbf{Q})$ in terms of spherical harmonics. At small values of Q the zero-order coefficient only contributes to the expansion of $I(\mathbf{Q})$. The method starts from the assumption that the particle is spherical, and harmonics of successively higher order are then introduced to obtain closer and closer agreement with the observed $I(\mathbf{Q})$.

In the indirect Fourier transformation (IFT) method of analysis (Glatter, 1977, 1980) no specific assumptions are made regarding a structural model. The method can only be applied to dilute particle systems, but it allows an estimate of the radius of gyration which is generally superior to the estimate from the Guinier approximation.

The type of analysis adopted depends on the nature of the sample under investigation. Biologists using irregular and rather ill-defined scattering samples will tend to use the Glatter or Stuhrmann analysis. Soft-matter scientists with better-defined samples will adopt model-fitted form factors and interaction potentials, whilst materials scientists will subject their data to Fourier inversion in order to extract density correlation functions.

10.4 Some applications of SANS

The small-angle scattering of neutrons has been widely used for the study of structural problems in materials science, molecular biology and polymer science. Here we can only give a flavour of work in these fields.

A specialist application in physics is the study of the structure of superconductors. In superconductors the repulsive forces between the magnetic flux lines causes the lines to order into a flux-line lattice (FLL) and there have been many studies of these FLLs by small-angle neutron scattering. Huxley *et al.* (2000) used the D22 instrument at the ILL to show that there is a fundamental change in symmetry of the FLL with temperature in the superconductor UPt_3. The same instrument and another instrument at the Paul Scherrer Institute, Switzerland, were employed by Gilardi *et al.* (2002) to show a field-induced crossover from a triangular to a square FLL in superconducting derivative of La_2CuO_4. A plethora of interesting phenomena has been discovered by SANS in the superconductor $CeCoIn_5$ (Eskildsen *et al.*, 2003; DeBeer-Schmitt *et al.*, 2006), including a square FLL, a reorientation transition and a field-dependent coherence length below the critical temperature T_c. These SANS observations present a formidable challenge to the theorists.

The vast majority of SANS experiments have been carried out on polymeric and biological systems, where the method of *contrast variation* is of particular importance. We have seen that for particles in a water solvent, the contrast in scattering length density between particle and solvent is readily altered by adjusting the D_2O/H_2O ratio. When the hydrogen in the water is replaced by deuterium, some of the protons of the particle (but not all) are also replaced by deuterons. As a rule, H bound to N or O will exchange, whereas H bound to C will not. Hence the scattering power of the particle itself will vary with the H/D content, as well as the scattering power of the solvent.

Figure 10.8 gives the values of the scattering length densities $\bar{b}$ for some of the most common molecules found in biological systems, plotted as a function of the D_2O content of the solvent in which they are immersed. For water alone $\bar{b}$ varies between -0.562×10^{-14} cm $\cdot$ Å^{-3} for H_2O and $+6.404 \times 10^{-14}$ cm $\cdot$ Å^{-3} for D_2O. At the crossing point of the water line with that of a given molecule the contrast $\bar{\rho}$ is zero: this is called the 'matching point' (sometimes referred to as the *isopycnic point*), where the scattering density of the solvent is equal to that of the particle and the small-angle scattering intensity falls to zero.

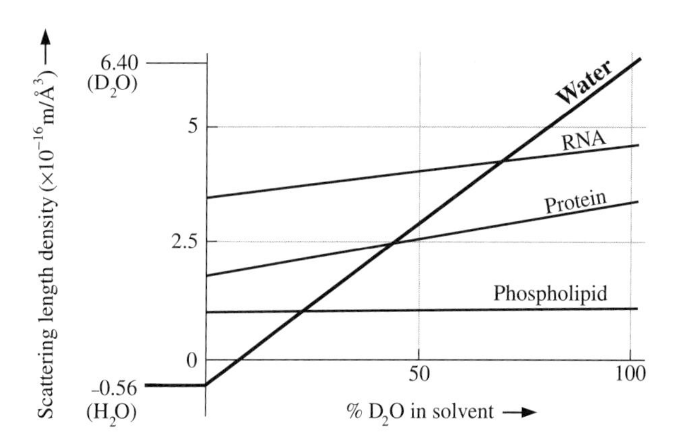

Fig. 10.8 Scattering length density of various molecular components, plotted as a function of the H/D ratio of the water in which they are immersed.

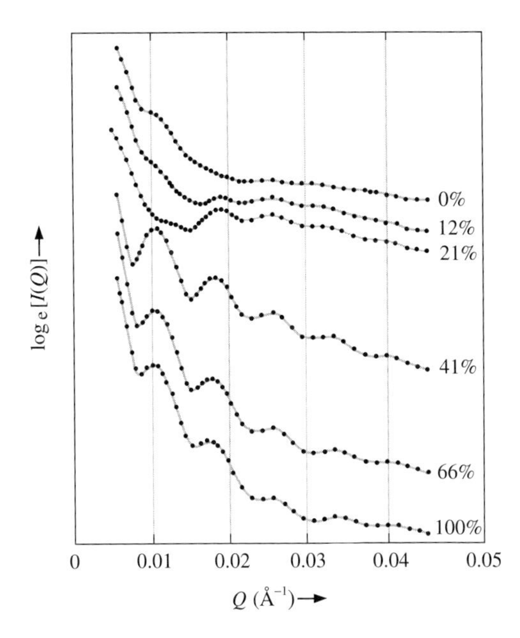

Fig. 10.9 Scattering curves at six different contrasts for the influenza virus. The percentage figures indicate the degree of deuteration. (*After* Cusack *et al.*, 1985.)

Thus contrast variation allows certain components of the sample to be rendered invisible and enhances the scattering from other components. An example is in the study of the influenza virus (Cusack *et al.*, 1985). The virus contains a lipid bilayer, in addition to RNA and protein. Scattering curves at six different contrasts (see Fig. 10.9) were consistent with a structural model of spherical virus particles 1,200 Å in diameter. An inner sphere of RNA, with a radius of 400 Å, is surrounded by a lipid bilayer between 40 and 50 Å thick and then by an outer coat of protein on the surface of the virus.

Schurtenberger *et al.* (1998) used SANS to analyse the solution structure of proteins and they determined the basic features of the structure of proliferating cell nuclear antigen (PCNA). The function of PCNA in DNA replication and repair is to form a sliding clamp in the machinery of DNA replication. The measurements supported a trimeric ring-like structure, in agreement with the determination of the crystal structure of yeast PCNA. The radius of gyration and shape of the subunits of a multi-subunit enzyme have been determined by Callow *et al.* (2007) using SANS and contrast variation.

Small-angle neutron scattering has been used since the late 1970s for the study of polymer conformation and morphology. The SANS study of polymers is adequately reviewed in the books of Higgins and Benôit (1994) and Roe (2000) and the technique has been applied to the examination of many polymer systems. These include amorphous polymers such as polystyrene (PS) and PMMA, crystalline polymers such as melt-grown crystals of polyethylene, polymers in solution and networks of cross-linked polymers. The rapid flow of entangled polymers through an abrupt contraction has been studied by Graham *et al.* (2006).

A general method of determining the structure of micelles was proposed by Hayter and Penfold (1983) and this has been followed by many investigations of *micelles* and *surfactants* in solution. Surfactants are of domestic and industrial importance and a recent study of mixed surfactants is described by Penfold *et al.* (2004).

A technological application of SANS is in the prospecting for oil for the petroleum industry. Oil shales contain voids, which are either empty or full of hydrocarbons. Hydrogen has a negative scattering length for neutrons, and so the mean scattering length of a void tends to be nearly or exactly zero, whether the void is filled or not. Thus SANS locates all the voids, whereas X-ray small-angle scattering senses only the empty voids: by combining both measurements an assessment can be made of the potential of the oil shale as a source of petroleum.

References

Boothroyd, A. T. (1989) *J. Appl. Crystallogr.* **22** 252–255. "*The effect of gravity on the resolution of small-angle neutron scattering.*"

Callow, P., Sukhodub, A., Taylor, J. E. and Kneale, G. G. (2007) *J. Mol. Biol.* **369** 177–185. "*Shape and subunit organization of the DNA methyltransferase M. Ahdl by small-angle neutron scattering.*"

Cotton, J. P., Decker, D., Benoit, H., Farnoux, B., Higgins, J. S., Jannink, G., Ober, R., Picot, C. and des Cloizeaux, J. (1974) *Macromolecules* **7** 863–872. "*Conformation of polymer chain in the bulk.*"

Cusack, S., Ruigrok, R. W. H., Krygsman, P. C. J. and Mellema, J. E. (1985) *J. Mol. Biol.* **186** 565–582. "*Structure and composition of influenza virus: a small-angle neutron scattering study.*"

DeBeer-Schmidt, L., Dewhurst, C. D., Hoogenboom, B. W., Petrovic, C. and Eskildsen, M. R. (2006) *Phys. Rev. Lett.* **97** 127001-1. "*Field dependent coherence length in the superclean, high-κ superconductor $CeCoIn_5$.*"

Debye, P. and Bueche, B. (1952) *J. Chem. Phys.* **20** 1337–1338. "*Distribution of segments in a coiling polymer molecule.*"

Eskildsen, M. R., Dewhurst, C. D. and Hoogenboom, B. W. (2003) *Phys. Rev. Lett.* **90** 187001-1. "*Hexagonal and square flux line lattices in $CeCoIn_5$.*"

Foret, M., Pelous, J. and Vacher, R. (1992) *J. de Phys. I* **2** 791–799. "*Small-angle neutron scattering in model porous systems: a study of fractal aggregates of silica spheres.*"

Gilardi, R., Mesot, J., Drew, A., Divakar, U., Lee, S. L., Forgan, E. M., Zaharko, O., Conder, K., Aswal, V. K., Dewhurst, C. D., Cubitt, R., Momono, M. and Oda, M. (2002) *Phys. Rev. Lett.* **88** 217003-1. "*Direct evidence for an intrinsic square vortex lattice in the overdoped high-T_c superconductor $La_{1.83}Sr_{0.17}CuO_{4+\delta}$.*"

Glatter, O. (1977) *J. Appl. Crystallogr.* **10** 415–421. "*A new method for the evaluation of small-angle scattering data.*"

Glatter, O. (1980) *J. Appl. Crystallogr.* **13** 577–584. "*Evaluation of small-angle scattering data from lamellar and cylindrical particles by the indirect transformation method.*"

Graham, R. S., Bent, J., Hutchings, L. R., Richards, R. W., Groves, D. J., Embery, J., Nicholson, T. M., McLeish, C. B., Likhtman, A. E., Harlen, O. G., Read, D. J., Gough, T., Spares, R., Coates, P. D. and Grillo, I. (2006) *Macromolecules* **39** 2700–2709. "*Measuring and predicting the dynamics of linear monodisperse entangled polymers in rapid flow through an abrupt contraction. A small-angle neutron scattering study.*"

Guinier, A. and Fournet, G. (1955) "*Small-angle scattering of X-rays.*" John Wiley, New York.

Hayter, J. B. and Penfold, J. (1983) *Colloid Polymer Sci.* **261** 1022–1030. "*Determination of micelle structure and charge by neutron small-angle scattering.*"

Higgins, J. S. and Benôit, H. C. (1994) "*Polymers and Neutron Scattering.*" University Press, Oxford.

Huxley, A., Rodière, P., McK. Paul, D., van Dijk, N., Cubitt, R. and Flouquet, J. (2000) *Nature* **406** 160–164. "*Realignment of the flux-line lattice by a change of symmetry of superconductivity in* UPt_3."

Jacrot, B. (1976) *Rep. Prog. Phys.* **39** 911–953. "*The study of biological structures by neutron scattering from solution.*"

King, S. M. (1999) "*Modern techniques for polymer characterisation.*" Chapter 7, edited by Pethrick, R. A. and Dawkins, J. V. New York, John Wiley.

Penfold, J., Staples, E., Tucker, I. and Thomas, R. K. (2004) *Langmuir* **20** 1269–1283. "*Surface and solution behaviour of the mixed dialkyl chain cationic and non-ionic surfactants.*"

Porod, G. *Kolloid Zeitschrift.* (1951) **124** 83–114. "*Die Röntgenkleinwinkelstreuung von dichtgepackten kolloiden Systemen.*"

Radulescu, A., Kentzinger, E., Stellbrink, J., Dohmen, L., Alefeld, B., Rücker, U., Heiderich, M., Schwahn, D., Brückel, T. and Richter, D. (2005) *Neutron News* **16** 18–21. "*The new small-angle neutron scattering instrument based on focusing-mirror optics.*"

Roe, R. J. (2000) "*Methods of neutron and X-ray scattering in polymer science.*" University Press, Oxford.

Schurtenberger, P., Egelhaaf, S. U., Hindges, R., Maga, G., Jonsson, Z. O., May, R. P., Glatter, O. and Hübscher, U. (1998) *J. Mol. Biol.* **275** 123–132. "*The solution structure of functionally active human proliferating cell nuclear antigen determined by small-angle nuclear scattering.*"

Stuhrmann, H. B. (1970) *Acta Crystallogr. A* **26** 297–306. "*Interpretation of small-angle scattering functions of dilute solutions and gases. A representation of the structures related to a one-particle scattering function.*"

Wignall, G. D., Christen, D. K. and Ramakrishnan, V. (1988) *J. Appl. Crystallogr.* **21** 438–451. "*Instrumental resolution effects in small-angle neutron scattering.*"

Zimm, B. H. *J. Chem. Phys.* (1948) **16** 1093–1099. "*The scattering of light and the radial distribution function of high polymer solutions.*"

Neutron optics

11

Neutron optics refers to phenomena which arise from the interference between the incident and the scattered neutron waves. This is in contrast to the topics in earlier chapters of this book, where we have been concerned with interference between the scattered waves alone. The wave-like property of the neutron was first demonstrated experimentally by Halban and Preiswerk (1936) and by Mitchell and Powers (1936) using 'white' neutrons from a radium–beryllium source, and more convincingly by Zeilinger *et al.* (1988) using monochromatic neutrons from a reactor source. The subject of neutron optics itself began with the work of Fermi and Zinn (1946), who demonstrated the mirror reflection of a collimated beam of thermal neutrons. Since then nearly all the classical optical phenomena of reflection, refraction and interference have been shown to occur with neutrons.

The mirror reflection of neutrons has led to the construction of guide tubes, bottles for containing ultracold neutrons (UCNs) and refractometers for determining the scattering lengths of the elements. Neutron interferometry provides an alternative method of accurately measuring neutron scattering lengths. These and related topics are discussed in this chapter. For a more comprehensive and very readable treatment of neutron optics reference should be made to the book by Sears (1989). An experimental survey of neutron optics is given in the article by Kaiser and Rauch (1999) and neutron interferometry is covered in the book by Rauch and Werner (2000).

11.1 Mirror reflection of neutrons

When a plane wave in vacuum enters a medium at a glancing angle θ, the angle θ in the medium is changed to θ' in accordance with Snell's law of refraction

$$\cos\theta = n\cos\theta' \tag{11.1}$$

where n is the refractive index of the medium (see Fig. 5.2). We can derive an expression for the refractive index of neutrons, in terms of the neutron wavelength, the mean scattering length of the atoms in the medium and their density. This derivation is essentially the same as that first given by Fermi (1950).

Figure 11.1 shows a plane wave striking a plate of thickness t at normal incidence. If the direction of propagation is along the z-axis, the wave

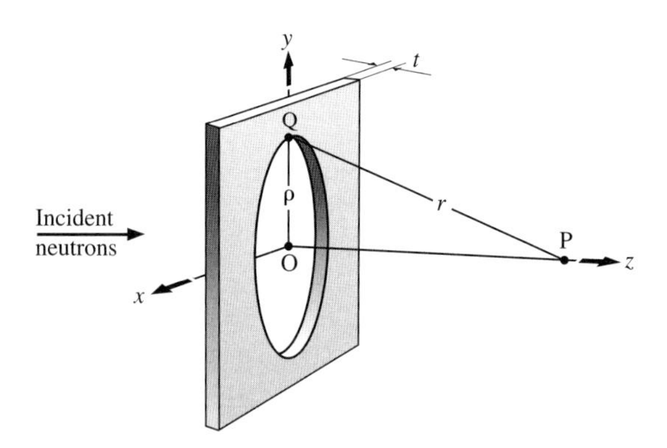

Fig. 11.1 Thin slab of material with the incident neutrons normal to the surface.

function of the incident neutrons is

$$\psi = \exp(ikz) \tag{11.2}$$

where $k = 2\pi/\lambda$. On passing through the plate the wave number k changes to nk, and the phase is changed by $(nk - k)t$. Thus, the wave function on emerging from the plate is

$$\psi' = \exp[\mathrm{i}\,(n-1)\,kt] \tag{11.3}$$

The emerging wave can also be considered as the superposition of the incident wave and the sum of the waves scattered by all the individual nuclei in the plate. In Fig. 11.1, we consider the scattering from an annulus of radius ρ, width $\mathrm{d}\rho$ and thickness t. This annulus contains $2\pi\rho\mathrm{d}\rho \cdot Nt$ atoms where N is the number of atoms per unit volume. Thus, the amplitude of the spherical waves scattered from the annulus at Q to the point P is

$$-\frac{\bar{b}}{r}\exp(ikr) \cdot 2\pi\rho\mathrm{d}\rho \cdot Nt \tag{11.4}$$

where r is the distance PQ and $\bar{b}$ is the mean coherent scattering length of the atoms in the plate. We have assumed that the plate is so thin that there is relatively little difference in phase between the spherical waves scattered from the front and back of the plate (*i.e.* $kt \ll 1$). Also $\rho\mathrm{d}\rho = r\mathrm{d}r$ and so the total scattering at P is

$$\psi_{\mathrm{sc}} = -\int_z^\infty \bar{b} \cdot 2\pi Nt \cdot \exp(ikr)\,\mathrm{d}r \tag{11.5}$$

The emerging wave is the sum of the initial and scattered waves:

$$\psi' = \psi + \psi_{\mathrm{sc}} = \exp(\mathrm{i}kz) - \bar{b} \cdot 2\pi Nt \int_z^\infty \exp(\mathrm{i}kr)\,\mathrm{d}r \tag{11.6}$$

To evaluate the integral in eqn. (11.6) we introduce a factor $\exp(-\beta^2 r)$ in the integrand and then proceed to the limit $\beta \to 0$ when this factor becomes unity. Hence we have

$$\begin{aligned} \int_z^\infty \exp(\mathrm{i}kr) \cdot \exp(-\beta^2 r)\,\mathrm{d}r &= \frac{1}{(\mathrm{i}k - \beta^2)} \left[\exp(\mathrm{i}k - \beta^2)\, r\right]_z^\infty \\ &= \frac{-\exp(\mathrm{i}k - \beta^2)}{(\mathrm{i}k - \beta^2)} \\ &= \frac{\mathrm{i}}{k} \exp(\mathrm{i}kz) \quad \text{as } \beta^2 \to 0 \end{aligned} \tag{11.7}$$

Inserting (11.7) into (11.6) and combining with (11.3) gives the final expression (recalling that $kt \ll 1$):

$$n = 1 - \frac{2\pi N\bar{b}}{k^2} \tag{11.8}$$

or

$$n = 1 - \frac{\lambda^2 N\bar{b}}{2\pi} \tag{11.9}$$

An alternative derivation of eqn. (11.9) is based on the concept that, in passing from a vacuum to a bulk material, the kinetic energy of a neutron is reduced by its interaction with a constant potential of the medium. For a single nucleus at the position **r**, the Fermi pseudo-potential is

$$V_\mathrm{F}(\mathbf{r}) = \frac{2\pi\hbar^2}{m_n} b\delta(\mathbf{r}) \tag{11.10}$$

(see Section 2.3) and so the mean potential V_F seen by neutrons whose wavelengths are very long in comparison with internuclear distances is

$$V_\mathrm{F} = \frac{2\pi\hbar^2}{m_n} N\bar{b} \tag{11.11}$$

Thus, we can write

$$E_\mathrm{t} = E_\mathrm{i} - V_\mathrm{F} \tag{11.12}$$

where E_t and E_i are the energies of the transmitted and incident beams, respectively. Putting $\mathrm{E} = \hbar^2 k^2 / 2m_\mathrm{n}$ (with $\hbar k$ the neutron momentum) gives

$$k_\mathrm{t}^2 = k_\mathrm{i}^2 - 4\pi N\bar{b} \tag{11.13}$$

The macroscopic refractive index n is $k_\mathrm{t}/k_\mathrm{i}$ and so

$$n^2 = \frac{k_\mathrm{t}^2}{k_\mathrm{i}^2} = 1 - \frac{\lambda^2 N\bar{b}}{\pi} \tag{11.14}$$

or

$$n = \left(1 - \frac{\lambda^2 N\bar{b}}{\pi}\right)^{1/2} \tag{11.15}$$

where λ is the neutron wavelength. Equation (11.9) is valid to first order, since only single scattering was considered, whereas eqn. (11.15) holds to higher order. The two expressions are almost the same if the refractive index is close to unity.

$1 - n$ is of the order of 10^{-6} for a neutron wavelength $\lambda \approx 1$ Å. Provided $\bar{b}$ is positive, the refractive index of the medium is very slightly less than unity and total external reflection can occur at the surface of the medium. The critical glancing angle θ_c for total reflection is given by putting $\theta' = 0$ in eqn. (11.1):

$$n = \cos\theta_c \approx 1 - \frac{1}{2}\theta_c^2 \tag{11.16}$$

for $\theta_c \ll 1$. Hence from eqn. (11.9) we have

$$\theta_c = \lambda\sqrt{\frac{N\bar{b}}{\pi}} \tag{11.17}$$

Table 5.1 lists the values of the critical angle at a wavelength of 1 Å for some selected elements. Total external reflection cannot occur for those few elements (hydrogen, titanium, manganese, *etc.*) possessing a negative value of $\bar{b}$.

The mirror reflection of neutrons has led to the construction of guide tubes (see Section 5.1), of bottles for containing UCNs (Section 11.2) and of devices for *polarizing* a neutron beam (Section 9.3). Mirror reflection is also used in the operation of refractometers for determining the scattering lengths of the elements.

11.1.1 Gravity refractometer

This is an instrument which was used to obtain absolute values of neutron scattering amplitudes by measuring the total reflection of neutrons. The original concept of the gravity refractometer was suggested by Maier-Leibnitz, and it was then fully developed as an experimental technique by Koester (Koester, 1965, 1977; Koester and Nistler, 1975).

Neutrons falling freely through a vertical height H gain an energy of $m_n gH$, and for $H = 1$ m this energy is the same order of magnitude as the potential energy V_F of neutrons interacting with matter. As long as the energy of free fall is less than V_F, the neutrons are unable to penetrate a horizontal plane surface of the sample and are totally reflected. We have already seen that the potential V_F is $(2\pi\hbar^2/m_n)\, N\bar{b}$ and so penetration will occur when the drop in height H exceeds the critical height H_c given by

$$m_n g H_c = \frac{2\pi\hbar^2}{m_n} N\bar{b} \tag{11.18}$$

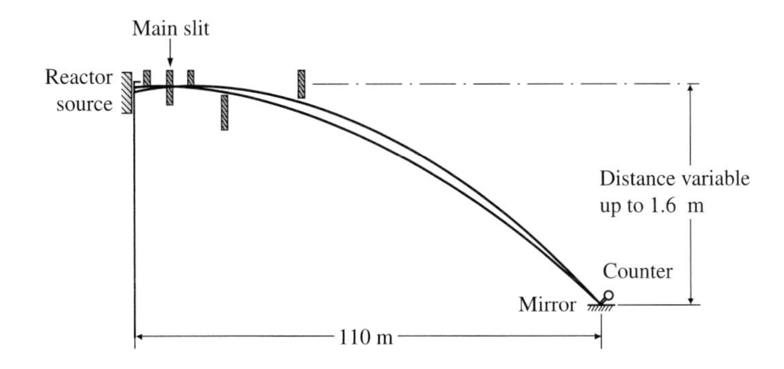

Fig. 11.2 Diagram of the neutron gravity refractometer designed by Koester (1965). The neutrons follow a parabolic path in an evacuated tube before striking the sample acting as a mirror.

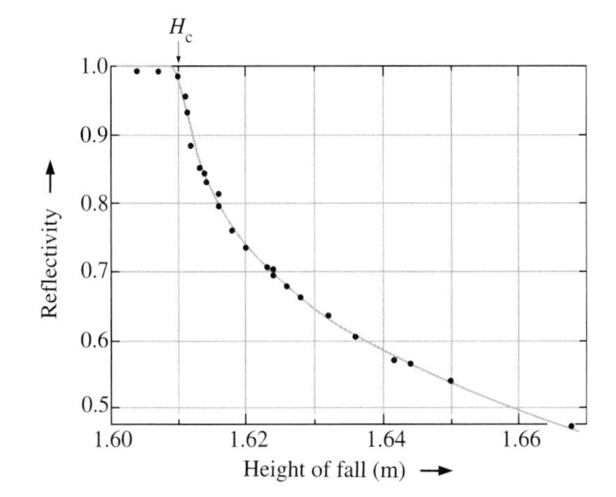

Fig. 11.3 Reflectivity of heavy water as a function of the vertical drop of neutrons. Total reflection occurs at the critical height H_c. (*After* Koester, 1977.)

This equation shows that there is direct proportionality between the mean scattering length $\bar{b}$ and the critical height of fall H_c, and all the other quantities in eqn. (11.18) are known constants. The wavelength (or velocity) does not enter into the expression, but the horizontal distance at which mirror reflection occurs is directly proportional to wavelength and observations are usually made at a wavelength of about 15 Å. Equation (11.18) forms the basis for the accurate determination of the scattering length $\bar{b}$ on an absolute scale.

A diagram of the gravity refractometer built by Koester (1965) at the FRM reactor in Munich is shown in Fig. 11.2. A neutron beam emerges in a horizontal direction from a collimator in the reactor; the neutrons then travel in an evacuated tube of about 110 m in length, describing a parabolic path with a fall in height of up to 1.6 m. At the end of their flight they strike the liquid sample, which is spread as a thin layer on a horizontal glass plate. The glass plate supporting the liquid can be adjusted vertically to a precision better than 0.01 mm, and H_c is then determined to an accuracy of a few parts in 10^5.

A typical curve of reflectivity versus height of fall for heavy water is shown in Fig. 11.3. The theoretical curve has a sharp cut-off at the critical height H_c, but the observed curve is rounded off by the influence of absorption and incoherent scattering. These factors are readily taken into account in calculating the reflectivity for a horizontal mirror, and very good agreement with the experimental points in Fig. 11.3 is achieved by choosing $H_c = 1.61\,\mathrm{m}$. The accuracy of measuring scattering lengths approaches two parts in 10^4, after applying corrections for centrifugal forces and for the *Coriolis effect.* This precision is only possible with liquid samples for which the surface can be kept strictly horizontal by gravity. By making measurements on various liquids of different organic substances containing the elements carbon, hydrogen and chlorine, Koester and Nistler (1975) deduced the following values for the bound coherent scattering amplitudes of hydrogen, carbon and chlorine:

$$\begin{aligned} b_{\mathrm{H}} &= -0.3741 \times 10^{-12}\,\mathrm{cm} \\ b_{\mathrm{C}} &= 0.6648 \times 10^{-12}\,\mathrm{cm} \\ b_{\mathrm{Cl}} &= 0.9579 \times 10^{-12}\,\mathrm{cm} \end{aligned}$$

An extensive compilation of scattering lengths, including measurements made by mirror reflection, has been given by Sears (1992) and a more recent compilation by Rauch and Waschkowski (2000) (see Appendix 2).

11.2 Ultracold neutrons

When neutrons of energy E are incident on a medium of potential V_{F}, the neutrons are totally reflected at all angles of incidence when E is less than V_{F}, and so the possibility then exists of storing the neutrons in a container or 'neutron bottle'. These neutrons have very long wavelengths, in excess of 600 Å, and are known as ultracold neutrons, or UCNs. Experiments with bottled UCNs have been used to investigate the properties of the neutron itself and to address several fundamental questions in particle physics and astrophysics.

Thus in the standard *Big Bang model* of cosmology one can calculate the ratio of the masses of helium to hydrogen which are formed during the first few minutes after the start of the universe. Precise measurements of the neutron lifetime have shown that the number of particle families is fixed at three, and this in turn indicates that the helium to hydrogen ratio in the universe is about 1:4 (Dubbers, 1994). Other UCN experiments have been stimulated by the search for an electric dipole moment (EDM) of the neutron. In the Big Bang model one expects the creation of an equal number of particles and antiparticles, and yet the number of antiparticles is known to be far less (by a factor of 10^4) than the number of particles. A possible explanation of this discrepancy is CP violation, that is, violation of both space parity and time reversal symmetry in the forces operating in particle creation. If CP violation occurs, the neutron should have an EDM of order 10^{-25}–10^{-27} e cm. The most recent result giving the experimental limit on the magnitude of the neutron EDM was reported by Baker *et al.* (2006), who measured the Larmor precession frequency of UCNs in a weak magnetic field: the

Table 11.1 Scattering potential and corresponding critical velocity for some selected materials.

Material	V_F (10^{-7} eV)	v_c (msec^{-1})
Be	2.47	6.9
Al	0.55	3.2
Si	0.56	3.3
Fused silica	0.98	4.3
Cu	1.67	5.7
H_2O	0.15	1.7
D_2O	1.67	5.7

absence of any frequency shift with an electric field applied alternately parallel and antiparallel to the magnetic field gave an upper limit to the neutron EDM of 2.9×10^{-26} e cm. In the next decade, we expect further improvements in the precision of these neutron measurements.

To examine the properties of materials for UCN experiments, we return to eqn. (11.11) expressing the space-averaged potential as

$$V_{\mathrm{F}} = \frac{2\pi\hbar^2 N\bar{b}}{m_{\mathrm{n}}} \tag{11.19}$$

where $\bar{b}$ is the average scattering length and N is the number of atoms in unit volume. Total reflection occurs when E $(= 1/2 m_{\mathrm{n}} v^2) < V_{\mathrm{F}}$ or $v < v_{\mathrm{c}}$, where v_{c} is the critical velocity given by

$$v_{\mathrm{c}} = \frac{\hbar}{m_{\mathrm{n}}} \left(4\pi N\bar{b}\right)^{1/2} \tag{11.20}$$

Table 11.1 lists the potential V_{F} for various materials, together with the corresponding critical velocities as calculated from eqn. (11.20). Thus in the case of aluminium the critical velocity is 3.25 ms^{-1}, corresponding to a wavelength a little longer than 1,200 Å and a temperature of 0.64 mK.

A UCN storage vessel (neutron bottle) should have reflecting walls made from a material with a high potential V_{F} and a low absorption cross-section. Suitable candidates are beryllium and beryllium oxide, but they present difficulties arising from their toxicity. An attractive alternative is diamond-like carbon (DLC), which has low absorption and a potential of 2.7×10^{-7}eV, which is slightly higher than V_{F} for Be. Techniques for producing high-quality DLC coatings have been well developed in industry, and neutron bottles with these coatings have been made at the neutron spallation source of the Paul Scherrer Institute (PSI) in Switzerland (Atchison *et al.*, 2006).

The ultimate storage time of a bottled neutron is limited by the lifetime of the neutron itself, which is about 10^3 sec (see Section 2.2). The question now arises as to whether or not the neutron is absorbed in the wall of the container before it decays. At each collision with the wall the neutron penetrates to a depth which is of the order of magnitude of its wavelength λ. Hence the probability of its being absorbed is $N\sigma_{\mathrm{a}}\lambda$

where σ_a is the absorption cross-section. The number of collisions n_0 before the neutron is captured is $n_0 = 1/(N\sigma_a\lambda)$. For boron-free glass with $N = 2.7 \times 10^{22}\,\text{cm}^{-3}$ the critical velocity is $v_c = 4 \times 10\,\text{msec}^{-1}$ and at this velocity (assuming that σ_a is proportional to $1/v$) the number of collisions is 7.3×10^4. If the average dimension of the container is l, then the time between successive collisions with the wall is $t = l/v_c$ or 0.048 sec for $l = 20\,\text{cm}$, and the time before capture is nt or 3,400 sec. This time is sufficiently greater than the neutron lifetime of 1,000 sec for us to expect that the storage time in a glass bottle should be about the same as the lifetime.

In practice, the storage time is much shorter than this. The discrepancy has been attributed partly to surface roughness, but the main culprit for losses is believed to be inelastic scattering by hydrogen impurities and by small loosely attached surface particles. Some results of Pendlebury (2000) indicate that, in about one wall collision in 20,000, a neutron may suddenly double or triple its kinetic energy and escape. A model that fits these results is one where there are nanoparticles of order 200 Å in diameter skating around on the bottle surface and executing Brownian motion. Such particles are of a size to scatter UCNs coherently and thus have a huge neutron cross-section. The origin and nature of these particles remains a mystery.

UCNs constitute a very small fraction of the neutrons in the Maxwellian spectrum of a reactor, even at low temperatures. Cold sources operate in the temperature range 20–30 K, and at 20 K the UCN fraction in the emergent flux is approximately $(V_F/k_BT)^2$ or $\approx 10^{-9}$ assuming $V_F = 10^{-7}\,\text{eV}$. In the reactor at the ILL very cold neutrons from the D_2 cold source are conducted vertically upwards through highly polished guide tubes (see Fig. 11.4) and are then slowed down to about 5 msec^{-1} by reflection from the retreating blades of a turbine. The maximum UCN density available for experiments is 50 cm^{-3}, and this corresponds to 100,000 neutrons in a bottle with a volume of 20 litres. The UCNs can be stored for hundreds of seconds before the bottle is refilled.

UCNs cannot be extracted directly from the cold source of a nuclear reactor via a conventional beam tube, as they will be either absorbed by the wall of the beam tube or totally reflected before emerging from the tube. This problem can be overcome by inserting a 'converter' in the form of a thin sheet of moderator material at the end of the beam tube and close to the reactor core. This material should have a small absorption cross-section σ_a and a large inelastic cross-section σ_{inel}. σ_{inel} must be large in order to repopulate, by means of repeated collisions, the neutrons lost from the low-energy tail of the Maxwellian distribution after entering the beam tube. The ideal choice of converter would be hydrogen with $V_F < 0$ (to avoid total reflection) or deuterium. In the UCN source at the ILL the liquid deuterium of the vertical cold source plays the role of converter. The thick walls of the flask containing D_2 separate it from the main D_2O moderator, while the vacuum in the vertical extraction guide is separated from the D_2 by a 1 mm aluminium wall. In newer developments, promising a 100-fold increase in the UCN

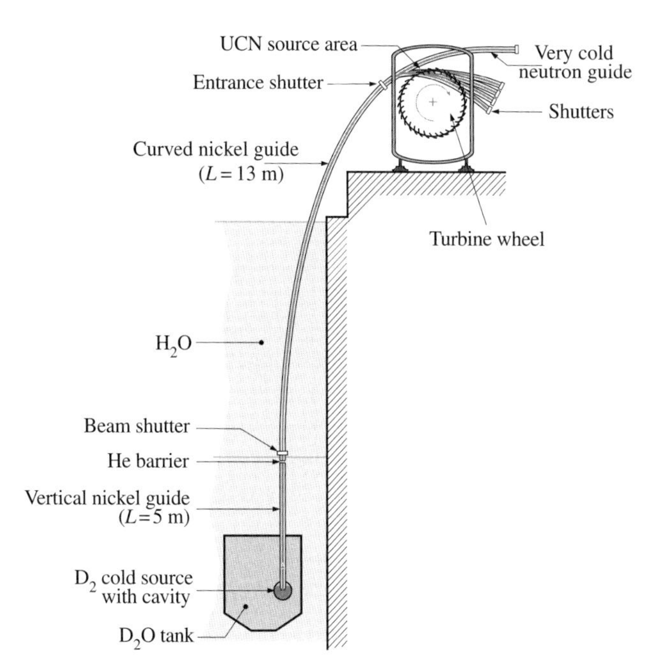

Fig. 11.4 The production of UCNs at the ILL, Grenoble. Very cold neutrons emerge from the liquid deuterium cold source of the reactor and are transported vertically, decelerating under gravity, via highly polished guide tubes to the top floor of the reactor building. Here they are slowed down by a Steyerl turbine and are then led off to the experiments via horizontal guide tubes.

density, liquid D_2 at 22 K will be changed to solid D_2 at 5 K or to liquid He at 0.5 K.

Several methods can be used to slow down cold neutrons to the UCN range of 0–5 msec^{-1} after emerging from the converter with velocities in the range 10–100 msec^{-1}. We have seen that neutrons are slowed by Doppler shifting after undergoing successive reflections from the receding polished and rotating blades of a 'Steyerl turbine' (Steyerl, 1989). In another method gravity slows down the neutrons in a vertical neutron guide tube of height H, where the gravitational loss in energy is $m_{\text{n}}gH$ (Steyerl, 1969). (1 cm of height is equivalent to 1 neV of energy.) The separation of UCNs from the background of faster neutrons is achieved by the simple expedient of introducing bends in the guide tube.

For a review of the field of UCNs the reader is referred to the article by Pendlebury (1993). Earlier reviews are given by Luschikov (1977) and Golub and Pendlebury (1979). The book entitled *Ultra-Cold Neutrons* by Golub *et al.* (1991), written by experimental physicists, complements the more theoretical book *The Physics of Ultracold Neutrons* by Ignatovich (1990).

11.3 Dynamical neutron diffraction

For most crystals a perfectly ordered crystalline array of atoms exists only within very small regions. These regions are known as mosaic blocks; their linear dimensions are no more than 10^4 Å and the crystal is referred to as a *mosaic crystal*. The diffraction properties of mosaic crystals are described by the *kinematic theory* of diffraction used in Chapters 7 and

8. In adopting the mosaic model, it is justifiable to ignore the possibility that the diffracted waves can be diffracted back into the direction of the incident beam. A few crystals, such as large dislocation-free crystals of silicon, have no mosaic structure at all, and so there are precise phase relationships between incident and scattered waves throughout the entire crystal. It is then necessary to replace the simple kinematic theory by the *dynamical theory* of diffraction, in which there is a dynamic exchange of energy between direct and scattered beams as the radiation is propagated throughout the crystal.

The dynamical theory was first developed for X-ray diffraction by Paul Ewald in three seminal papers published in *Annalen der Physik* (Ewald, 1916a, 1916b, 1917). Ewald's work began with the publication in 1912 of his doctorate thesis, which attempted to express the laws of crystal optics in terms of a discrete lattice rather than in terms of a continuum. Following the suggestion by von Laue in the same year that X-rays might be diffracted by crystals, and its subsequent experimental confirmation by Friedrich and Knipping, Ewald extended his treatment to the much shorter wavelengths of X-rays. In this section we shall give a brief outline of the theory: a fuller account of the dynamical theory for neutrons is given in the review by Rauch and Petrascheck (1978). We shall then discuss two types of neutron experiment covering applications of the theory. The first type concerns experiments in neutron interferometry, where beams of neutrons are split and then brought together again, allowing measurements to be made, for example, of the neutron scattering lengths of samples inserted in the interferometer. The second type, known as Pendellösung effects, can also be used to obtain accurate scattering lengths of large perfect crystals.

11.3.1 Basic equations of dynamical diffraction theory

The neutron wave field inside the crystal is derived by solving the time-independent Schrödinger equation:

$$\left[-\left(\frac{\hbar^2}{2m_\mathrm{n}}\right)\nabla^2 + V(\mathbf{r})\right]\psi(\mathbf{r}) = E\psi(\mathbf{r}) \tag{11.21}$$

where $V(\mathbf{r}) = V(\mathbf{r} + \mathbf{R}_\mathrm{n})$ is the periodic potential describing the interaction between neutrons and the nuclei of the crystal, $\mathbf{R}_\mathrm{n}$ is a lattice vector and E is the kinetic energy of the incident neutrons. E is equal to $\hbar^2 k_0^2/(2m_n)$ where $\mathbf{k}_0$ is the wave vector of the neutron wave outside the crystal.

The periodic nature of $V(\mathbf{r})$ suggests a Bloch function as the solution of eqn. (11.21). Hence

$$\psi_j(\mathbf{r}) = u_j(\mathbf{r})\exp(i\mathbf{k}_j.\mathbf{r}) \tag{11.22}$$

where the index j labels the different Bloch waves, of amplitude $u(\mathbf{r})$ and wave vector $\mathbf{k}$ inside the crystal. Both $u(\mathbf{r})$ and $V(\mathbf{r})$ are periodic

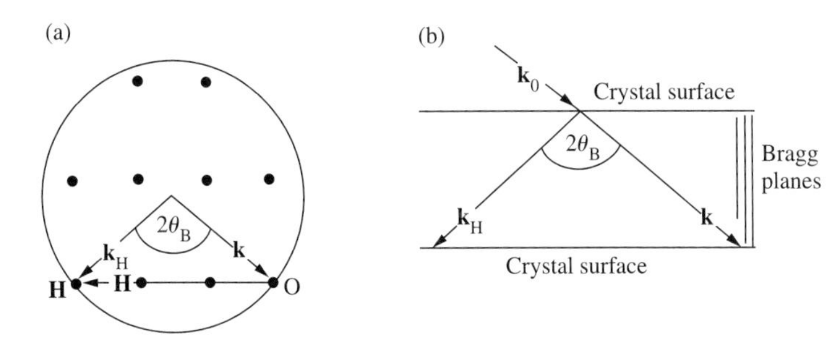

Fig. 11.5 (a) Ewald construction for two-beam case. $\mathbf{k}$ is the wave vector of the direct beam and $\mathbf{k}_{\mathrm{H}}$ that of the diffracted beam. (b) Corresponding diagram in direct space for Laue geometry (*i.e.* with reflecting planes normal to crystal surface).

lattice functions, and so each can be expanded in a Fourier series. Thus

$$u(\mathbf{r}) = \sum_{\mathbf{H}} u(\mathbf{H}) \exp(i\mathbf{H} \cdot \mathbf{r}) \tag{11.23}$$

and

$$V(\mathbf{r}) = \sum_{\mathbf{H}} V(\mathbf{H}) \exp(i\mathbf{H} \cdot \mathbf{r}) \tag{11.24}$$

where $\mathbf{H}$ is a reciprocal lattice vector.

Combining eqns (11.22), (11.23) and (11.24) with eqn. (11.21) gives

$$\left[\left(\frac{\hbar^2}{2m_n}\right)(\mathbf{k}+\mathbf{H})^2 - E\right] u(\mathbf{H}) + \sum_{\mathbf{H}'} V(\mathbf{H}-\mathbf{H}') u(\mathbf{H}') = 0 \tag{11.25}$$

This is a homogeneous system of linear equations in $u(\mathbf{H})$. They are the basic equations of Dynamical Diffraction Theory. The equations cannot be solved exactly, and so approximations are necessary.

In the *two-beam approximation* Bragg scattering takes place when just two reciprocal lattice points, *viz.* (000) and (hkl), are coincident with the Ewald sphere. These are the points O and H in Fig. 11.5(a). Hence only two beams are excited in the crystal, $\mathbf{k}$ and $\mathbf{k}_{\mathrm{H}}$, and $u(\mathbf{H}')$ in eqn. (11.25) is zero except for $\mathbf{H}' = 0$ and $\mathbf{H}' = \mathbf{H}$. These equations then reduce to

$$\left[\left(\frac{\hbar^2}{2m_n}\right) k^2 - E + V(\mathbf{0})\right] u(\mathbf{0}) + V(-\mathbf{H})\, u(\mathbf{H}) = 0 \tag{11.26}$$

and

$$V(\mathbf{H})\, u(\mathbf{0}) + \left[\left(\frac{\hbar^2}{2m_n}\right)(\mathbf{k}+\mathbf{H})^2 - E + V(\mathbf{0})\right] u(\mathbf{H}) = 0 \tag{11.27}$$

The condition for a non-trivial solution of (11.26) and (11.27) is that the secular determinant is zero:

$$\begin{vmatrix} 2\varepsilon + V(\mathbf{O})/\mathrm{E} & \mathrm{V}(\mathbf{H})/\mathrm{E} \\ \mathrm{V}(\mathbf{H}/\mathrm{E}) & 2\varepsilon + \alpha + V(\mathbf{O})/E \end{vmatrix} = 0 \tag{11.28}$$

In eqn. (11.28), ε is the so-called *Anregungsfehler* defined by

$$k^2 = k_0^2(1 + 2\varepsilon) \tag{11.29}$$

with $\mathbf{k}_0$ the wave vector outside the crystal. Typical values for $V(\mathbf{0})/E$ are in the range 10^{-5}–10^{-6} and ε is of a similar order of magnitude. Thus the wave vector of the direct beam is changed only very slightly on entering the crystal. α in eqn. (11.28) is the deviation of the scattered wave $\mathbf{k}_{\mathrm{H}}$ from the Bragg direction:

$$\alpha = (\theta_{\mathrm{B}} - \theta)\sin 2\theta_{\mathrm{B}} \tag{11.30}$$

Equation (11.28) is a quadratic equation whose solution to the first order in ε is

$$\varepsilon_{1,2} = \frac{1}{4}\left\{-\alpha - 2\frac{V(\mathbf{O})}{E} \pm \left[\alpha^2 + \frac{4V(\mathbf{H})\,V(-\mathbf{H})}{E^2}\right]^{1/2}\right\} \tag{11.31}$$

The direct beam is split into *two* beams inside the crystal with wave vectors of magnitudes

$$k_1 \approx k_0\,(1+\varepsilon_1) \tag{11.32}$$

and

$$k_2 \approx k_0\,(1+\varepsilon_2) \tag{11.33}$$

where ε_1 corresponds to the positive sign in (11.31) and ε_2 to the negative sign. Similarly, the diffracted beam $\mathbf{H}$ is split into two beams. Finally, the vector $\mathbf{k}$ lies on a pair of dispersion surfaces inside the crystal, which defines the relation between the magnitude of the wave vectors

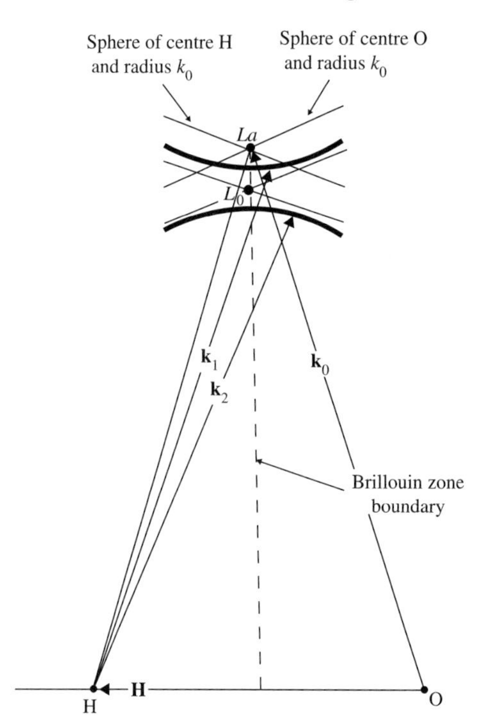

Fig. 11.6 Dispersion surfaces, shown as heavy lines, of the wave vectors $\mathbf{k}_1$ and $\mathbf{k}_2$. *La* is known as the Laue point and *Lo* as the Lorentz point.

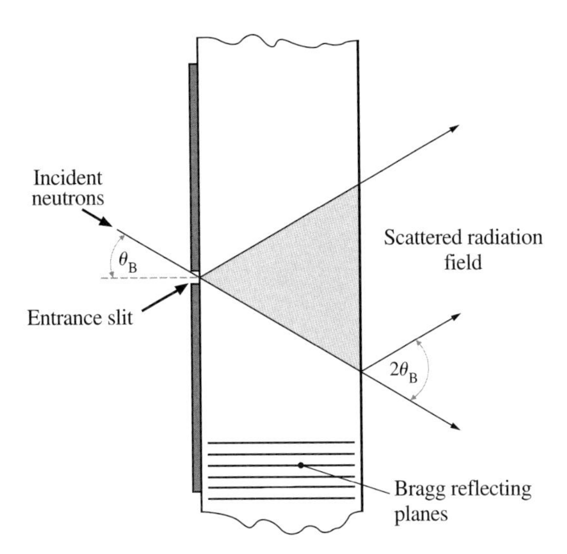

Fig. 11.7 Perfect single crystal set in Bragg reflection with Laue diffraction geometry. For both the forward scattering and Bragg scattering directions two wave fields are transported through the crystal. The shaded region is filled with radiation.

(or energy) and their direction (or momentum). These dispersion surfaces are approximately hyperboloids (see Fig. 11.6).

At the Bragg angle of scattering we see that the transmitted beam inside the crystal is split into two wave fields, whose wave vectors are respectively slightly shorter and slightly longer than the incoming wave vector. (The difference between the two wave vectors depends on the strength of the interaction with the scattering atoms.) One of the wave fields has nodes at the atomic planes and the other has nodes between the atomic planes. Similarly, there are two wave fields in the direction of Bragg scattering, so that, in all, there are four wave fields propagated within the crystal. On emerging from the exit face of the crystal, the wave fields recombine to produce a single field along the transmitted direction and a single field along the scattered direction (Fig. 11.7).

11.3.2 Neutron interferometry

In experiments with a neutron interferometer the forward-scattered fields (the O-beam) are separated from the Bragg-scattered fields (the H-beam) and then brought together again to produce interference fringes. The first perfect-crystal neutron interferometer was constructed by Rauch *et al.* (1974), following the development of a similar instrument for X-rays by Bonse and Hart (1965, 1966). The conventional form of the interferometer is illustrated in Fig. 11.8. Three parallel slices (S, M and A) are cut from a perfect monolithic crystal of silicon; the (220) planes of the crystal are perpendicular to the faces of the slices, allowing the slices to operate in Laue transmission geometry. The incident beam is split into the O-beam and the H-beam at the first slice S, to be separated by several cm at the next slice M. M effectively acts as a mirror, reflecting the beams back thus allowing them to recombine at the analyser slice A.

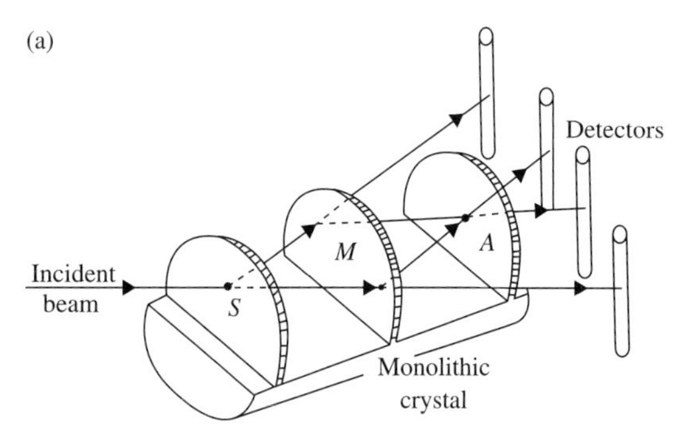

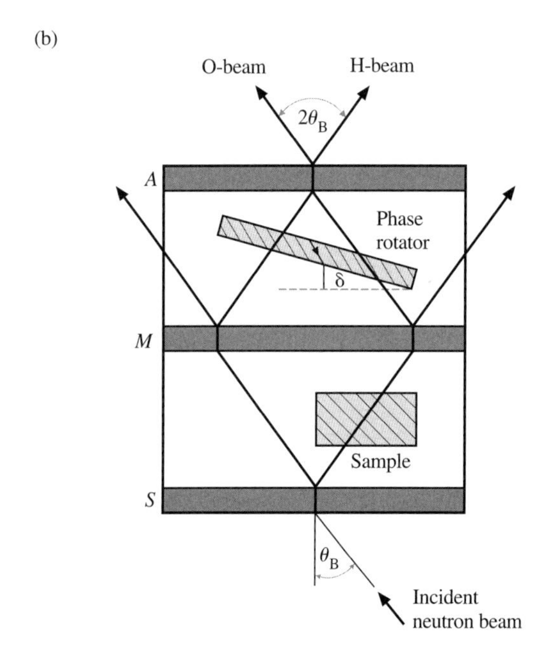

Fig. 11.8 (a) The Bonse–Hart perfect-crystal interferometer. S is the first crystal, M is the mirror crystal and A is the analyser crystal. S, M and A are all cut from the same monolithic crystal of silicon. (b) Measurement of the refractive index of a sample.

An important application of neutron interferometry is in the accurate determination of atomic coherent scattering lengths, many of which have been measured in this way for the elements in the Periodic Table. The forward-scattered and Bragg-scattered beams in Fig. 11.8 traverse widely separated paths before being brought together again at the analyser. Hence, a sample can be inserted between the M and S slices which intersects the H-beam only, introducing a phase change between the O- and H-beams given by

$$\phi = \frac{2\pi}{\lambda}(1-n)\,t \tag{11.34}$$

where t is the thickness of the sample and n its refractive index. To measure the refractive index a plane-parallel slab, typically aluminium, is inserted between the M and A slices that intersects both the O- and

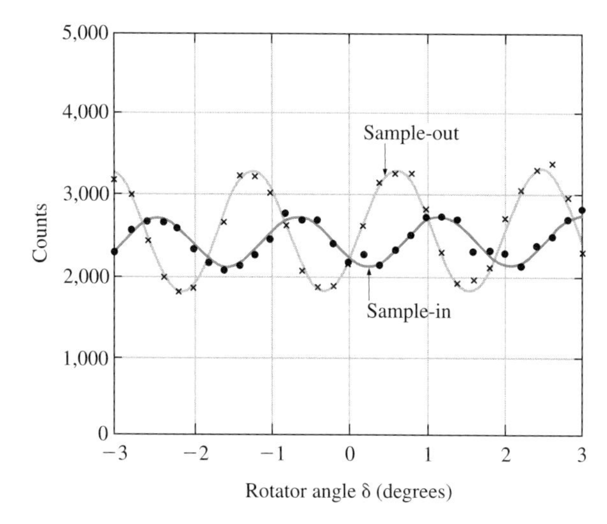

Fig. 11.9 Interference fringes used in determining the scattering length of ^{235}U. (*After* Kaiser *et al.*, 1986.)

H-beams [see Fig. 11.8(b)]. This slab, which can be rotated about an axis perpendicular to the scattering plane, acts as a phase rotator or phase shifter with the phase changing with the angle of rotation δ. Interference fringes are obtained by varying δ, and these fringes are observed first with the sample-in and then with the sample-out. The presence of the sample gives a measurable shift in the fringe pattern, leading to a direct measurement of the phase shift ϕ. The refractive index n is then derived using eqn. (11.34) and the mean scattering length $\bar{b}$ is obtained from eqn. (11.9).

Figure 11.9 shows results published by Kaiser *et al.* (1986) using a sample of uranium foil to derive the scattering length of the uranium isotope ^{235}U. The figure shows the phase shift between the sample-in and sample-out patterns. This uranium isotope has a large absorption cross-section, accounting for the reduction in fringe contrast on inserting the sample. The coherent scattering amplitude of ^{235}U was measured at $\lambda = 1.045$ Å as 10.39 ± 0.03 fm, an accuracy which was primarily limited by the error in measuring the thickness t of the uranium foil.

A similar technique can be used for gaseous samples. A pressure vessel containing the gas is placed in the sample position of Fig. 11.8(b) and fringes are produced by varying the pressure in the gas vessel, thereby altering the refractive index by changing the number density N of atoms. The absolute magnitude of the scattering length can be calculated directly from the fringe spacing, and its sign can be determined by noting in which direction the pattern shifts when the phase shifter is rotated by a small angle δ. Interference fringes measured by Kaiser *et al.* (1979) for various gases are shown in Fig. 11.10. For each gas the fringes for both the O-beam and the H-beam are shown: the two patterns are exactly 180° out of phase, as demanded by the dynamical diffraction theory.

The neutron interferometer has also been used in the famous $2\pi - 4\pi$ experiment (Rauch *et al.*, 1975; Werner *et al.*, 1975). The origin of

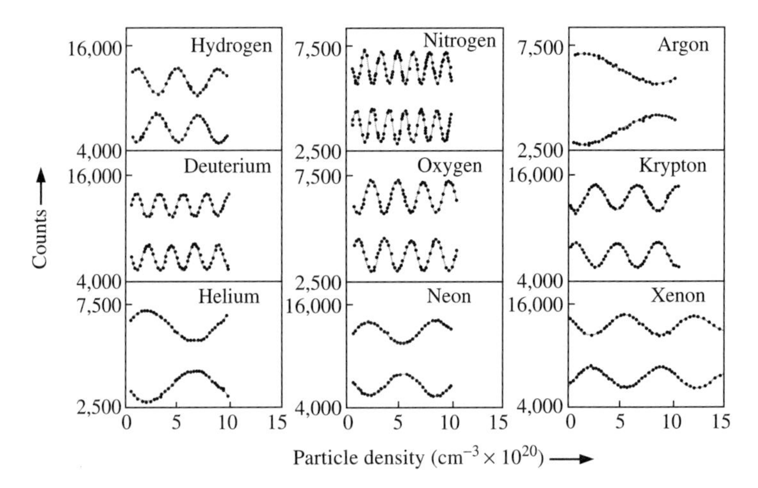

Fig. 11.10 Interference fringes for various gases. The upper patterns refer to the O-field and the lower patterns to the H-field. (*After* Kaiser *et al.*, 1979.)

this experiment lies in the well-known concept that the operator for rotation through an angle of 2π radians causes a reversal in sign of the wave function of a *fermion*. *Bosons* (integral spin particles) return to their original state after turning through a full circle, but fermions (half-integral spin particles) such as neutrons do not. Intensity oscillations of the O- and H-beam were observed while varying the magnetic field along the flight path of the neutrons. In a magnetic field, the neutrons perform Larmor precession and the period of the oscillations showed clearly that the identical wave function was reproduced after a rotation of 4π rather than 2π.

11.3.3 Pendellösung effects

The name Pendellösung ('pendulum solution') was coined by Ewald, who pointed out the analogy between the interaction of coupled pendulums and the interaction of the wave fields inside a crystal. Consider two identical pendulums which are coupled together by a weak spring and can undergo oscillations without damping. If one pendulum is displaced and set into oscillation, its amplitude gradually falls as energy is transferred to the other pendulum. After a while the first pendulum is at rest and the other has a maximum amplitude of oscillation. The process is then reversed and the whole cycle repeated with energy being swapped continuously from one pendulum to the other. In a similar way there is a periodic beating of the radiation inside a perfect crystal with the neutron flux being exchanged, back and forth, between the direct and diffracted beams as the depth of penetration into the crystal increases. In the case of coupled pendulums the rate of energy transfer is proportional to the stiffness of the spring; for a crystal the rate is proportional to the strength of interaction between the two wave fields.

Figure 11.11 is a diagram of a perfect crystal of uniform thickness t set in symmetric Laue diffraction, with the Bragg reflecting planes normal

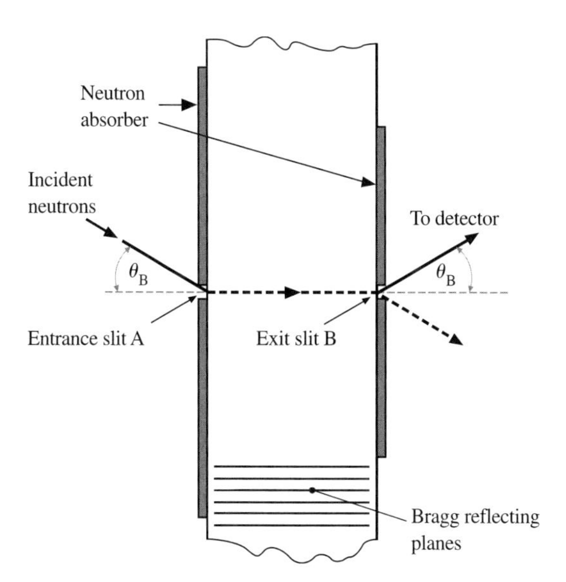

Fig. 11.11 Perfect single crystal in Laue diffraction geometry. The geometry is the same as in Fig. 11.7 apart from the addition of the narrow exit slit B.

to the surface of the crystal. At the entrance slit ($t=0$) the diffracted wave field has zero amplitude; the diffracted amplitude then rises steadily as the depth of penetration increases, reaching a maximum at a depth ($t=\Delta/2$) and falling to zero again at $t=\Delta$. Δ is known as the Pendellösung length. The energy flow is along the Bragg planes, and when the radiation arrives at the exit slit B (which is directly opposite to the entrance slit A in Fig. 11.11), the wave fields inside the crystal recombine into two components travelling in the forward Bragg direction. According to the dynamical theory Δ is given by

$$\Delta = \frac{\pi \cos \theta_{\mathrm{B}}}{F_{hkl} N_{\mathrm{c}} \lambda} \tag{11.35}$$

where N_c is the number of unit cells per unit volume and F_{hkl} is the crystal structure factor per unit cell. Thus, the Pendellösung length varies inversely as the strength of scattering by the Bragg planes. Δ also depends on the neutron wavelength λ, so that the intensity oscillations can be observed either by varying the thickness t at a fixed wavelength or by varying the wavelength at a fixed thickness of the crystal. The first method was adopted by Sippel *et al.* (1965) and the second by Shull (1968).

Figure 11.12 shows the experimental results of Sippel, who was the first to observe Pendellösung fringes with neutrons. The intensity of the 220 reflection of silicon was measured for a plane-parallel disc examined in symmetrical (Laue) transmission. Measurements were repeated for different thicknesses t obtained by progressively etching the silicon crystal. The experimental points, fitted to the full line of the dynamical theory, led to an estimate of the Pendellösung length Δ. The absolute value of the scattering length b_{Si} of silicon was then obtained by substituting

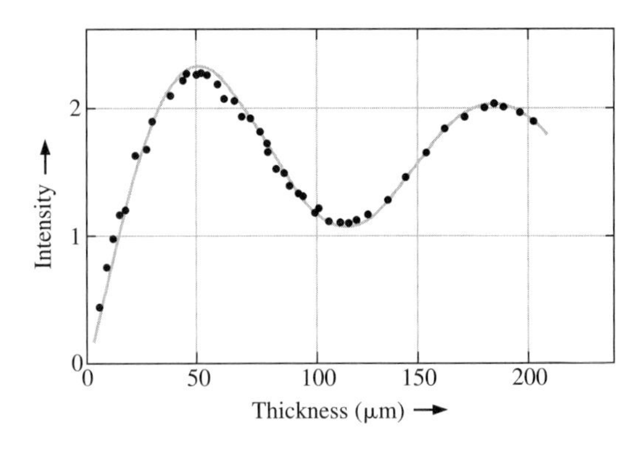

Fig. 11.12 Pendellösung fringes observed in a perfect silicon crystal. The points refer to intensity measurements made as the thickness of the crystal was gradually reduced by etching. (*After* Sippel *et al.*, 1965.)

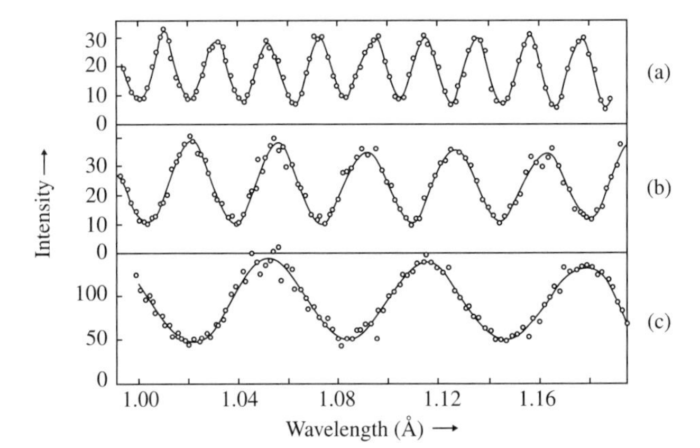

Fig. 11.13 Pendellösung oscillations in parallel-sided crystals of silicon of different thicknesses $t =$ (a) 10 mm, (b) 5.9 mm and (c) 3.3 mm. (*After* Shull, 1968.)

into eqn. (11.35), assuming that $F_{220} = 8b_{\mathrm{Si}}$ without including a Debye–Waller temperature factor.

In Shull's experiment Pendellösung fringes in silicon were observed by measuring the intensity as a function of wavelength. His results are shown in Fig. 11.13. The three curves refer to slices of three different thicknesses. As the wavelength was changed, the incident angle was altered in order to maintain the Bragg reflecting condition. After the Debye–Waller correction was made for thermal motion, b_{Si} was derived to an accuracy of four decimal places. However, this is not a general method of measuring nuclear scattering lengths, as it only applies to materials which have low neutron absorption and which are available as large perfect crystals.

In this chapter, we have referred briefly to various aspects of neutron optics, including experimental work on the reflection, refraction, diffraction and interference of neutrons. For those interested in reading further we recommend the book by Sears (1989) entitled *Neutron Optics*, the book *De Broglie Wave Optics* by Kaiser and Rauch (1979) and the book *Neutron Interferometry: Lessons in Experimental Quantum Mechanics* by Rauch and Werner (2000).

References

Atchison, F., Blau, B., Daum, M., Fierlinger, P., Geltenbort, P., Henneck, R., Heule, S., Kasprzak, M., Kirch, K., Kohlik, K., Kuzniak, M., Meier, M., Meyer, C.-F., Pichlmaier, A., Plonka, C., Schmidt-Wellenburg, P., Schultrich, B., Stucky, Th., Weihnacht, V. and Zimmer, O. (2006) *Phys. Rev. C* **74** 055501–055509. "*Storage of ultra-cold neutrons in a volume coated with diamond-like carbon.*"

Baker, C. A., Doyle, D.D., Geltenbort, P., Green, K., van der Grinten, M.G.D., Harris, P. G., Iaydjiev, P., Ivanov, S. N., May, D. J. R., Pendlebury, J. M., Richardson, J. D., Shiers, D. and Smith, K. F. (2006) *Phys. Rev. Lett.* **97** 131801. "*Improved experimental limit on the electric dipole moment of the neutron.*"

Bonse, U. and Hart, M. (1965) *Appl. Phys. Lett.* **7** 99–100. "*An X-ray interferometer with long separated interfering beam paths.*"

Bonse, U. and Hart, M. (1966) *Z. Physik.* **194** 1–17. "*An X-ray interferometer with Bragg case beam splitting and beam recombination.*"

Dubbers, D. (1994) *Neutron News* **5** 21–24. "*The neutron and some basic questions.*"

Ewald, P. P. (1916a) *Annalen der Physik* **49** 1–38. "*Zur Begründung der Kristalloptik. I. Dispersionstheorie.*"

Ewald, P. P. (1916b) *Annalen der Physik* **49** 117–143. "*Zur Begründung der Kristalloptik. II. Theorie der Reflexion und Brechung.*"

Ewald, P. P. (1917) *Annalen der Physik* **54** 519–597. "*Zur Begründung der Kristalloptik. III. Die Kristalloptik der Röntgenstrahlen.*"

Fermi, E. (1950) "*Notes on nuclear physics.*" Edited by Orear, J., Rosenfeld, A. H. and Schluter, R. A. University of Chicago Press.

Fermi, E. and Zinn, W. H. (1946) *Phys. Rev.* **70** 103. "*Reflection of neutrons on mirrors.*"

Golub, R. and Pendlebury, J. M. (1979) *Rep. Progr. Phys.* **42** 439–501. "*Ultra-cold neutrons.*"

Golub, R., Richardson, D. and Lamoreaux, S. K. (1991) "*Ultra-cold neutrons.*" Adam-Hilger, London.

Halban, H. and Preiswerk, P. (1936) *Comptes Rendus Acad. Sci.* **203** 73–75. "*Preuve expérimentale de la Diffraction des Neutrons.*"

Ignatovich, V. K. (1990) "*The physics of ultracold neutrons.*" Oxford University Press, Oxford.

Kaiser, H. and Rauch, H. (1999) *De Broglie wave optics: neutrons, atoms and molecules* in "*Optics of waves and particles.*" (edited by H. Niedrig). W. de Gruyter, Berlin.

Kaiser, H., Rauch, H., Badurek, G., Bauspiess, W. and Bonse, U. (1979) *Z. Phys.* A **291** 231–238. "*Measurement of coherent neutron scattering lengths of gases.*"

Kaiser, H., Arif, M., Werner, S. A. and Willis, J. O. (1986) *Physica B* **136** 134–136 "*Precision measurement of the scattering amplitude of* ^{235}U."

Koester, L. (1965) *Z. Phys.* **182** 328–336. "*Absolutmessung der kohärenten Streulänge von Quecksilber mit dem Neutronen-Schwerkraft-Refraktometer am FRM.*"

Koester, L. (1977) "*Neutron physics*" in "*Springer tracts in modern physics.*" Vol. 80, pp. 1–55. Springer-Verlag, New York.

Koester. L. and Nistler, W. (1975) *Z. Phys. A* **272** 189–196. "*New determination of the neutron-proton scattering amplitude and precise measurements of the scattering amplitudes of carbon, chlorine, fluorine and bromine.*"

Luschikov, V. I. (1977) *Physics Today* **June** 42–51. "*Ultracold Neutrons.*"

Mitchell, D. P. and Powers, P. N. (1936) *Phys. Rev.* **50** 486–487. "*Bragg reflection of slow neutrons.*"

Pendlebury, J. M. (1993) *Ann. Rev. Nucl. Part. Sci.* **43** 687–727. "*Fundamental physics with ultracold neutrons.*"

Pendlebury, J. M. (2000). Private communication.

Rauch, H. and Petrascheck, D. (1978) *Topics in Current Physics.* Volume **6**, pp. 303–351. "*Dynamical neutron diffraction and its application.*" Springer-Verlag, Berlin.

Rauch, H. and Waschkowski, W. (2000) "*Landolt- Börnstein.*" Volume **1/16A**, Chapter 6. Springer-Verlag, Berlin.

Rauch, H. and Werner, S. A. (2000) "*Neutron interferometry: Lessons in experimental quantum mechanics.*" Oxford University Press, Oxford.

Rauch, H., Treimer, W. and Bonse, U. (1974) *Phys. Lett. A* **47** 369–371. "*Test of a single crystal neutron interferometer.*"

Rauch, H., Zeilinger, A., Badurek, G., Wilfing, A., Bauspiess, W. and Bonse, U. (1975) *Phys. Lett.* **54A** 425–427. "*Verification of coherent spinor rotation of fermions.*"

Sears, V. F. (1989) "*Neutron optics: An introduction to the theory of neutron optical phenomena and their applications.*" Oxford University Press, Oxford.

Sears, V. F. (1992) *Neutron News* **3** 26–37. "*Neutron scattering lengths and cross-sections.*"

Shull, C. G. (1968) *Phys. Rev. Lett.* **21** 1585–1589. "*Observation of Pendellösung fringe structure in neutron diffraction.*"

Sippel, D., Kleinstück, K. and Schulze, G. E. R. (1965) *Phys. Lett.* **14** 174–175. "*Pendellösung Interferenzen mit thermischen Neutronen an Si- Einkristallen.*"

Steyerl, A. (1969) *Phys. Lett.* **29**B 33–35. "*Measurements of total cross-sections for very slow neutrons with velocities from 100m/sec to 5m/sec.*"

Steyerl, A. (1989) *Physica B* **156–157** 528–533. "*Ultracold neutrons: production and experiments.*"

Werner, S. A., Colella, R., Overhauser, A. W. and Eagen, C. F. (1975) *Phys. Rev. Lett.* **35** 1053–1055. "*Observation of the phase shift of a neutron due to precession in a magnetic field.*"

Zeilinger, A., Gähler, R., Shull, C. G., Treimer, W. and Mampe, W. (1988) *Rev. Modern Phys.* **60** 1067–1073. "*Single- and double-slit diffraction of neutrons.*"

Neutron reflectometry

12

Neutron reflectometry, in which neutrons are reflected from planar structures, is a powerful method for investigating properties of interfaces. It is one of the neutron techniques in most demand today and neutron reflectometers are amongst the first instruments to be commissioned at the new pulsed sources SNS and J-PARC. The technique has been applied to a wide variety of materials, including surfactants adsorbed at solid/liquid interfaces, Langmuir–Blodgett and polymer films, lipid bilayers in biological cell membranes, and magnetic ultrathin films and multilayers. By measuring the reflected specular intensity as a function of the incident angle or wavelength, the scattering profile normal to the surface is obtained. Non-specular scattering yields information on the lateral structure of the interface. The critical glancing angles for the total reflection of neutrons are very small (see Section 5.1) and measurements are made near grazing incidence with highly collimated beams. For the wavelength range 1–7 Å the critical angle of incidence is in the range 0.25–1.8°.

There are several reasons why neutrons have an advantage over X-rays for these experiments, even though the enormous X-ray fluxes available with synchrotron radiation vastly exceed the fluxes available for neutron work. Neutrons are not destructive and can examine *buried* interfaces. For soft-matter studies, neutrons are strongly scattered by light atoms such as H, C, O and N, which are the constituents of organic and biological materials. Isotopic substitution produces contrasts in the scattering density allowing the use of *contrast variation.* In particular, by using H/D isotopic substitution, the refractive index distribution can be manipulated at the surface or interface whilst leaving the chemistry unaltered, and this deuteration can serve either to highlight or to mask selected features in the sample. Finally, magnetic materials have a refractive index which depends on the neutron spin, and the reflection of spin-polarized neutrons provides a method for studying the magnetism of thin films.

The importance of neutron specular reflectometry is evidenced by the large number of reflectometers installed at reactor and pulsed neutron sources during the past two decades. The field was opened up in a seminal paper by Hayter *et al.* (1981). It then rapidly expanded, and the early work was reviewed in articles by Penfold and Thomas (1990), Lu *et al.* (1996) and Lu and Thomas (1998). A more recent review is by Penfold (2002). Studies of magnetic structures by polarized neutron reflectivity are described by Zabel and Theis-Bröhl (2003).

12.1 Theory of neutron reflectivity

If a neutron beam is incident on a flat smooth surface, the neutrons interact at the surface with a potential V_F given by eqn. (11.11) of Chapter 11:

$$V_\mathrm{F} = \frac{2\pi\hbar^2}{m_\mathrm{n}} N\bar{b} \tag{12.1}$$

where N is the number density of the atoms in the surface and $\bar{b}$ is their average coherent scattering length.

The potential acts perpendicular to the surface and it changes the normal component of the wave vector $\mathbf{k}_\mathrm{i}$ of the incident wave. The kinetic energy of this normal component is

$$E_{\mathrm{i}\perp} = \frac{(\hbar k_\mathrm{i} \sin\theta)^2}{2m_\mathrm{n}} \tag{12.2}$$

where k_i is the wave number of the incident neutron and θ is the grazing angle of incidence. Total reflection occurs when $E_{\mathrm{i}\perp} = V_\mathrm{F}$, or $\theta = \theta_\mathrm{c}$ with θ_c denoting the critical glancing angle. For elastic scattering the momentum transfer Q is normal to the surface and equal to $2k_\mathrm{i}\sin\theta$. From eqns. (12.1) and (12.2) the critical value of Q is

$$Q_\mathrm{c} = (16\pi N\bar{b})^{1/2} \tag{12.3}$$

The refractive index for neutrons is very close to unity and so $\sin\theta_\mathrm{c} \approx \theta_\mathrm{c}$. Putting $k_\mathrm{i} = 2\pi/\lambda$ where λ is the neutron wavelength, we can write

$$\theta_\mathrm{c} \approx \lambda \left(\frac{N\bar{b}}{\pi}\right)^{1/2} \tag{12.4}$$

The measurement of the intensity of the totally reflected beam at $\theta = \theta_\mathrm{c}$ provides an absolute value for the reflectivity of the surface. The technique of neutron reflectometry is concerned with measurements at grazing angles above this critical angle, where the reflection is not total and the neutrons can either be reflected or transmitted into the bulk of the material. To calculate the reflection at the interface for $\theta > \theta_\mathrm{c}$ we need a quantum mechanical treatment.

Consider the reflection of an incident beam from a surface normal to the z-axis and possessing a step-wise potential gradient V_F. For this simple step the Schrödinger equation along the z-direction is

$$\left[-\left(\frac{\hbar^2}{2m_\mathrm{n}}\right)\frac{\partial^2\psi_z}{\partial^2 z} + V_\mathrm{F}\right]\psi_z = E_\perp\psi_z \tag{12.5}$$

where $E_\perp$ is the energy of the reflected beam along z and is given by

$$E_\perp = \frac{\hbar^2 k_\perp^2}{2m_\mathrm{n}} = \frac{\hbar^2}{2m_\mathrm{n}}(k_f^2 - k_\parallel^2) \tag{12.6}$$

k_f is $2\pi/\lambda$ and the symbols $\perp$ and $\|$ refer to the directions perpendicular and parallel to the surface. The solution of (12.5) above the surface is

$$\psi_z = \exp(ik_{i\perp}z) + r\exp(-ik_{i\perp}z) \tag{12.7}$$

where r is the probability amplitude of reflection, and the solution below the surface is

$$\psi_z = t\exp(ik_{i\perp}z) \tag{12.8}$$

where t is the probability amplitude of transmission. Continuity conditions for the wave function and its derivatives lead to the Fresnel expressions of classical optics:

$$r = \frac{k_{i\perp} - k_{t\perp}}{k_{i\perp} + k_{t\perp}} \quad \text{and} \quad t = \frac{2k_{t\perp}}{k_{i\perp} + k_{t\perp}} \tag{12.9}$$

The reflectivity R, which is the measured intensity, is the square of the quantum mechanical amplitude r:

$$R = r^2 = \left[\frac{Q - (Q^2 - Q_c^2)^{1/2}}{Q + (Q^2 - Q_c^2)^{1/2}}\right]^2 \tag{12.10}$$

When $Q \gg Q_c$ (12.10) reduces to Fresnel's law:

$$R(Q) \approx \frac{16\pi^2}{Q^4}\left(N\bar{b}\right)^2 \tag{12.11}$$

Thus, the reflectivity decays rapidly with the inverse fourth power of the scattering vector Q, as illustrated in Fig. 12.1.

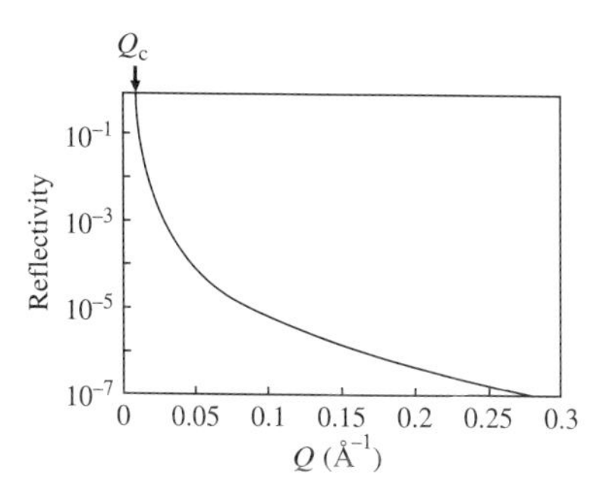

Fig. 12.1 The calculated reflectivity of a silicon/air interface as a function of Q.

Equation (12.8) is the wave function below the surface when Q exceeds Q_c. When $Q < Q_c$, that is, in the region of total reflection, there is a solution within the material of the form

$$\psi_z = t\exp\left[-\frac{1}{2}\left(Q_c^2 - Q^2\right)^{1/2} z\right] \tag{12.12}$$

This equation shows that the neutrons penetrate the surface to a characteristic depth of $\left(Q_c^2 - Q^2\right)^{-1/2}$, even though the potential barrier is higher than the neutron energy perpendicular to the surface. An *evanescent wave* of neutrons propagates along the surface for a very short time before being ejected from the bulk material into the specular direction. This phenomenon is important in understanding the reflection from very thin films. If $Q = 0$ the penetration depth is $1/Q_c$, and this is about 100 Å for reflection from a surface of silicon, rising rapidly to infinity as Q approaches Q_c. Now suppose that a 100 Å film of another material is deposited on the silicon with a Q_c value higher than that of silicon and giving a penetration depth in excess of 100 Å. An incident beam will tunnel through the thin film before it is reflected from the underlying silicon. (This tunnelling is reminiscent of the behaviour of alpha particles ejected from an atomic nucleus.)

When there is a thin uniform layer sandwiched between two bulk phases, the neutron beam is reflected at both the first and the second interfaces, giving rise to interference of the two beams and to the appearance of fringes in the reflectivity profile. The calculation of the reflectivity can be carried out by standard optical methods, using Fresnel coefficients for the transmitted and reflected amplitudes at each interface. Provided that Q is at least twice as large as Q_c, a useful expression for the reflectivity of a system consisting of a layer (phase 2) of thickness t sandwiched between two bulk phases (1 and 3) is

$$R = \frac{16\pi^2}{Q^4} \Big[\left(N\bar{b}_1 - N\bar{b}_2\right)^2 + \left(N\bar{b}_3 - N\bar{b}_2\right)^2 + 2\left(N\bar{b}_1 - N\bar{b}_2\right)\left(N\bar{b}_2 - N\bar{b}_3\right)\cos\left(Qt\right) \Big] \quad (12.13)$$

The formula (12.13) shows that the reflectivity curve consists of a series of fringes superimposed on a curve decaying rapidly as $1/Q^4$. The separation of the fringes is $2\pi/t$. A reflectivity profile of this form is shown in Fig. 12.2.

The calculation of the reflectivity profile for a system with many layers can be carried out by the so-called *optical matrix method* (Born and Wolf, 1999). The scattering system is divided into an arbitrary number of layers, each with its own characteristic thickness and scattering density. Using the condition that the wave function and its derivative are continuous at an interface, the transmission and reflection are determined for each layer by a characteristic matrix; the reflectivity profile is then obtained by multiplying together the matrices of all the layers. The various techniques for inverting the observed profile to extract the thickness and scattering length density $N\bar{b}$ for each layer is described by Zhou and Chen (1995).

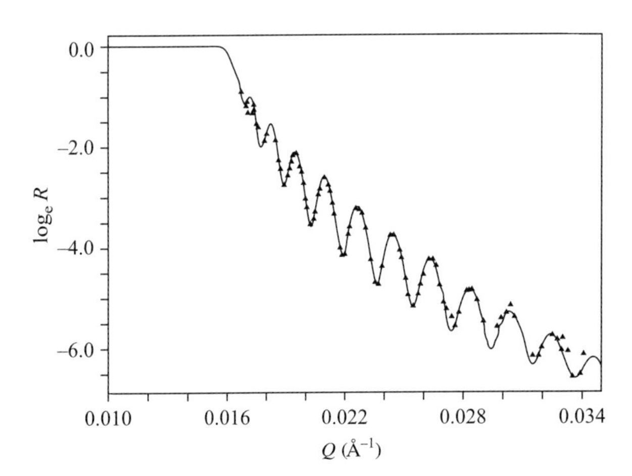

Fig. 12.2 Reflectivity from a silver nitride film deposited on a crystalline silicon substrate. The experimental points are shown as triangles. The continuous curve was calculated from eqn. (12.13) using a layer thickness of 2,430 Å. (*After* Ashworth *et al.*, 1989.)

12.1.1 Non-specular scattering

Non-specular (or diffuse) scattering occurs when the scattering density is not uniform within the planar surface. Because the scattering angles at grazing incidence are very small, the extent of the coherent incident wave front may exceed the length scale of in-plane fluctuations. Intensity is lost from the specular reflection and is redistributed in the rest of reciprocal space. This is analogous to the effect of the thermal motion of the atoms on the reflection of radiation from a crystal. The intensities of the Bragg reflections are diminished by the lattice vibrations, as expressed by the Debye–Waller factor, and thermal diffuse scattering is seen in directions not allowed by Bragg's law: similarly, the specular intensity is reduced by an exponential factor of the form

$$\exp\left(-\frac{Q^2\sigma^2}{2}\right) \tag{12.14}$$

where σ is a characteristic length of the in-plane fluctuation.

While specular reflection, in which the angle of the incoming beam is equal to the angle of the reflected beam, gives information in a direction perpendicular to the interface, the lateral structure of the interface is probed by non-specular scattering. The theory of X-ray and neutron scattering from rough surfaces is described by Sinha *et al.* (1988) and by Pynn *et al.* (1992, 1999). A practical example in biophysics is the off-specular scattering from highly oriented multilamellar phospholipid membranes (Münster *et al.*, 1999), opening up the possibility of determining the lateral structure of membranes on length scales from a few Å to several μm.

12.2 Spin-polarized neutron reflectometry

Polarized neutron reflectometry (PNR) measures the reflected neutron intensity as a function of the perpendicular momentum and spin eigenstate. It is used to determine the magnitude and orientation of atomic magnetic moments in thin films and multilayer media. There are over a score of reflectometers at different institutes producing data on magnetic surfaces, interfaces and multilayers. Reviews of the field have been given by Felcher (1993, 1999), who is a pioneer of the PNR technique; a more recent review, with particular reference to the study of nanomagnetism, is given by Fitzsimmons *et al.* (2004).

We shall give a rudimentary description of the theory of neutron reflection from magnetic multilayers: a more complete treatment is given by Blundell and Bland (1992). In a magnetic sample there is an additional potential whose magnitude is related to the magnetic flux density **B** in the material. This magnetic potential is given by

$$V_{\mathrm{mag}} = -\boldsymbol{\mu}\cdot\mathbf{B} \tag{12.15}$$

where μ is the magnetic dipole moment of the neutron. It can be expressed in the form of a magnetic scattering length given by

$$b_{\mathrm{m}} = 1.913e^2 S/m_{\mathrm{e}} \tag{12.16}$$

where S is the spin of the magnetic atom in the direction perpendicular to the momentum transfer $\mathbf{Q}$, and e and m_{e} are the charge and mass, respectively, of the electron. The total coherent scattering length is then

$$b_{\mathrm{total}} = b_{\mathrm{nuclear}} \pm b_{\mathrm{m}} \tag{12.17}$$

where the sign $\pm$ corresponds to the beam being polarized parallel or antiparallel to the magnetization direction of the sample. Magnetic reflection from a thin magnetic layer can only occur if there is a component of magnetization within the plane of the layer, and for maximum interaction of the neutron magnetic moment with the magnetization of a stratified sample, the polarization direction of the incident beam must be oriented within the plane of the layer.

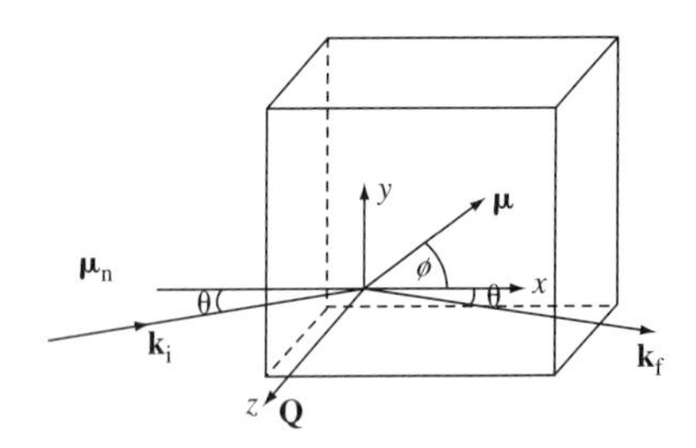

Fig. 12.3 Scattering geometry for measuring polarized spin reflectivity of a thin film in the $x-y$ plane. The neutron spin moment μ_n is along y and the scattering vector $\mathbf{Q}$ is along z. The magnetic moment of the film μ is within the $x-y$ plane and at an angle ϕ to the x-axis. θ is the angle of incidence and reflection.

Figure 12.3 illustrates the scattering geometry in an experiment for measuring the specular reflection of a polarized neutron beam from a thin film. The film is in the $x-y$ plane and the scattering vector $\mathbf{Q}$ is along the z-direction normal to the film. A supermirror in the incident beam polarizes the beam along the $+y$ direction in the spin-up state (+). The beam can be changed to the spin-down state (−) by activating a π-spin flipper between the sample and supermirror. Components of the magnetization along z cannot reflect the in-plane polarized beam, but they can cause spin flip so that a fraction of the (+) state is changed to the (−) state by reflection. This fraction is measured using a second supermirror and π-spin flipper in the reflected beam. Thus four reflectivities are measured. If both flippers are deactivated the non-spin flip scattering cross-section is obtained for the (+ +) states, and if both flippers are activated the non-spin flip cross-section for the (− −) states is measured: the spin-flip cross-sections (+ −) and (− +) are measured by activating one flipper but not the other.

The sensitivity of the technique is illustrated by measurements of the flipping ratio (see Section 9.5.2), which is the ratio of the intensity of specular reflection for neutrons polarized parallel and antiparallel to the magnetization. Figure 12.4 shows the calculated flipping ratio for the specular reflection by ferromagnetic nickel of neutrons of 5 Å wavelength. There are two critical angles in the material corresponding to the two scattering amplitudes in eqn. (12.15): θ_{c}^{+} is the critical angle for the neutron state with the neutron spin parallel to the magnetization direction and θ_{c}^{-} is the critical angle for the antiparallel state. The reflectivity is unity for both states, when the glancing angle θ_0 is less than θ_{c}^{-}. When θ_0 is between θ_{c}^{+} and θ_{c}^{+}, there is a sharp rise in the flipping ratio, since only the positively polarized neutrons are totally reflected. For θ_0 greater than θ_{c}^{+} the flipping ratio suddenly decreases towards an asymptotic value of nearly 2.

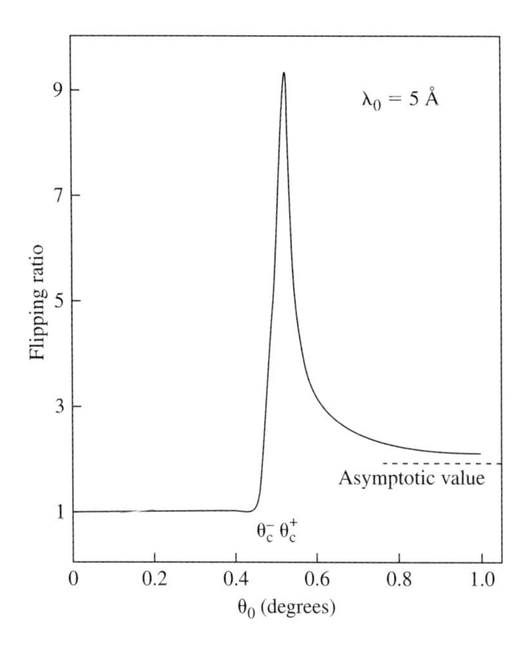

Fig. 12.4 Flipping ratio for the specular reflection of 5 Å neutrons by ferromagnetic nickel, plotted against the glancing angle θ_0. (*After* Felcher, 1981.)

12.3 Experimental methods

The measurement of the reflectivity as a function of Q can be achieved either by using a single monochromatic wavelength and varying the grazing angle of incidence, or by using a fixed angle of incidence and a white beam of neutrons which are sorted into specific wavelengths by time-of-flight. The latter approach is adopted with pulsed neutron sources and the data are measured on instruments such as SURF or CRISP at ISIS. The D17 reflectometer at the ILL is the corresponding instrument for use with monochromatic radiation and measurements are made by scanning the grazing angle whilst moving the detector through twice this angle ($\theta/2\theta$ scan).

The resolution $\Delta Q/Q$ of pulsed-source instruments, typically from 2% to 5%, is determined mainly by the spread $\Delta\theta$ in the angle of the incident beam and is constant for the entire reflectivity profile. The profile is measured simultaneously over a wide range of Q and over the same area of sample. The resolution of reactor instruments varies with angle through the term $\cot\theta\ \Delta\theta$ (see Section 8.2), and the area of illumination of the sample varies with θ and so a geometrical correction must be applied. At long wavelengths the reactor instrument tends to have a higher flux than a reflectometer on a pulsed source, and so is more suited for measurements at low Q.

Figure 12.5 is a schematic diagram of the Critical Surface Reflection Spectrometer, CRISP, which has operated at the ISIS pulsed neutron source for over 20 years (Penfold *et al.*, 1987). The instrument views the 20 K hydrogen moderator, which transmits a wide range of wavelengths from 0.5 Å to 13 Å to the instrument. The incident beam is inclined at an angle of 1.5° below the horizontal for examining liquid surfaces; solid

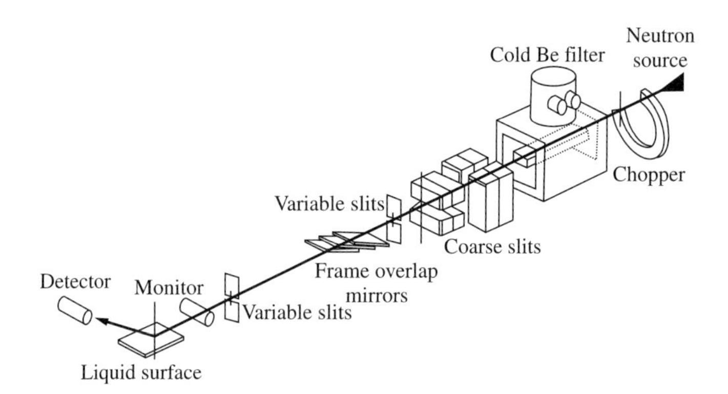

Fig. 12.5 The CRISP reflectometer. (*After* Penfold and Thomas, 1990).

surfaces can be studied over a range of angles from 0.2° upwards. Background noise is suppressed by attenuating the fast neutrons in the moderator at the start of each time frame. Frame overlap is suppressed with frame-overlap mirrors which reflect neutrons of wavelengths exceeding 13 Å out of the main beam. SURF is a 'second generation' reflectometer at ISIS which is optimized for the study of the surface chemistry of soft matter (Penfold *et al.*, 1997).

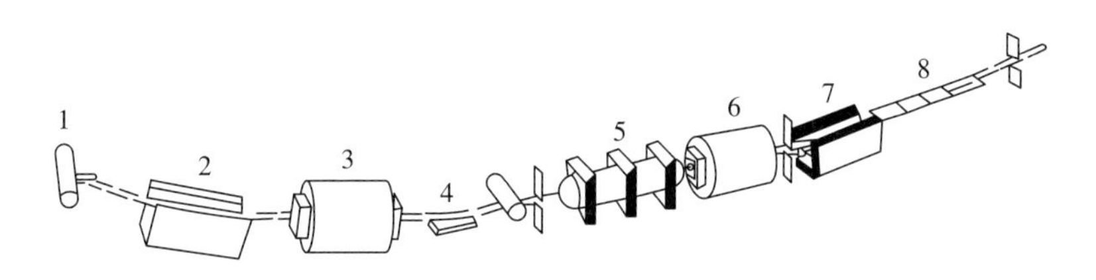

Fig. 12.6 The CRISP reflectometer operating in polarization analysis mode. 1: Detector; 2: polarization mirror; 3: spin flipper; 4: sample; 5: guide field; 6: spin flipper; 7: polarizing supermirror; 8: frame overlap mirrors.

CRISP can also be used in the polarized beam mode for studying magnetic materials, where the polarization is provided by a supermirror of cobalt–titanium (Fig. 12.6). The initially unpolarized beam is polarized vertically by a polarization mirror and this is followed by a π-spin flipper. The four combinations of the initial and final polarization states can be measured by inserting a second π flipper and polarizer in the reflected beam. A small vertical guide field ensures that the neutron spins are aligned along a vertical axis between the initial and final polarizing mirrors. Whilst the actions of the spin flippers and polarizing mirrors are not 100% perfect, this can be allowed for by repeating the four measurements with a non-magnetic sample such as graphite.

12.4 Examples of reflectivity studies

Neutrons are not destructive and can probe buried interfaces, such as liquid/liquid or solid/liquid, as well as solid/air and liquid/air interfaces. They are particularly useful for soft-matter studies since they are strongly scattered by light atoms such as H, C and O, which occur in most organic and biological materials. Moreover, the nuclei of different isotopes of the same element have different scattering cross-sections and, in the case of protons and deuterons, scatter with opposite phase. By using *contrast variation* different parts of the interface may be highlighted. For biophysical studies, the required quantity of sample is very small ($<10^{-6}$ g) and it is possible to work with expensive or rare macromolecules.

The technique has been used for studies in surface chemistry (surfactants, lipids, mixtures adsorbed at liquid/fluid and solid/fluid interfaces), surface magnetism (ultrathin Fe films, spin valve structures, magnetic multilayers, superconductors) and solid films (Langmuir–Blodgett films, thin solid films, polymer films). We can only refer to a few representative cases.

12.4.1 Surfactant monolayers

Surfactants are surface wetting agents, such as ordinary commercial detergents, which lower the surface tension of a liquid by being adsorbed at the liquid–air interface and allowing easier spreading. They are usually organic molecular compounds containing both hydrophilic 'tails' and hydrophobic 'heads', and so they are soluble in both organic solvents and water. The most common biological example of a surfactant is the coating of the small air sacs of the lungs serving as the site of gas exchange.

The large difference between the scattering lengths of hydrogen and deuterium leads to wide differences in the scattering length density of protonated and deuterated materials. By adding D_2O to H_2O in the appropriate amount, it is possible to make the scattering length density zero. There is then no reflection at the air–water interface and the water is deemed null-reflecting water (NRW). Lu *et al.* (1995) examined the structure of the hydrocarbon chain in a surfactant monolayer at the air/water interface of the system hexadecyl-trimethyl-ammonium bromide ($C_{16}TAB$). The molecule was isotopically labelled into ethyl or ethylene fragments (C_4D_8 or C_4D_9) and into butyl or butylene fragments (C_2D_4 or $C_2\,D_5$). The reflection profiles were then measured for a series of compounds in which null reflecting water was used to isolate the scattering from the individual fragments. The analysis gave the separation between the different fragments of the chain and between the fragments and the solvent; the part of the chain nearest to the alkyl head group was shown to be closer to the surface than to the outer parts. This method for probing the average structure along the surface can be used in situations where the layer is too complex to be investigated by other methods.

Critical assessments of the results obtained by neutron reflectometry on surfactant layers at the air–water interface, and their comparison with results available from other methods, are published in comprehensive articles by Lu *et al.* (2000) and Taylor *et al.* (2007).

12.4.2 Magnetic films

Polarized neutron reflectometry has been widely used for studying ferromagnetic materials in ultrathin films and multilayer structures. This interest has been driven by two developments. First, advanced deposition techniques, such as sputtering and molecular beam epitaxy, can be used to grow magnetic multilayers with atomic-plane precision and to produce materials with tailor-made physical properties. This development has been accompanied by the discovery in 1988 of Giant Magneto Resistance (GMR) between two thin layers of ferromagnetic metal separated by a layer of non-magnetic metal. In the GMR effect there is a large electronic resistance when the magnetic moments of the two ferromagnetic layers are aligned antiparallel, whereas the electronic resistance is low when the moments are parallel. A panoply of commercial applications have followed from these developments. They exploit the spin rather than the charge of the electron and include magnetic devices for high-density data storage and reading heads ('spin valves') for computer hard discs.

The unidirectional anisotropy known as 'exchange bias' is important in the operation of spin valves. The hysteresis loop of a ferromagnet describes the dependence of the magnetization M on the externally applied field H. The loop is centred at the origin, *i.e.* $M(H) = -M(-H)$, and so time reversal symmetry is satisfied. A system consisting of a ferromagnet and an antiferromagnet in atomic proximity can exhibit a different behaviour, with the loop offset from the applied field. Steadman *et al.* (2002) used spin-polarized reflectometry to examine exchange bias in spin-engineered double superlattices, in which sputtering produces a ferromagnetically coupled superlattice grown on to an antiferromagnetically coupled superlattice. The PNR experiment showed that the exchange bias can be explained in terms of a spiral spin structure in the antiferromagnetic layer of the double superlattice.

The surface magnetization of a ferromagnetic commercial alloy in the form of a metallic glass ribbon has been studied by Ivison *et al.* (1989). Figure 12.7 shows the reflectivity plotted on a log scale as a function of the neutron wavelength. The analysis of the data gave a bulk moment

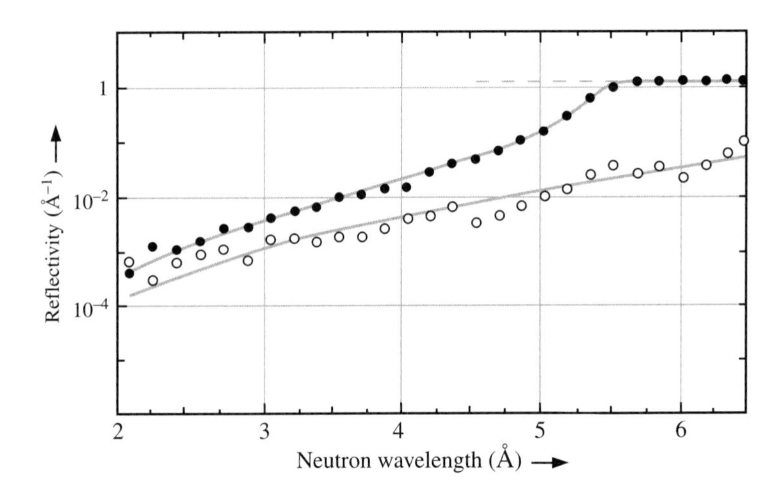

Fig. 12.7 Neutron reflectivity for a ferromagnetic metallic glass. Full circles: spin parallel to applied field. Open circles: spin antiparallel to applied field. (*After* Ivison *et al.*, 1989.)

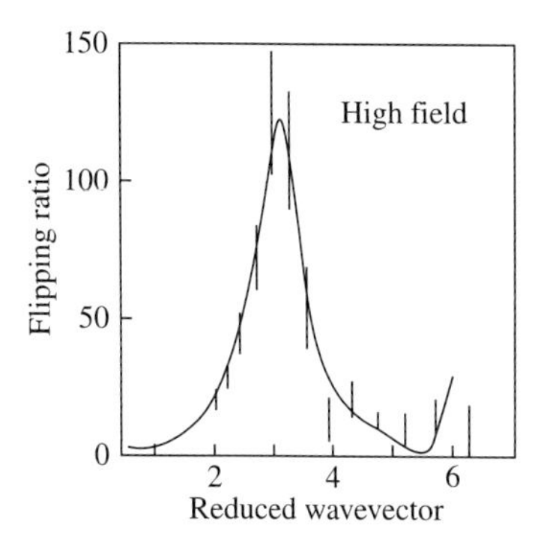

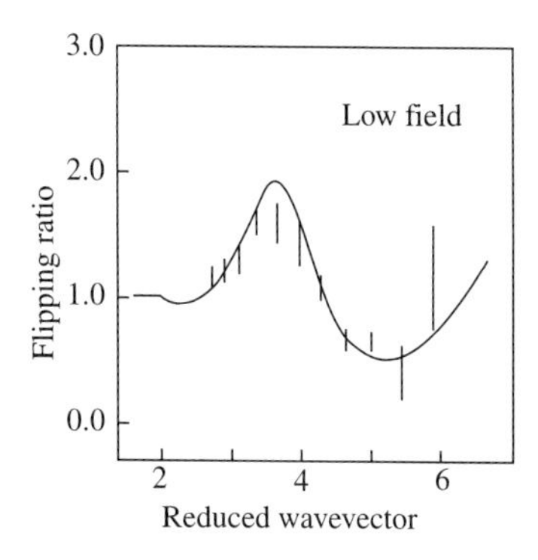

Fig. 12.8 The dependence of the flipping ratio on the reduced wave vector, as measured in an Fe/Cr/Fe multilayer by the PNR technique. In (a) the high field of 2,000 Gauss produces a parallel alignment of the moments in the Fe films. In (b) the low field of 50 Gauss leads to an antiparallel alignment. (*After* Dalgleish and Langridge, 2004.)

per iron atom of 1.7 Bohr magnetons which is 13% larger than that expected from bulk magnetization measurements.

A final example is the study of the GMR effect in an ultrathin Cr film sandwiched between two Fe layers. Epitaxial films of Fe/Cr/Fe couple ferromagetically at a high magnetic field, but at low fields they couple antiferromagnetically. Figure 12.8 shows measurements of the flipping ratio for these two cases, showing that there is a dramatic change in the magnitude of the flipping ratio in passing from a high to a low magnetic field. This kind of PNR experiment has been used to investigate the properties of many magnetic nanostructures; progress and prospects in this field are summarized by Dalgliesh and Langridge (2004).

12.4.3 Lipid bilayers

Membranes are ubiquitous in living material and carry out highly specialized functions. They surround both cells and organelles within cells and represent the surface through which interaction occurs with the outside world. Cell membranes consist mostly of lipids and proteins although sterols are other important constituents. Lipids are amphiphilic molecules consisting of a hydrophilic head and hydrophobic double chains. They form a continuous bilayer, which acts as a barrier to water-soluble molecules and provides the framework for the incorporation of membrane proteins. A knowledge of the structural properties of lipids leads to an understanding of their function at membrane surfaces.

Phospholipids pass through a variety of different phases as the temperature is changed. In going from high to low temperature, they undergo a fluid to gel phase transition at a temperature T_{m} (melting temperature). While in the fluid phase, L_{α}, lipid chains have a liquid-like conformation and are mobile in the lateral direction, in the gel phase, L_{β}, the chains are stiffer. For some lipids, just below T_{m}, there is the so-called *ripple phase*, P_{β}, where the lamellae are deformed by a periodic modulation.

Supported lipid bilayers can be deposited on solid substrates by the Langmuir–Blodgett or Langmuir–Schaeffer techniques. Fragneto *et al.* (2001, 2003) have prepared stable and reproducible double bilayers, in which the second bilayer floats a few Å above the first one adsorbed on a solid substrate. Reflectivity studies were performed on a highly hydrated and fluctuating bilayer, where the composition of each leaflet could be chosen separately. By fitting the data with a model including thermal fluctuations, it was possible to determine the value of the bilayer tension and rigidity modulus. The bending modulus was about 300 $k_B T$ in the gel phase and 10–30 $k_B T$ in the fluid phase.

Further examples of the applications of neutron reflectivity in biophysics are described by Fragneto-Cusani (2001). These include the adsorption of two proteins (β-casein and β-lactoglobulin) on hydrophobic silicon, and the interaction of a peptide with phospholipid bilayers deposited on silicon.

References

Ashworth, C. D., Messoloras, S., Stewart, R. J., Wilkes, J. G., Baldwin, I. S. and Penfold, J. (1989) *Phil. Mag. Lett.* **60** 57–65. "*Neutron reflectivity from a silicon nitride layer on a silicon substrate.*"

Blundell, S. J. and Bland, J. A. C. (1992) *Phys. Rev. B* **46** 3391–3400. "*Polarised neutron reflection as a probe of magnetic films and multilayers.*"

Born, M. and Wolf, E. (1999) "*Principles of Optics.*" Cambridge University Press: Cambridge.

Dalgliesh, R. M. and Langridge, S. (2004) *Notiziari Neutroni e Luce di Sincrotone* **9** 3–22. "*Magnetic nanostructures studied by polarised neutron reflectometry.*"

Felcher, G. P. (1981) *Phys. Rev. B* **24** 1595–1598. "*Neutron reflection as a probe of surface magnetism.*"

Felcher, G. P. (1993) *Physica B* **192** 137–149. "*Magnetic depth profiling studies by polarised neutron reflection.*"

Felcher, G. P. (1999) *Physica B* **267–268** 154–161. "*Polarised neutron reflectometry—a historical perspective.*"

Fitzsimmons, M. R., Bader, S. D., Borchers, J. A., Felcher, G. P., Furdyna, J. K., Hoffmann, A., Kortright, J. B., Schuller, I. K., Schulthess, T. C., Sinha, S. K., Toney, M. F., Weller, D. and Wolf, S. (2004) *J. Magnetism Magnetic Materials* **271** 103–146. "*Neutron scattering studies of nanomagnetism and artificially structured materials.*"

Fragneto, G., Charitat, T., Graner, F., Mecke, K., Perino-Gallice, L. and Bellet-Amalric, E. (2001) *Europhys. Lett.* **53** 100–106. " *A fluid floating bilayer.*"

Fragneto, G., Charitat, T., Bellet-Amalric, E., Cubitt, R. and Graner, F. (2003) *Langmuir* **19** 7695–7702. "*Swelling of phospholipid floating bilayers.*"

Fragneto-Cusani, G. (2001) *J. Phys. Condens. Matter* **13** 4973–4989. "*Neutron reflectivity at the solid/liquid interface: examples in biophysics.*"

Hayter, J. B., Highfield, R. R., Pullman, B. J., Thomas, R. K., McMullen, A. I. and Penfold, J. (1981) *J. Chem. Soc. Faraday Trans.* **77** 1437–1448. "*Critical reflection of neutrons. A new technique for investigating interfacial phenomena.*"

Ivison, P. K., Cowlam, N., Gibbs, M. R. J., Penfold, J. and Shackleton, C. (1989) *J. Phys. Condens. Matter* **1** 3655–3662. "*A direct measurement of the surface magnetisation of a ferromagnetic metallic glass.*"

Lu, J. R. and Thomas, R. K. (1998) *J. Chem. Soc. Faraday Trans.* **94** 995–1018. "*Neutron reflection from wet surfaces.*"

Lu, J. R., Li, Z. X., Smallwood, J. and Thomas, R. K. (1995) *J. Phys. Chem.* **99** 8233–8243. "*Detailed structure of the hydrocarbon chain in a surfactant monolayer at the air/water interface: neutron reflection from hexadecycltrimethyl-ammonium bromide.*"

Lu, J. R., Lee, E. M. and Thomas, R. K. (1996) *Acta Crystallogr. A* **52** 11–41. "*The analysis and interpretation of neutron and X-ray specular reflection.*"

Lu, J. R., Thomas, R. K. and Penfold, J. (2000) *Adv. Colloid Interf. Sci.* **84** 143–304. "*Surfactant layers at the air-water interface: structure and composition.*"

Münster, C., Salditt, T., Vogel, M., Siebrecht, R. and Peisl, J. (1999) *Europhys. Lett.* **46** 486–492. "*Non-specular neutron scattering from aligned phospholipid membranes.*"

Penfold, J. (2002) *Curr. Opin. Colloid Interf. Sci.* **7** 139–147. "*Neutron reflectivity and soft condensed matter.*"

Penfold, J. and Thomas, R. K. (1990) *J. Phys. Condens. Matter* **2** 1369–1412. "*The application of the specular reflection of neutrons to the study of surfaces and interfaces.*"

Penfold, J., Ward, R. C. and Williams, W. G. (1987) *J. Phys. E* **20** 1411–1417. "*A time-of-flight neutron reflectometer for surface and interfacial studies.*"

Penfold, J., Richardson, R. M., Zarbakhsh, A., Webster, J. R. P., Bucknall, D. G., Rennie, A. R., Jones, R. A. L., Cosgrove, T., Thomas, R. K., Higgins, J. S., Fletcher, P. D. I., Dickinson, E., Roser, S. J., McLure, I. A., Hillman, A. R., Richards, R. W., Staples, E. J., Burgess, A. N., Simister, E. A. and White, J. W. (1997) *J. Chem. Soc. Faraday Trans.* **93** 3899–3917. "*Recent advances in the study of chemical surfaces and interfaces by specular neutron reflection.*"

Pynn, R. (1992) *Phys. Rev. B* **45** 602. "*Neutron scattering by rough surfaces at grazing incidence.*"

Pynn, R., Baker, S. M., Smith, G. and Fitzsimmons, M. (1999) *J. Neutron Res.* **7** 139–158. "*Off-specular scattering in neutron reflectometry.*"

Sinha, S. K., Sirota, E. B. and Garoff, S. (1988) *Phys. Rev. B* **38** 2297–2311. "*X-ray and neutron scattering from rough surfaces.*"

Steadman, P., Ali, M., Hindmarch, A. T., Marrows, C. H., Hickey, B. J., Langridge, S., Dalgleish, R. M. and Foster, S. (2002) *Phys. Rev. Lett.* **89** 077201. "*Exchange bias in spin-engineered double superlattices.*"

Taylor, D. J. F., Thomas, R. K. and Penfold, J. (2007) *Adv. Colloid Interf. Sci.* **132** 69–110. "*Polymer/surfactant interactions at the air-water interface.*"

Zabel, H. and Theis-Bröhl, K. (2003) *J. Phys. C* **15** S505–S517. "*Polarised neutron reflectivity and scattering studies of magnetic heterostructures.*"

Zhou, X. and Chen, S. (1995) *Phys. Rep.* **257** 223–348. "*Theoretical foundation of X-ray and neutron reflectometry.*"

Liquids, glasses and amorphous materials

13

In this chapter, we shall discuss the application of neutron scattering to the study of liquids, glasses and amorphous materials. Our principal concern will be the understanding of the 'structure', or the spatial distribution, of the atoms in these disordered systems. The dynamics of the systems can be investigated by inelastic neutron scattering, but we shall only discuss inelastic scattering processes in so far as they affect the interpretation of data relating to structure.

The atomic structure in a liquid is intermediate between the highly perfect, long-range order in a crystalline solid and the complete disorder in a gas. In a typical model for a crystal the atoms occupy well-defined sites on a crystal lattice and interact by harmonic forces. In the model for a perfect gas the atoms undergo random, uncorrelated motions. There is no correspondingly simple model for a liquid: a liquid possesses a local structural motif but this does not repeat itself indefinitely, and at large distances of separation the atomic positions are completely uncorrelated with one another. We discuss first the theory of scattering from such a partially ordered system.

13.1 Diffraction theory

The amplitude of scattering from a collection of N atoms at positions $\mathbf{r}_j$ is

$$A(\mathbf{Q}) = \sum_{j=1}^{N} b_j \exp(i\mathbf{Q}.\mathbf{r}_j) \tag{13.1}$$

where b_j is the scattering factor of atom j and $i = \sqrt{-1}$. The scattered intensity, $I(\mathbf{Q})$, is given by $|A(\mathbf{Q})|^2$ or $A(\mathbf{Q})A*(\mathbf{Q})$:

$$I(\mathbf{Q}) = \sum_{j,k}^{N} b_j b_k \exp[i\mathbf{Q}.(\mathbf{r}_j - \mathbf{r}_k)] \tag{13.2}$$

This equation shows that the diffraction experiment gives information about the separation of pairs of atoms j, k but no direct information about their absolute positions. We no longer have the luxury offered by single-crystal diffraction (Chapter 7), in which the regular arrangement

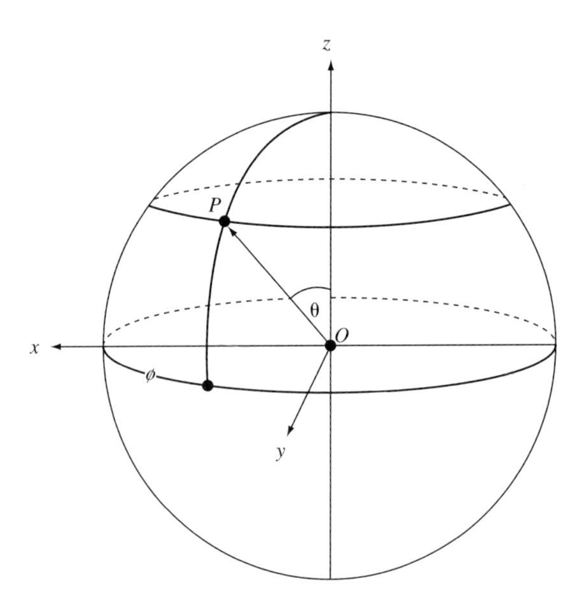

Fig. 13.1 Polar coordinates θ, ϕ of vector OP.

of atoms on a periodic lattice, extending throughout the entire crystal, allows us to derive the individual atomic positions $\mathbf{r}_j$, $\mathbf{r}_k$.

For most powders, glasses and liquids, $I(\mathbf{Q})$ depends on the magnitude of $\mathbf{Q}$ and not on its direction. This is equivalent to saying that the points in reciprocal space are distributed uniformly in shells of radius $Q\,(=|\mathbf{Q}|)$. Thus, the term $\exp[\mathrm{i}\mathbf{Q}\cdot(\mathbf{r}_j - \mathbf{r}_k)]$ in eqn. (13.2) must be averaged over all relative orientations of the vectors $\mathbf{Q}$ and $\mathbf{r}_j - \mathbf{r}_k$. Introducing polar coordinates θ and ϕ (see Fig. 13.1), the average value of $\exp[\mathrm{i}\mathbf{Q}\cdot(\mathbf{r}_j - \mathbf{r}_k)]$ is

$$\begin{aligned}\langle\exp[\mathrm{i}\mathbf{Q}\cdot(\mathbf{r}_j - \mathbf{r}_k)]\rangle &= \frac{\int \exp[\mathrm{i}\mathbf{Q}\cdot(\mathbf{r}_j - \mathbf{r}_k)]\,\mathrm{d}v}{\int \mathrm{d}v} \\ &= \int_0^{2\pi}\mathrm{d}\phi\int_0^{\pi}\exp\left(\mathrm{i}Q\cdot r_{jk}\cdot\cos\theta\right)\sin\theta\,\mathrm{d}\theta/4\pi \\ &= \frac{\sin(Qr_{jk})}{Qr_{jk}} \end{aligned} \tag{13.3}$$

where r_{jk} is the magnitude of $\mathbf{r}_j - \mathbf{r}_k$ and $< \cdots >$ indicates average value.

The next step is to replace the sum over atom pairs, *jk*, by the so-called *radial distribution function* $g(r)$. For a monatomic system this function represents the local density of atoms at a distance r from an atom at the origin ($r = 0$) divided by the average density of atoms throughout the entire system. The atomic positions fluctuate at any instant of time, and so we assume that the averages are taken over a time interval exceeding 1 ns. If r is much greater than the range of interatomic forces, $g(r)$ is unity, and for r less than the size of the atom $g(r)$ is zero. This is in contrast with the case of a gas, where the combined volume of the atoms

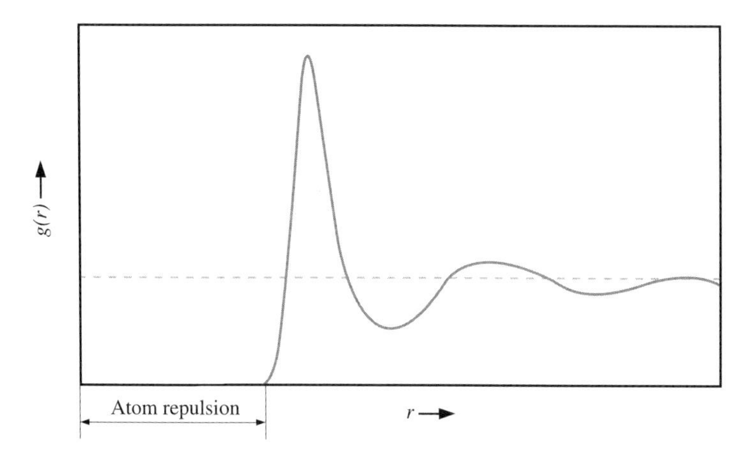

Fig. 13.2 The static radial distribution function $g(r)$ for a liquid. The first peak corresponds to nearest-neighbour atoms, and the second peak to next-nearest-neighbour atoms.

is negligible compared with the available space which they occupy, and $g(r)$ is unity throughout all space.

The form of the static pair distribution function $g(r)$ for a simple monatomic liquid is shown in Fig. 13.2. Beyond the closest distance of approach of the atoms $g(r)$ increases rapidly to the first maximum, which corresponds to the most probable separation of neighbouring atoms. The first minimum gives an indication of the spatial extent of the short-range interactions. Weaker oscillations occur at larger values of r, and at very large separations $g(r)$ settles down to the constant value of unity. Clearly, the precise form of the distribution function $g(r)$ depends on the forces between atoms, and theories of the liquid state relate $g(r)$ to the interatomic pair potential $u(r)$. In the simplest theories, three-body interactions are neglected, and the potential $u(r)$ is treated as a basic quantity from which nearly all other properties of the liquid can be calculated.

$I(Q)$ in eqn. (13.2) contains a sum over the same atom ($j = k$) as well as over different atoms ($j \neq k$). For a monatomic system the summation over identical atoms is $N\langle b^2\rangle$ where $\langle b^2\rangle$ indicates the square of the scattering length of the atom, averaged over its spin and isotope states. Thus we have

$$I(Q) = N\langle b^2\rangle + \sum_{j\neq k}^{N} \langle b_j\rangle \langle b_k\rangle \exp\left[\mathrm{i}\mathbf{Q}\cdot(\mathbf{r}_j - \mathbf{r}_k)\right] \qquad (13.4)$$

The averaging terms in eqn. (13.4) can be varied by isotopic substitution.

The volume of the spherical shell of radius r and thickness $\mathrm{d}r$ is $4\pi\mathrm{r}^2\mathrm{d}r$ and the number of atoms in this shell is $4\pi\mathrm{r}^2\mathrm{d}r \cdot \rho\mathrm{g}(r)$ where ρ is the average density. Hence, using eqn. (13.3) the above expression becomes

$$I(Q) = N\langle b^2\rangle + 4\pi\rho\,\langle b\rangle^2 \int_0^\infty r^2 g\,(r) \frac{\sin(Qr)}{Qr}\,\mathrm{d}r \qquad (13.5)$$

Now the integral $\int_0^\infty r^2(\sin(Qr)/Qr\mathrm{d}r)$ is zero unless $Q=0$. The intensity at $2\theta = 0$ is experimentally inaccessible: subtracting it from eqn. (13.5) we obtain the final expression for the observed intensity:

$$I(Q) = N\langle b^2\rangle \left[1 + 4\pi\rho/Q \int_0^\infty r\left[g(r) - 1\right]\sin(Qr)\,\mathrm{d}r\right] \tag{13.6}$$

This equation is customarily written in the form

$$I(Q) = N\langle b^2\rangle\, S(Q) \tag{13.7}$$

where $S(Q)$ is the so-called *liquid structure factor* and is given by

$$S(Q) = 1 + \frac{4\pi\rho}{Q}\int_0^\infty r\left[g(r) - 1\right]\sin(Qr)\,\mathrm{d}r \tag{13.8}$$

The term 'structure factor' is very different from the structure factor of a crystalline lattice (see Chapter 7): it has only a few peaks and it expresses intensity rather than amplitude. The Fourier inversion of (13.8) to $g(r)$ is given by

$$g(r) = 1 + \frac{1}{2\pi^2\rho}\int_0^\infty Q^2\left[S(Q) - 1\right]\frac{\sin(Qr)}{Qr}\,\mathrm{d}Q \tag{13.9}$$

For a two-component system AB there is a separate structure-factor term, $S_{\alpha\beta}(Q)$, for each distinct type of atom pair α, β. α and β can take the values A or B, and so there are three independent partial structure factors S_{AA}, S_{AB} and S_{BB}. [For a liquid with n types of atom there are $n(n+1)/2$ site–site radial distribution functions: for a three-component system there are six functions and for $n = 4$ there are 10 functions to be determined.] $S_{\alpha\beta}(Q)$ is related to the pair distribution functions $g_{\alpha\beta}(r)$, representing the average distribution of type β atoms as observed by a type α atom at the origin, and is given by the equation:

$$S_{\alpha\beta}(Q) = 1 + 4\pi\rho/Q\int_0^\infty r\left[g_{\alpha\beta}(r) - 1\right]\sin(Qr)\,\mathrm{d}r \tag{13.10}$$

The structure factor $S(Q)$ in (13.8) is then replaced by the 'total structure factor' $D(Q)$:

$$D(Q) = \sum_{\alpha} c_\alpha\left\langle b_\alpha^2\right\rangle + \sum_{\alpha,\beta} c_\alpha\left\langle b_\alpha\right\rangle c_\beta\left\langle b_\beta\right\rangle\left(S_{\alpha\beta}(Q) - 1\right) \tag{13.11}$$

where c_α is the atomic fraction of α, b_α its scattering length and the brackets $\langle\rangle$ indicate average value over spins and isotopes. The first term in (13.11) is the 'single atom' or 'self' term, while the second is the 'distinct' or 'interference' term which contains the information about structure. The self term is associated with the incoherent component and the distinct term with the coherent component of scattering.

The measured differential cross-section $\mathrm{d}\sigma/\mathrm{d}\Omega$ is proportional to $D(Q)$ and can be split in the same way into self and distinct terms:

$$\frac{\mathrm{d}\sigma}{\mathrm{d}\Omega} = \left(\frac{\mathrm{d}\sigma}{\mathrm{d}\Omega}\right)_s + \left(\frac{\mathrm{d}\sigma}{\mathrm{d}\Omega}\right)_d \tag{13.12}$$

where

$$\left(\frac{\mathrm{d}\sigma}{\mathrm{d}\Omega}\right)_s = \sum_{\alpha} c_\alpha b_\alpha^2 \tag{13.13}$$

and

$$\left(\frac{\mathrm{d}\sigma}{\mathrm{d}\Omega}\right)_d = \sum_{\alpha,\beta} c_\alpha b_\alpha c_\beta b_\beta \left(S_{\alpha\beta}\left(Q\right) - 1\right) \tag{13.14}$$

In a neutron diffraction experiment on a liquid there is strictly no elastic scattering because the atomic nucleus of the sample recoils under neutron impact, resulting in an exchange of energy between the neutron and the scattering system. Thus, the structure factor is not *directly* accessible by diffraction measurements. The measurements are made at fixed scattering angle (reactor source) or at fixed time-of-flight (pulsed source) but not at fixed Q. The detector integrates the counts recorded in $(\mathbf{Q},\, \omega)$ space, and only if we could access all of $(\mathbf{Q},\, \omega)$ space could we measure the structure factor. The necessary corrections for inelastic effects are considered in Section 13.2, but first we shall refer to the van Hove dynamic structure factor $S_{(Q,\,\omega)}$ and the corresponding correlation function $G(\mathbf{r}, t)$.

13.1.1 Van Hove correlation functions

The time-dependent correlation function $G(\mathbf{r}, t)$ was introduced by van Hove (1954). For a classical system, $G(\mathbf{r}, t)$ is the probability that, given there is an atom at the origin ($\mathbf{r} = 0$) at the initial time ($t = 0$), the same or any other atom is found at a position $\mathbf{r}$ at a later time t. It is convenient to separate $G(\mathbf{r}, t)$ into two parts:

$$G(\mathbf{r}, t) = G_s\left(\mathbf{r}, t\right) + G_d\left(\mathbf{r}, t\right) \tag{13.15}$$

$G_s(\mathbf{r}, t)$ refers to the probability that the same atom is observed at $\mathbf{r}$ at a time t later and is the *self* pair correlation function; $G_d(\mathbf{r}, t)$ is the probability that a different atom is at $(\mathbf{r},\, t)$ and is the *distinct* pair correlation function. The self correlation function describes the motion of just one atom, whereas the distinct correlation function describes the motion of many atoms. We would expect, therefore, that $G_s(\mathbf{r}, t)$ is a simpler type of function than $G_d(\mathbf{r}, t)$.

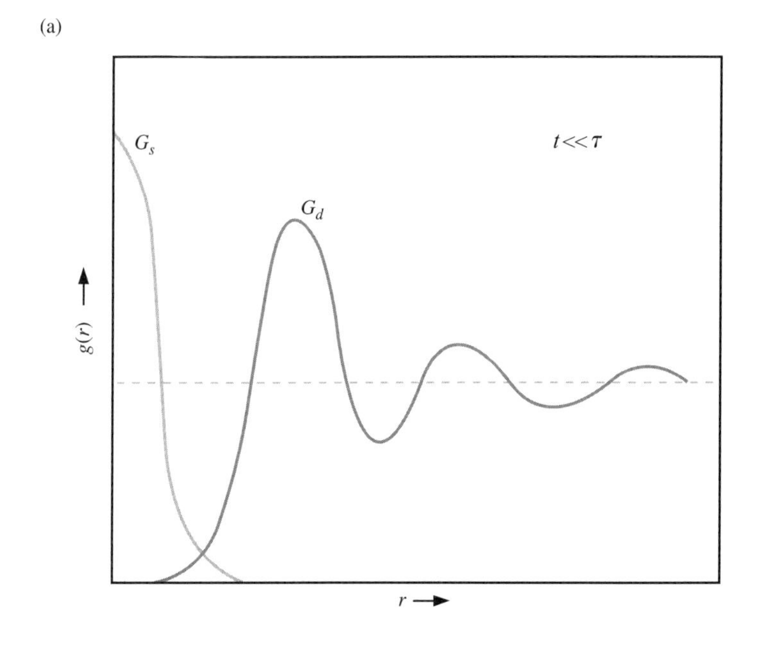

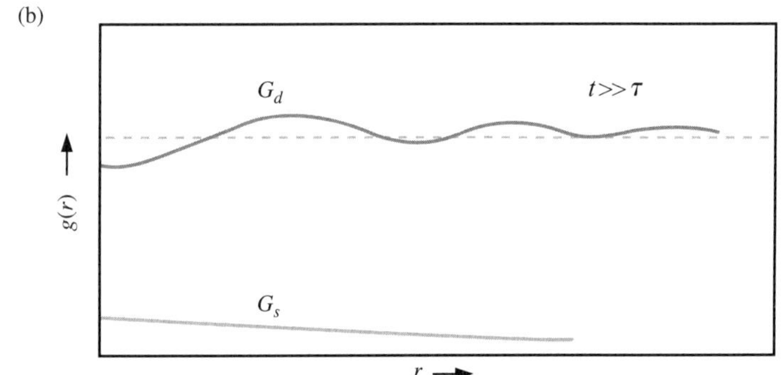

Fig. 13.3 The components G_s, G_d of van Hove's space–time correlation function. Curves (a) relate to very short times ($<10^{-13}$ sec) and curves (b) to much longer times.

As time develops the correlation functions change, and their behaviour can be shown diagrammatically (Fig. 13.3) for the two extreme cases in which the time is very short or very long compared with the relaxation time τ (of the order of 10^{-13} sec) required to damp out the disturbance arising from the displacement of an individual atom. For $t \ll \tau$ [Fig. 13.3(a)] the atom at the origin has not had sufficient time to move far away, and G_s falls to zero within a few interatomic distances. G_d determines the positions of all the other atoms; it is zero at $r = 0$ and is identical to $g(r)$ at $t = 0$. For $t \gg \tau$ [Fig. 13.3(b)] the atom originally at the origin has diffused away, although it is still slightly more likely to be near the origin than further away: see curve G_s. The curve for G_d shows that the oscillations have been largely damped out, and they settle down to the value of the instantaneous distribution function $g(r)$ at $r = \infty$.

Van Hove showed that the coherent double differential scattering cross-section for a monatomic system of N nuclei is given by

$$\left(\frac{\mathrm{d}^2\sigma}{\mathrm{d}\Omega\mathrm{d}\omega}\right) = \left(\frac{\mathrm{d}^2\sigma}{\mathrm{d}\Omega\mathrm{d}\omega}\right)_s + \left(\frac{\mathrm{d}^2\sigma}{\mathrm{d}\Omega\mathrm{d}\omega}\right)_d \tag{13.16}$$

where

$$\left(\frac{\mathrm{d}^2\sigma}{\mathrm{d}\Omega\mathrm{d}\omega}\right)_s = \frac{k_\mathrm{f}}{k_\mathrm{i}}\left\langle b^2\right\rangle N S_s(\mathbf{Q},\omega) \tag{13.17}$$

and

$$\left(\frac{\mathrm{d}^2\sigma}{\mathrm{d}\Omega\mathrm{d}\omega}\right)_d = \frac{k_\mathrm{f}}{k_\mathrm{i}}\left\langle b\right\rangle^2 N S_d(\mathbf{Q},\omega) \tag{13.18}$$

The functions $S_s(\mathbf{Q},\omega)$ $S_d(\mathbf{Q},\omega)$ are related to the correlation functions G_s, G_d by

$$S_s(\mathbf{Q},\omega) = \frac{1}{2\pi}\iint G_s(\mathbf{r},t)\exp[i\,(\mathbf{Q}\cdot\mathbf{r}-\omega t)]\;\mathrm{d}\mathbf{r}\,\mathrm{d}t \tag{13.19}$$

and

$$S_d(\mathbf{Q},\omega) = \frac{1}{2\pi}\iint G_d(\mathbf{r},t)\exp[\mathrm{i}\,(\mathbf{Q}\cdot\mathbf{r}-\omega t)]\;\mathrm{d}\mathbf{r}\,\mathrm{d}t \tag{13.20}$$

Our discussion has been given in terms of the self and distinct contributions to the double differential scattering cross-section, rather than to the more usual coherent and incoherent components used in Section 2.4.

The static structure factors $S_\alpha(Q)$ and $S_{\alpha\beta}(Q)$ are related to the dynamic structure factors $S_\alpha(Q,\omega)$ and $S_{\alpha\beta}(Q,\omega)$ by the expressions

$$S_\alpha(Q) = \int_{-\infty}^{\infty} S_\alpha(Q,\omega)\mathrm{d}\omega \tag{13.21}$$

and

$$S_{\alpha\beta}(Q) = \int_{-\infty}^{\infty} S_{\alpha\beta}(Q,\omega)\mathrm{d}\omega \tag{13.22}$$

where the integrals are evaluated at constant Q. The aim of the diffraction experiment is to determine the static structure factor $S(Q)$. It is not practical to carry out the appropriate intensity measurements at all energy transfers but Placzek (1952) has shown that, for nuclei much more massive than the neutron, the inelasticity correction is independent of the detailed dynamics and is related only to the nuclear mass, to the incident neutron energy and to the temperature of the sample.

13.2 Analysis of diffraction data

The basic steps in data analysis consist of (i) measuring the intensity data over as wide a range of Q as possible, (ii) normalizing the data to the incident monitor, (iii) correcting for background, (iv) putting on an absolute scale by comparison with vanadium scattering, (v) correcting for multiple scattering, absorption and container scattering and (vi) subtracting the term for single-atom scattering.

It is possible to approach accuracies of 1% in measuring the intensities, but this requires extreme care in applying corrections to the raw data. Most of these corrections are well understood and are routinely applied in a normal investigation. The absence of long-range order in liquids and amorphous materials implies that the scattered neutron intensity is distributed over all points in reciprocal space; consequently, the intensity at any specific point Q tends to be rather weak and the effects of absorption (neutron capture), self-shielding and multiple scattering acquire greater significance than in measuring Bragg intensities. Absorption corrections are available analytically for flat-plate geometry, but all other cases require numerical solution.

Cylindrical sample containers are essential for work at high pressure or high temperature. On the other hand, there are strong arguments for using flat-plate sample containers, as the multiple scattering can be calculated easily and the experimental data are insensitive to sample positioning errors. To ensure that multiple scattering will not introduce too large a systematic error, the sample size should be so restricted that no more than 20% of the incident beam is scattered. An important practical advantage in using neutrons rather than X-rays is that the absorption coefficient for neutrons of many materials is small. This means that samples can be relatively bulky, with the neutron beam being scattered from a depth of 10 mm or more, and the measurements are representative of the entire irradiated volume. For the same reason, the container for a liquid can be made mechanically strong, allowing measurements to be made on materials at extreme pressures and temperatures.

The most difficult correction to the data is the calculation of the effect of inelasticity. If the intensity measurements are made by scattering neutrons of incident energy E_i through a given angle 2θ, all neutrons scattered at that angle are recorded in the detector irrespective of any change of energy on scattering. This change in energy $\hbar\omega$ is accompanied by a change in the momentum transfer $\hbar Q$ between the neutron and the sample, which is given by the expression

$$\frac{\hbar^2 Q^2}{2m_\mathrm{n}} = 2E_\mathrm{i} + \hbar\omega - 2E_\mathrm{i}\left(1 + \hbar\omega/E_\mathrm{i}\right)^{1/2}\cos 2\theta \qquad (13.23)$$

Thus in the conventional diffraction experiment an exceedingly complex cross-section is measured, and it is only when the incident energy E_i vastly exceeds the energy transfer that the structure factor $S(Q)$ is directly determined. This condition, $E_\mathrm{i} \gg \hbar\omega$, is known as the *static approximation*, because dynamical information is lost as in the corresponding X-ray experiment. Placzek (1952) has shown that the measured cross-section at constant scattering angle is proportional to

$$\int_{-E_{\mathrm{i}}/\hbar}^{\infty} \frac{k_{\mathrm{f}}}{k_{\mathrm{i}}} S\left(Q, \omega\right) \varepsilon\left(\omega\right) \mathrm{d}\omega \tag{13.24}$$

where $\varepsilon(\omega)$ is the energy dependence of the detector. Starting from this formula, and expanding the integrand in energy moments of $S(Q, \omega)$, Placzek showed that for neutrons of high incident energy and for atoms much heavier than neutrons, the first few terms of the expansion are sufficient to give a good estimate of the cross-section. If the energy of the incident neutrons is much greater than the energy change on scattering, the magnitude of the Placzek correction is relatively small: this favours the choice of epithermal neutrons with incident energies in excess of several electron volts, as can be obtained with a spallation neutron source. The correction is notably larger at low neutron energies, high temperatures and small nuclear masses, and it is particularly significant for liquids containing light atoms such as hydrogen or deuterium, and for molecules with high energies of vibrational motion. There is an extensive literature for the cases where the scattering system is only partially excited by the incident neutrons; lengthy algebra is involved, and the correct formalism to use is still in dispute. The early literature is summarized in a review by Egelstaff (1987) who discusses different experimental conditions and gives detailed formulae for the scattering from atoms and molecules.

Some calculations of the Placzek correction for vanadium at 293 K and at a fixed incident wavelength are shown in Fig. 13.4. The correction applies only to the self-scattering term which is then subtracted from the

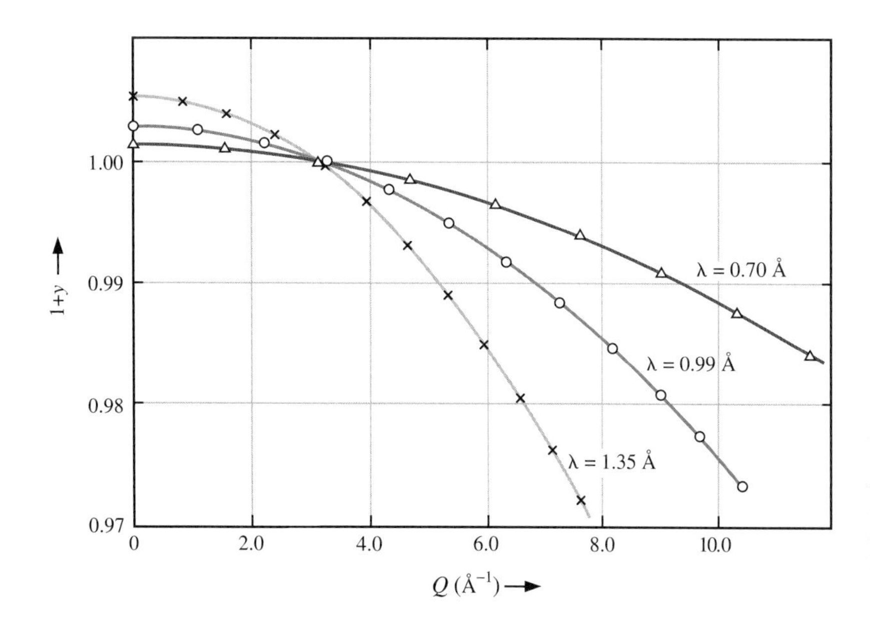

Fig. 13.4 Placzek corrections for vanadium at 293 K using a $1/v$ detector. The intensity must be divided by $1 + y$ where y is the magnitude of the correction. λ is the wavelength of the incident neutrons. The correction is less at higher incident energies. (*After* Page, 1973.)

total scattering leaving the distinct or interference term. The departure from the static approximation gives a characteristic droop in Fig. 13.4 in the intensities measured at high Q. Vanadium is an atom of medium mass and the correction is only a few per cent. For lighter elements, the Placzek correction fails and there are various empirical ways of removing the self term.

The theory of the Placzek correction for time-of-flight data has been considered by Sinclair and Wright (1974) and by Grimley *et al.* (1990). These authors recommend a short flight path from the sample to the detector, together with a liquid-nitrogen cooled moderator. By using a cold moderator, such that the Maxwellian contribution to the incident spectrum is removed beyond the wavelength range of interest, the scattering occurs predominantly from neutrons in the 'slowing down' epithermal portion of the spectrum whose functional form of $1/E$ is well characterized. They also recommend that the measurements should be made at as low a scattering angle 2θ as possible.

The final step in the data analysis is the Fourier inversion of the measured structure factors. Computer routines are available to carry out this Fourier transform but inevitably spurious structure will appear in the calculated $g(r)$ owing to the finite range of Q and the statistical noise in the data. A number of approaches have been tried in order to overcome this problem, including the Reverse Monte Carlo method (McGreevy, 2001). If the fluctuations in $r[g(r) - 1]$ with increasing r are made compatible with the observed width of the peaks in $S(Q)$, the effects of noise and data truncation are considerably reduced.

13.3 Diffractometers for liquid and amorphous samples

The D4 instrument at the Institut Laue-Langevin is a two-axis diffractometer dedicated to the study of liquid and amorphous samples. It is equipped with nine two-dimensional microstrip detectors, each of which is located in a ^{3}He chamber (Fig. 13.5). It uses short-wavelength neutrons from the hot source, ensuring that the inelasticity corrections are reduced to a minimum. The monochromator take-off angle $2\theta_M$ is quite small (20–25°) whereas the scattering angle 2θ from the specimen is 1.5–140°: this arrangement gives maximum flux at the sample and maximum accessible range of Q, without undue loss of resolution. (The D1A powder diffractometer described in Chapter 8 operates at a much higher $2\theta_M$ and higher resolution $\Delta Q/Q$.) The high-transmission monitor between monochromator and sample is used for normalizing the intensity measurements.

The corresponding instrument for time-of-flight measurements is typified by the SANDALS diffractometer at the ISIS Laboratory (Fig. 13.6). There are 660 zinc sulphide detectors (assembled in 33 modules each containing 20 detectors) covering an angular range from 4° to 39° in scattering angle 2θ. The combination of an intense pulsed neutron source and

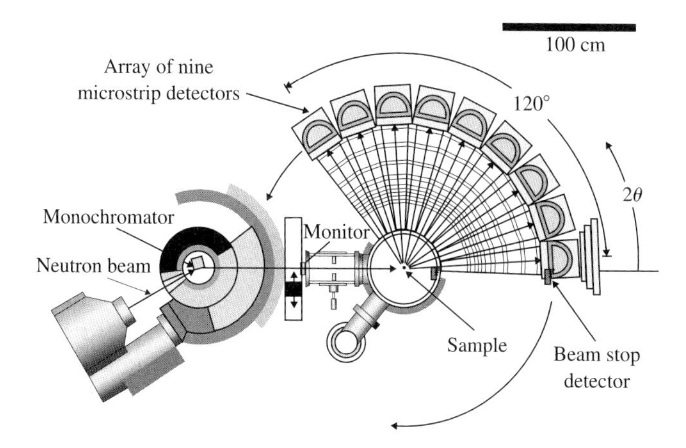

Fig. 13.5 Layout of the D4 diffractometer at the ILL.

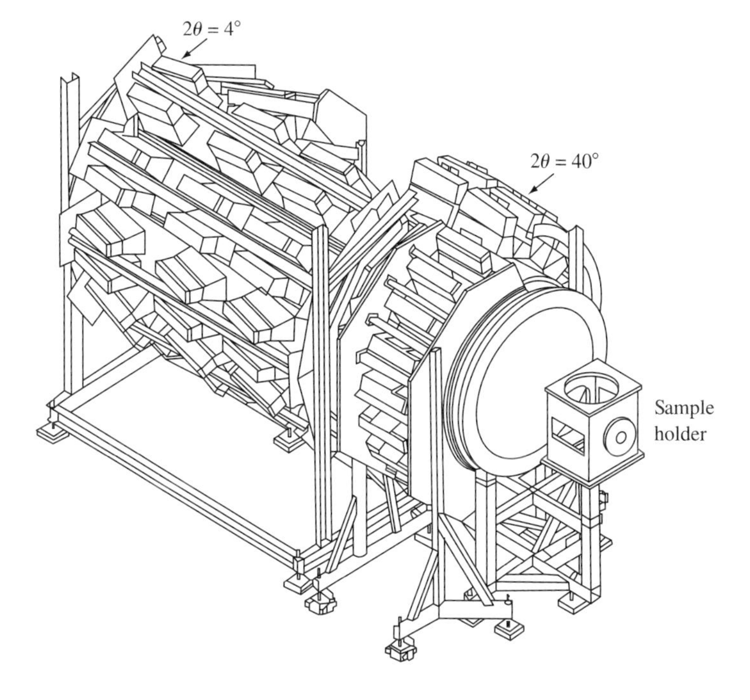

Fig. 13.6 The SANDALS diffractometer at the ISIS Laboratory. The scattered flight paths are 4 m at $2\theta = 4°$ and 0.8 m at $2\theta = 40°$.

a large number of detectors at low angles makes SANDALS particularly suitable for measuring static structure factors on samples containing light atoms such as hydrogen or deuterium. $S(Q)$ can be measured over a very wide range (0.10–$50\,\text{Å}^{-1}$) of momentum transfer Q.

13.4 Atomic structure of liquids

13.4.1 Simple liquids

The rare gases and liquid metals are the so-called *simple liquids*, for which there are many reported measurements of the structure factors

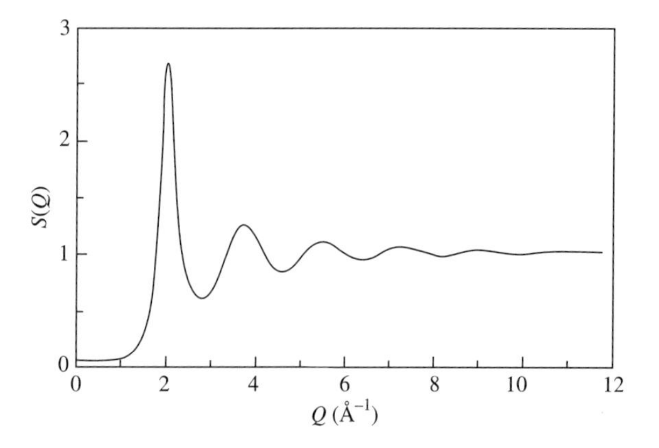

Fig. 13.7 Fully corrected experimental values of the structure factor $S(Q)$ for liquid argon at 85 K. The smooth curve through the experimental points was obtained from a molecular dynamics calculation. (*After* Yarnell *et al.*, 1973.)

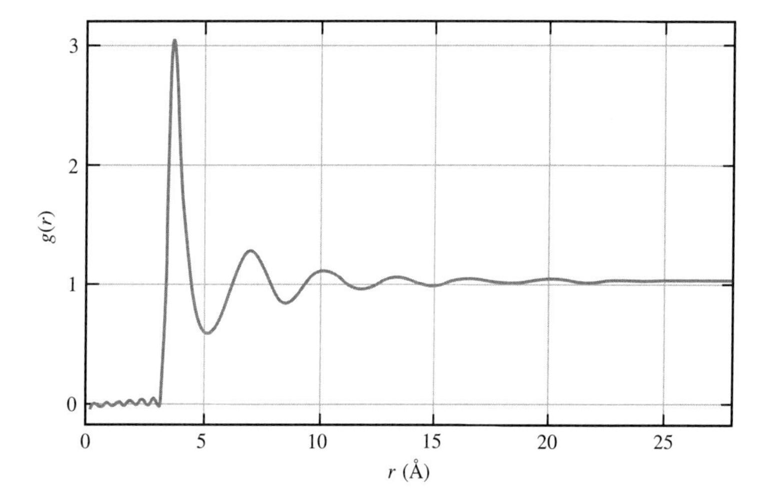

Fig. 13.8 Radial distribution function $g(r)$ for argon at 85 K. This curve is the Fourier transform of $S(Q)$ in Fig. 13.7. (*After* Yarnell *et al.*, 1973.)

$S(Q)$. An example of neutron scattering data on argon at 85 K, fully corrected for instrumental resolution and for the effects considered in Section 13.2, is shown in Fig. 13.7. The structure factor consists of a principal peak followed by an oscillatory curve which decreases in amplitude with increasing Q. The asymptotic limit of $S(Q)$ is close to unity, and this provides a check that the corrections have been carried out properly. The Fourier transform of $S(Q)$ for argon gives the radial distribution function $g(r)$ shown in Fig. 13.8. The data in Fig. 13.7 were extended artificially to values of Q beyond the range of measurement, so as to reduce the amplitude of the spurious oscillations in $g(r)$ at small values of r. One method of evaluating the experimental results is to compare the measured $g(r)$ with the theoretical curves for various models of the interaction between nearest neighbours in the liquid. In the 'hard sphere' model the potential $u(r)$ between two atoms is zero for separations r greater than a fixed value σ and is infinite for separations less

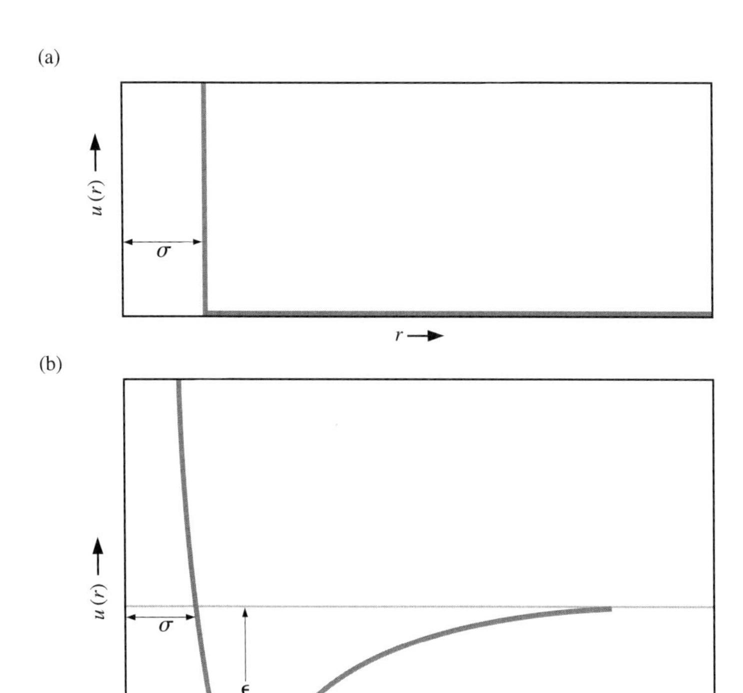

Fig. 13.9 Spherically symmetric potentials: (a) hard-sphere and (b) Lennard-Jones.

than σ:

$$u(r) = \infty \quad \text{for } r < \sigma$$
$$= 0 \quad \text{for } r > \sigma \tag{13.25}$$

[see Fig. 13.9(a)]. A more realistic model is the Lennard-Jones potential defined by

$$u(r) = 4\varepsilon\left[\left(\frac{\sigma}{r}\right)^{12} - \left(\frac{\sigma}{r}\right)^{6}\right] \tag{13.26}$$

and sometimes referred to as the 6–12 potential. This potential is illustrated in Fig. 13.9(b), which also shows the definitions of the parameters ε and σ. The Lennard-Jones model allows for partial overlap of adjacent atoms, and so avoids the assumption that the atoms are infinitely rigid. The softer core gives a rising edge which is less steep than with the hard-sphere model and a greater dampening of the oscillations in $g(r)$ on taking the Fourier transform of the intensity data.

In the Monte Carlo method of calculating $g(r)$ several thousands of atoms are allowed to interact through a suitable potential $u(r)$, and configurations are then generated with the probability of each configuration calculated from the assumed potential and the Boltzmann temperature factor. In the molecular dynamics method, the forces are calculated on each molecule in its initial configuration. The molecules are then moved by small amounts after applying the laws of motion to each molecule: the calculation is repeated many times by computer simulation and, after many collisions, the required distribution function $g(r)$ is extracted. The full line in Fig. 13.7 was calculated by the molecular dynamics method using a Lennard-Jones potential, and gives good agreement with the experimental data. A detailed description of computer simulation methods in liquids is given in the book by Allen and Tildesley (1989).

13.4.2 Binary liquids

The next stage in complexity after the simple liquid is a binary liquid, such as a binary alloy or a fully ionized molten salt. Instead of the single radial distribution function $g(r)$ we must now consider three separate functions $g_{\mathrm{AA}}(r)$, $g_{\mathrm{BB}}(r)$ and $g_{\mathrm{AB}}(r)$. Here A and B refer to the two kinds of atoms or ions, and $g_{\mathrm{AB}}(r)$, for example, represents the distribution of atoms of type B at a distance r as viewed by an atom of type A at $r = 0$. The structure factor $S(Q)$ is a linear combination of the three independent partial structure factors $S_{\mathrm{AA}}(Q)$, $S_{\mathrm{BB}}(Q)$ and $S_{\mathrm{AB}}(Q)$:

$$\begin{aligned} S(Q) = c_{\mathrm{A}} b_{\mathrm{A}}^2 \left[1 + c_{\mathrm{A}} \left(S_{\mathrm{AA}} - 1\right)\right] + 2 c_{\mathrm{A}} c_{\mathrm{B}} b_{\mathrm{A}} b_{\mathrm{B}} \left(S_{\mathrm{AB}} - 1\right) \\ + c_{\mathrm{B}} b_{\mathrm{B}}^2 \left[1 + c_{\mathrm{B}} \left(S_{\mathrm{BB}} - 1\right)\right] \end{aligned} \quad (13.27)$$

where $c_{\mathrm{A}}, c_{\mathrm{B}}$ are the concentrations of A, B in the sample and $b_{\mathrm{A}}, b_{\mathrm{B}}$ are the corresponding scattering lengths. Our aim is to determine the individual values of the three partial structure factors, which are defined in eqn. (13.10), and then to Fourier transform them to the radial distribution functions $g_{\alpha\beta}(r)$. Clearly, this requires three independent intensity measurements of $S(Q)$ and, in principle, this is possible by repeating the experiment for three different isotopic combinations of A and B.

Liquid alloys

The isotopic substitution technique was pioneered by Enderby *et al.* (1966) in a study of the alloy $\mathrm{Cu_6Sn_5}$. Measurements were made from three specimens consisting of natural tin alloyed with (i) natural copper, (ii) the copper isotope $^{63}\mathrm{Cu}$ and (iii) the copper isotope $^{65}\mathrm{Cu}$. The values of the corresponding scattering lengths are 0.76, 0.67 and 1.11 fermis, respectively. The corresponding partial structure factors were derived by solving three simultaneous equations of the form given in (13.27). The scattering lengths of natural Cu and $^{63}\mathrm{Cu}$ are too close to yield other than small intensity differences between the two samples, and so it was necessary to introduce additional mathematical constraints in the analysis.

Water

There have been many studies of water using isotopic substitution of hydrogen ^{1}H and deuterium ^{2}D. The scattering lengths of these isotopes are quite different and of opposite sign, and the intensities derived from various isotopic mixtures are very different. Nevertheless, diffraction from water presents some exceptionally challenging problems. Neutrons are scattered almost totally incoherently by light water, and so it is more difficult to obtain accurate structure factors from light water than from heavy water. A second problem is the limitation imposed by the Placzek correction, which is particularly difficult to evaluate for light atoms. A reliable set of radial distribution functions is necessary for understanding the properties of water at the atomic level, and there has been much progress in recent years in obtaining accurate data for water and molecular liquids.

Extensive intensity data are available for water and ice over a temperature range of 77–673 K and at pressures up to 400 MPa. Soper (2000) has developed a refinement procedure in which all three partial structure factors are fitted simultaneously to a three-dimensional ensemble of water molecules. The constraints on this structural model are that the density and intramolecular geometry are correct. Figure 13.10 shows the site distribution functions derived in this way for water at 298°K and at ambient pressure.

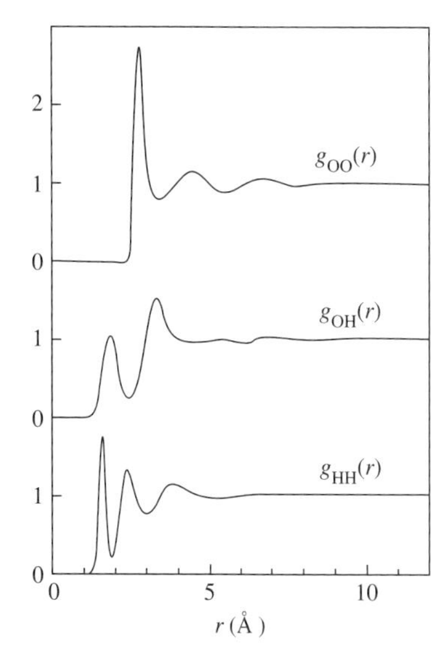

Fig. 13.10 Partial pair correlation functions of liquid water. (*After* Soper, 2000.)

Neutron diffraction studies of H_2O/D_2O mixtures at supercritical temperatures of 300°C and 400°C have been reported by Tromp *et al.* (1994). The three radial distribution functions $g_{HH}(r)$, $g_{OH}(r)$ and $g_{OO}(r)$ were determined after correcting for incoherent scattering and for Placzek effects using the technique described by Soper and Luzar (1992).

13.4.3 Glasses

Glasses are amorphous solids in which the underlying structure lacks any form of long-range order. On a macroscopic scale glasses are usually isotropic, and so a diffraction experiment can only yield a one-dimensional correlation function. For this reason modelling plays an important role in interpreting the diffraction data. The choice of suitable models is best made by employing results from both neutron and X-ray diffraction.

Neutron scattering from vitreous silica, using a time-of-flight diffractometer with a pulsed neutron source, has been reported by Grimley *et al.* (1990). The pulsed source is under-moderated, giving rise to a strong epithermal component in the incident flux, so that these short-wavelength neutrons allow measurements to be made to high values of Q. The time-of-flight data are shown in Fig. 13.11. The measurements were extended to Q values as high as 45 Å^{-1}, corresponding to a resolution of 0.12 Å in real space. Transforming the data leads to a composite correlation function, but the separation of the individual correlation functions is hampered by the fact that SiO_2 does not possess suitable isotopes with widely varying atomic scattering amplitudes.

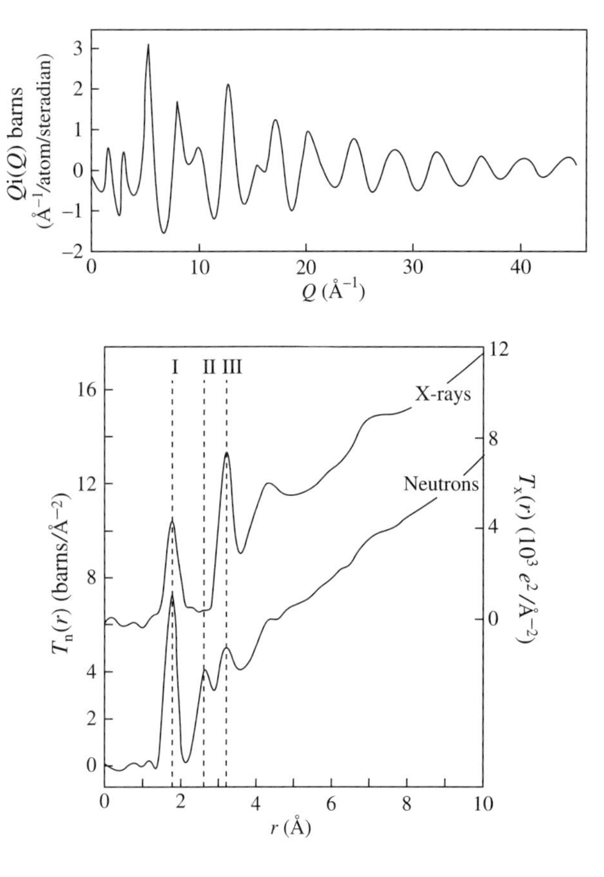

Fig. 13.11 Plot of the quantity $Q\,i(Q)$ versus Q for vitreous silica, as obtained from time-of-flight data. $i\,(Q)$ is the distinct, or interference, scattering. The experimental measurements are indicated by points, and the continuous line is a cubic spline fit. (*After* Grimley *et al.*, 1990.)

Fig. 13.12 Comparison of neutron and X-ray correlation functions for vitreous As_2O_3. The peak at I refers to As–O, at II to O–O and at III to As–As. (*After* Clare *et al.*, 1989.)

For many amorphous solids extra information can be obtained by combining neutron results with data from X-ray diffraction. An example of the complementarity of neutron and X-ray diffraction is shown in Fig. 13.12, which gives the neutron and X-ray total correlation functions for vitreous As_2O_3 as derived from the normalized diffraction patterns. In spite of the overlapping of the As–As, As–O and O–O distributions, the nearest-neighbour separations of the atoms can be deduced from the total correlation functions and are indicated by the vertical dashed lines in the figure. The O–O distance is clearly revealed by neutron diffraction; it is not observed in the X-ray diffraction pattern, because the atomic number of arsenic is much higher than that of oxygen and so the scattering from arsenic masks that from the lighter oxygen.

References

Allen, M. P. and Tildesley, J. (1989) "*Computer simulation of liquids.*" University Press, Oxford.

Clare, A. G., Wright, A. C., Sinclair, R. N., Galeener, F. L. and Geissberger, A. E. (1989) *J. Non-Crystalline Solids* **111** 123–138. "*A neutron diffraction investigation of the structure of vitreous As_2O_3.*"

Egelstaff, P. A. (1987) *Classical fluids.* in "*Methods of experimental physics*" Chapter 14, Volume **23**, Part B, pp. 405–469. Edited by Skold, K. and Price, D. L. Academic Press, London.

Enderby, J. E., North, D. M. and Egelstaff, P. A. (1966) *Phil. Mag.* **14** 961–970. "*The partial structure factors of liquid copper-tin.*"

Grimley, D. I., Wright, A. C. and Sinclair, R. N. (1990). *J. Non-crystalline Solids* **119** 49–64. "*Neutron scattering from vitreous silica.*"

McGreevy, R. L. (2001) *J. Phys. Condens. Matter* **13** R877–R913. "*Reverse Monte Carlo modelling.*"

Page, D. I. (1973) in "*Chemical applications of Neutron Scattering.*" Chapter 8, pp. 173–198. Edited by Willis, B. T. M. University Press, Oxford.

Placzek, G. (1952) *Phys. Rev.* **86** 377–388. "*The scattering of neutrons by systems of heavy nuclei.*"

Sinclair, R. N. and Wright, A. C. (1974) *Nucl. Instrum. Methods* **114** 451–457. "*Static approximation and neutron time-of-flight diffraction.*"

Soper, A. K. (2000) *Chem. Phys.* **258** 121–137. "*The radial distribution functions of water and ice from 220 to 673 K and at pressures up to 400 MPa.*"

Soper, A. K. and Luzar, A. K. (1992) *J. Chem. Phys.* **97** 1320–1331. "*A neutron diffraction study of dimethyl sulphoxide-water mixtures.*"

Tromp, R. H., Postorino, P., Neilson, G. W., Ricci, M. A. and Soper, A. K. (1994) *J. Chem. Phys.* **101** 6210–6215. "*Neutron diffraction studies of H_2O/D_2O at supercritical temperatures. A direct determination of $g_{HH}(r), g_{OH}(r)$ and $g_{OO}(r)$.*"

Van Hove, L. (1954) *Phys. Rev.* **95** 249–262. "*Correlations in space and time and Born approximation scattering in systems of interacting particles.*"

Yarnell, J. L., Katz, M. J., Wenzel, R. G. and Koenig, S. H. (1973) *Phys. Rev. A* **7** 2130–2144. "*Structure factor and radial distribution function for liquid argon at 85K.*"

Part III

Neutron spectroscopy

Coherent inelastic scattering from single crystals: study of phonons and magnons

14

Experiments in inelastic scattering require the measurement of the energy transferred from (or to) the scattered neutron to (or from) the scattering system. In this chapter, we shall be concerned with the use of coherent inelastic scattering in the study of the lattice and magnetic excitations in single crystals.

The experimental work of Brockhouse and collaborators in the 1950s and early 1960s, using inelastic neutron scattering, transformed our understanding of the lattice dynamics of crystalline solids. It was shown then that the inelastic scattering of neutrons from single crystals can yield both the frequency spectrum of the lattice vibrations and the dependence of the frequencies on the wave vectors of the vibrations. The essential features of the theory of lattice dynamics were established much earlier in the work of Born and von Kármán (1912, 1913), but experimental confirmation of the theory had to await the advent of neutron inelastic scattering.

14.1 The elements of lattice dynamics

Lattice-dynamical models form the basis for interpreting many physical properties, specific heat and thermal conductivity, which cannot be explained by a static model of the crystal. One of the first references to a lattice-dynamical model is found in Newton's *Principia*, which contains a discussion on the vibrations of a linear chain of equidistant mass points connected by springs. Indeed some of the important features of the dynamics of a three-dimensional crystal are revealed by considering the vibrational properties of a one-dimensional crystal treated as a linear chain of atoms. Figure 14.1 illustrates a linear chain of atoms in which the vibration is *longitudinal*, that is, where the atomic displacements are along the length of the chain. If the wavelength λ' of the vibration is very large compared with the spacing a between atoms, the wave propagates similar to a compressional wave in a continuum with the velocity

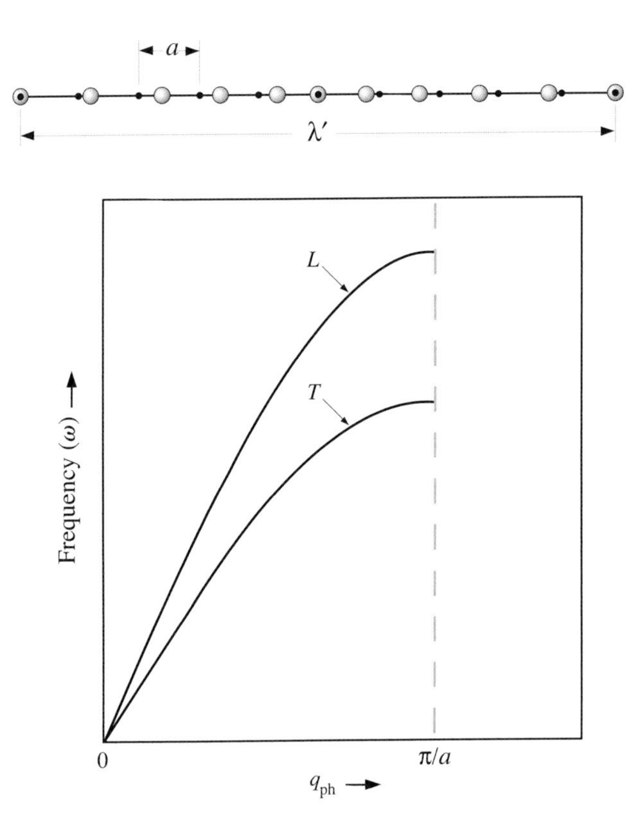

Fig. 14.1 One-dimensional representation of a longitudinal mode of vibration in a linear chain of atoms. The dots indicate undisplaced positions of the atoms and the open circles give their positions after displacement by the vibrational wave.

Fig. 14.2 Dispersion relation for a one-dimensional crystal showing the longitudinal (L) and transverse (T) branches. The *zone boundary* is at $q = \pi/a$, where a is the separation between adjacent atoms.

of sound v_S. The energy associated with the waves is quantized, and the corresponding quantum of energy is known as a *phonon.* The *phonon dispersion curve* gives the dependence of the (angular) frequency ω of a particular mode of vibration on its wave number q $(=2\pi/\lambda')$. We shall assume in the following that the forces between the atoms are *harmonic* and extend to nearest-neighbours only.

Figure 14.2 shows the dispersion curve (L) for these longitudinal waves. Near the origin at $q = 0$, where the wavelengths are much longer than the distance between the atoms, the oscillations are *acoustic waves* (although most of them have frequencies higher than can be perceived by ear). The initial slope $\mathrm{d}\omega/\mathrm{d}q$ in Fig. 14.2 is the macroscopic sound velocity v_S, but for larger values of q the dispersion relation is non-linear with the frequency rising to a maximum at the value $q = \pi/a$ and falling to zero again at $q = 2\pi/a$. The frequency is periodic with periodicity $2\pi/a$, which is the spacing of the one-dimensional reciprocal lattice. All possible frequencies are encompassed within this spacing, and the range $\pm\pi/a$ of the dispersion relation is known as the *first Brillouin zone.* Second Brillouin zone, third Brillouin zone, etc. are accessed by increasing q by integral values of $2\pi/a$, but the corresponding frequencies and patterns of atomic displacement are equivalent to those calculated for the first zone.

Next we allow the atoms in the one-dimensional chain to be displaced in any one of three perpendicular directions, one along the direction of propagation of the wave (giving *longitudinal* waves) and two at right angles to this direction (giving *transverse* waves). The force constants for the two transverse displacements are the same and so, for each value of q, there are three independent modes of vibration with two modes having the same frequency. There are three *branches* of the dispersion relations, but the transverse branch (curve T in Fig. 14.2) is doubly degenerate for a linear chain of atoms.

In a *diatomic* one-dimensional chain, the transverse waves of vibration exhibit further patterns of atomic displacement, as illustrated in Fig. 14.3. Neighbouring atoms tend to move against one another in the *optic* modes but together with one another in the acoustic modes. The corresponding dispersion relations are shown in Fig. 14.4. For the

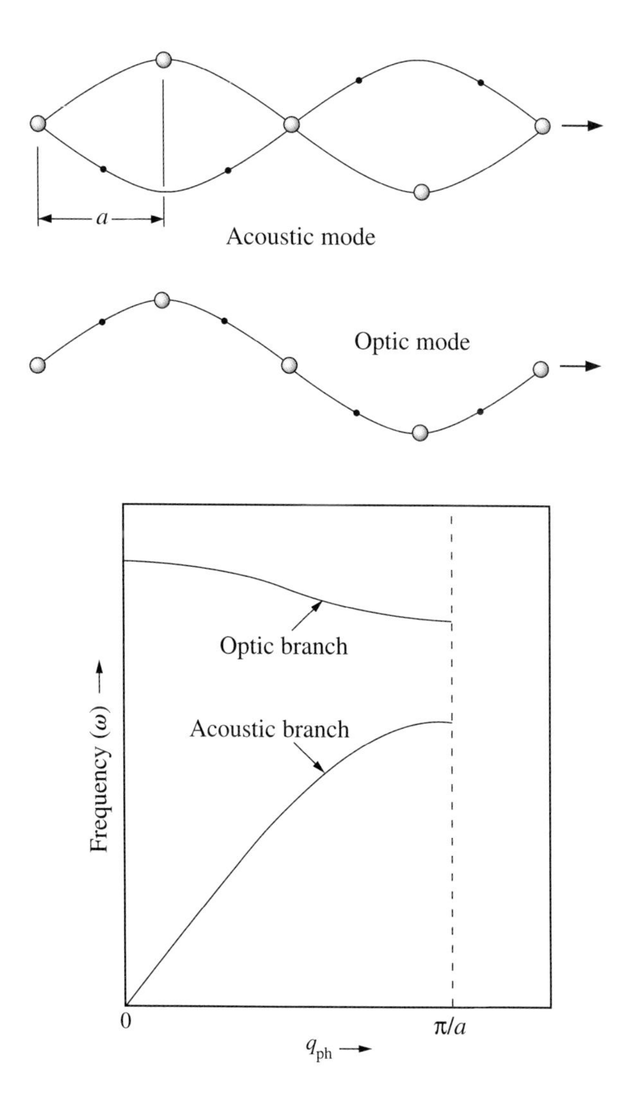

Fig. 14.3 Patterns of atomic displacement for transverse optic and transverse acoustic modes in a diatomic linear lattice. The arrows indicate the direction of propagation.

Fig. 14.4 The dispersion relations for a diatomic linear chain. The zone boundary is at $q = \pi/a$, where a is the atomic separation.

acoustic modes, the frequency goes to zero at the origin, whereas the frequency of the optic modes remains finite as $q \to 0$.

The three-dimensional theory of the lattice dynamics of crystals is based on the classical treatment of Born and von Kármán (1912, 1913). In this theory, the restoring force on an atom is determined not only by the displacement of the atom from its equilibrium position, but also by its displacement relative to its neighbours. The atomic motion is considered in terms of travelling waves extending throughout the whole crystal; these waves are the normal modes of vibration, in which each mode is characterized by a wave vector $\mathbf{q}$, an angular frequency $\omega(\mathbf{q})$ and certain polarization properties. For a crystal with n atoms in the unit cell, there are $3n$ branches of the dispersion curves relating the frequencies $\omega(\mathbf{q})$ with $\mathbf{q}$. Three of these branches have frequencies which tend to zero as the wave number goes to zero: these are the acoustic branches. The remaining $3n - 3$ branches are the optic branches. The frequencies depend on both the wave vector $\mathbf{q}$ and the particular branch j $(=1, 2, 3, \ldots, 3n)$ and so we label them $\omega_j(\mathbf{q})$.

It is because thermal neutrons have both a wavelength comparable with interatomic distances and an energy comparable with the energies of the normal modes of vibration, that they constitute such an effective probe for measuring $\omega_j(\mathbf{q})$ across the entire range of $\mathbf{q}$ in one Brillouin zone. Brockhouse and Stewart (1958) first reported the dispersion curves obtained in this way on single crystals of aluminium, and since then the neutron method has become a standard technique for obtaining detailed information about the lattice dynamics of single crystals.

Vibrational frequencies $\omega(\mathbf{q})$ can also be measured using X-rays, although X-ray frequencies are much larger than $\omega(\mathbf{q})$. In an experimental *tour de force*, using synchrotron X-rays and a specially constructed spectrometer with an energy resolution $\Delta E/E$ of one part in 10^6, Dorner *et al.* (1987) measured the phonon dispersion curves in beryllium for the longitudinal modes propagating in the [100] direction. More recently, the development of third-generation synchrotrons has made X-ray instruments competitive with neutron triple-axis spectrometers for phonon studies. The X-ray measurements are complicated, as they must achieve an energy resolution of 10^{-7} of the energy of the incident X-ray beam which is around 20 keV. The resulting X-ray beam is extremely small (often with a cross-sectional area of only $100\ \mu\text{m} \times 100\ \mu\text{m}$) and, therefore, measurements of small samples, or of samples containing isotopes that are highly absorbing to neutrons, are well suited to the X-ray technique. Provided a single crystal weighing more than 100 mg is available, the neutron technique is still to be preferred.

For 20 years after its publication in 1913, the Born–von Kármán treatment was eclipsed by the simpler theory of Debye (1912). In the Debye theory, the crystal is treated as a continuous medium and not as a discrete array of atoms. The theory gives a reasonable fit to the integral vibrational properties, such as the specific heat, of simple monatomic crystals. An even simpler model than Debye's is due to Einstein (1907), who considered the atoms in the crystal to be vibrating independently of each other and with the same frequency ω_E. By quantizing the energy of each atom in units of $\hbar\omega_E$, Einstein showed that the specific heat falls to zero at the temperature $T = 0$ and rises asymptotically to the Dulong and Petit value for $T \gg \hbar\omega_E/k_B$. Einstein's theory gives a more rapid fall-off of specific heat with decreasing temperature than is observed.

Deficiencies in the Debye theory were noted by Blackman (1937), who showed that they can be overcome by using the more rigorous Born–von Kármán analysis. Subsequently, Laval (1939), who carried out extensive X-ray studies on sylvine, aluminium and diamond, showed that the thermal diffuse scattering from these simple crystal structures can only be explained by the Born–von Kármán theory. The theory is routinely used in the interpretation of neutron data and is fully described in the book by Born and Huang (1954), and in more elementary terms in the books by Cochran (1973) and by Willis and Pryor (1975).

14.2 One-phonon coherent scattering of neutrons

In this section, we discuss the scattering of neutrons by lattice vibrations, in which one quantum of vibrational energy is exchanged between the neutron and the crystal. This is known as one-phonon scattering and is of primary interest in the study of the lattice dynamics of crystals. Neutrons can be scattered by more than one phonon, but multiphonon scattering simply gives rise to background scattering independent of $\mathbf{Q}$, and this is particularly troublesome with time-of-flight instruments, where a wide range of ($\mathbf{Q}$, ω) space is covered. Multiphonon scattering is of more significance in experiments exploiting inelastic incoherent scattering (see Chapter 15).

For one-phonon coherent scattering, the most important expressions are the following equations:

$$\mathbf{Q} \equiv \mathbf{k}_\mathrm{i} - \mathbf{k}_\mathrm{f} = \mathbf{H} \pm \mathbf{q} \tag{14.1}$$

and

$$\hbar\omega \equiv \frac{\hbar^2}{2m_\mathrm{n}} \left(k_\mathrm{i}^2 - k_\mathrm{f}^2\right) = \pm\hbar\omega_j\left(\mathbf{q}\right) \tag{14.2}$$

where $\mathbf{Q}$ is the scattering vector, $\mathbf{H}$ is a reciprocal lattice vector, and $\mathbf{k}_\mathrm{i}$ and $\mathbf{k}_\mathrm{f}$ are the initial and final wave vectors of the neutron, respectively. $\mathbf{q}$ is the phonon wave vector, and the right-hand side of eqn. (14.1) is written as $\mathbf{H} \pm \mathbf{q}$ because phonon dispersion curves are periodic in reciprocal space. These equations are illustrated pictorially in Fig. 14.5.

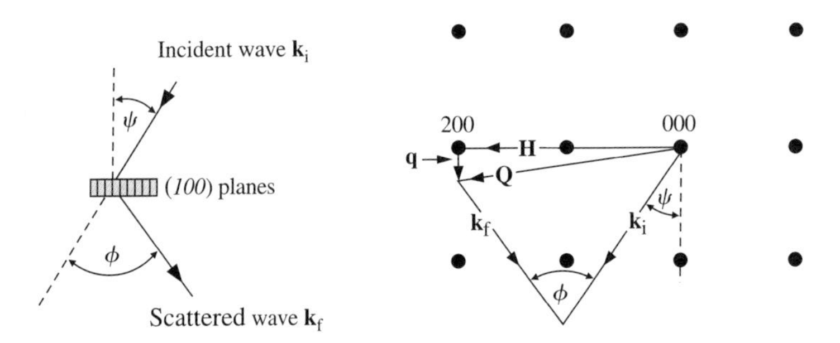

Fig. 14.5 Diagrams in real and reciprocal space illustrating the conservation laws for scattering by single phonons of wave vector $\mathbf{q}$. The angle ψ defines the direction of the incident beam. ϕ is the scattering angle, or the angle between the incident and scattered beams: it replaces the symbol 2θ used in Parts I and II for the case of elastic scattering.

The first equation can be regarded as an expression of the conservation of momentum in the scattering process, whereas the second equation expresses the conservation of energy. The positive sign on the right-hand side of the equations is considered when the neutron loses energy ($\hbar\omega > 0$) and creates a phonon in the crystal: this is known as *phonon emission* (or *phonon creation*). The negative sign is considered when the neutron gains energy ($\hbar\omega < 0$) and annihilates a phonon in the process known as *phonon absorption* (or *phonon annihilation*). In the first case, we also speak of neutron energy loss (or *down scattering*), and in the second of neutron energy gain (or *up scattering*).

The displacement of the atoms from their equilibrium positions results in scattering in all directions in reciprocal space subject to the restrictions imposed by the conservation laws [eqns (14.1) and (14.2)]. The *dynamic structure factor* $G_j(\mathbf{q})$ for this one-phonon scattering, representing the amplitude scattered by the lattice wave (or normal mode) with polarization vector $\mathbf{e}_j(\mathbf{q})$ and angular frequency $\omega_j(\mathbf{q})$, is given by

$$G_j(\mathbf{q},\mathbf{Q}) = \sum_l \left(\frac{1}{2m_l\omega_j(\mathbf{q})}\right)^{1/2} b_l\left[\mathbf{Q}\cdot\mathbf{e}_j(\mathbf{q})\right] \times \exp(\mathrm{i}\mathbf{Q}\cdot\mathbf{r}_l)\exp[-W_l(\mathbf{Q})] \qquad (14.3)$$

where $\mathbf{r}_l$ is the position of the lth atom in the unit cell, b_l is its scattering amplitude and m_l its mass. The second exponential in eqn. (14.3) is the Debye–Waller temperature factor of the atom. The equation refers to the coherent one-phonon inelastic scattering arising from the passage through the crystal of the wave ($j\mathbf{q}$) with wave vector $\mathbf{q}$ and belonging to the jth branch of the dispersion relations. If there is a nuclear spin or isotopic disorder in the sample, then there is also an incoherent inelastic scattering, and this type of one-phonon scattering is considered in Chapter 15.

The double differential scattering cross-section for coherent one-phonon scattering by the wave ($j\mathbf{q}$) is given by

$$\left(\frac{\mathrm{d}^2\sigma}{\mathrm{d}\Omega\mathrm{d}E_\mathrm{f}}\right)_{\mathrm{coh}} = \frac{k_\mathrm{f}}{k_\mathrm{i}}\cdot\frac{(2\pi)^3}{v_\mathrm{c}}\sum_{\mathrm{H}}\sum_{j\mathbf{q}}\left|G_j(\mathbf{q},\mathbf{Q})\right|^2 \times\left[n_j(\mathbf{q})\,\delta(\omega+\omega_j(\mathbf{q}))\,\delta(\mathbf{Q}+\mathbf{q}-\mathbf{H}) + (n_j(\mathbf{q})+1)\,\delta(\omega-\omega_j(\mathbf{q}))\,\delta(\mathbf{Q}-\mathbf{q}-\mathbf{H})\right] \qquad (14.4)$$

where the term $n_j(\mathbf{q})\,\delta(\omega+\omega_j(\mathbf{q}))\,\delta(\mathbf{Q}+\mathbf{q}-\mathbf{H})$ refers to phonon absorption and the term $(n_j(\mathbf{q})+1)\,\delta(\omega-\omega_j(\mathbf{q}))\,\delta(\mathbf{Q}-\mathbf{q}-\mathbf{H})$ refers to phonon emission. The delta functions embody the conservation relations in eqns (14.1) and (13.2). v_c is the volume of the unit cell and n_j is the number of phonons with energy $\hbar\omega_j$. Phonons (and magnons) obey Bose–Einstein statistics and so the Bose factor $n_j(\mathbf{q})$, giving the number of phonons with wave vector $\mathbf{q}$, is

$$n_j(\mathbf{q}) = \frac{1}{\exp[\hbar\omega_j(\mathbf{q})/k_{\mathrm{B}}T]-1} \tag{14.5}$$

The derivation of eqns (14.3) and (14.4) is given in standard texts, such as Squires (1978) and Lovesey (1984).

There are two important consequences arising from the form of these equations. First, on account of the term $\mathbf{Q}\cdot\mathbf{e}_j(\mathbf{q})$ in eqn. (14.3) the scattered intensity for the $(j\mathbf{q})$ mode vanishes when the scattering vector is normal to the polarization direction of the mode. This feature is useful in distinguishing between modes with different polarization properties. Second, there is always a greater probability that neutrons will be scattered with energy loss than with energy gain: at very low temperatures, where the Bose population factor $n_j(\mathbf{q})$ in eqn. (14.5) tends to zero, scattering takes place almost entirely by energy loss. This must be so, because as $T \to 0$ there are no phonons to be absorbed by the neutrons.

The cross-section for scattering by magnetic excitations (*magnons*) can be derived from that for one-phonon scattering after taking into account some important differences between phonon and magnon scattering. First, we must introduce the magnetic interaction vector (Section 9.2), so that for magnons there is no scattered intensity if the momentum transfer $\mathbf{Q}$ is parallel to the fluctuations, whereas for phonons the intensity vanishes if $\mathbf{Q}$ is perpendicular to the displacements. Second, the form factor $f(\mathbf{Q})$ for magnetic scattering decreases sharply with increasing Q and so the observations are restricted to relatively small vales of Q: on the other hand, for phonons the signal increases with Q^2 and measurements can be made at large Q. An important practical difference between phonons and magnons is that most studies of magnetic scattering are carried out below 100 K, where magnetic or electronic effects are often observed. Consequently, the Bose factor $n_j(\mathbf{q})$ is very small and magnetic experiments are feasible only with neutron energy loss.

14.2.1 The one-phonon scattering surface

Suppose that we illuminate a crystal with neutrons of energy and direction defined by $\mathbf{k}_{\mathrm{i}}$, and that the scattered neutrons $\mathbf{k}_{\mathrm{f}}$ emerge at a scattering angle ϕ. If the scattering is elastic, then $k_{\mathrm{i}} = k_{\mathrm{f}}$, and the locus of the end point of $\mathbf{k}_{\mathrm{f}}$, as ϕ varies and $\mathbf{k}_{\mathrm{i}}$ is kept fixed relative to the crystal, is a circle (see Fig. 14.6). In three dimensions the locus is the Ewald sphere, familiar in Bragg diffraction (see Section 7.1.3).

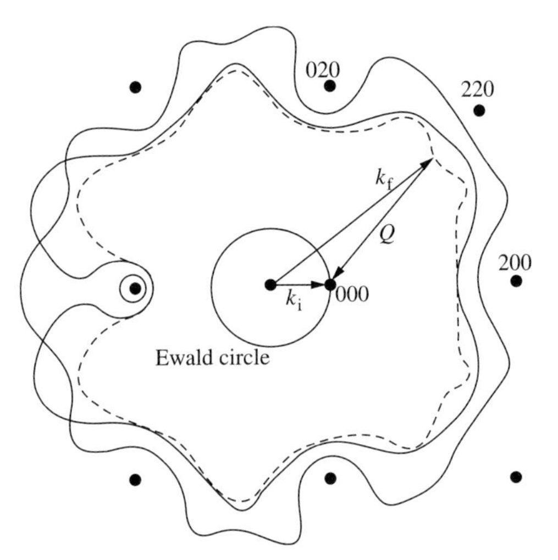

Fig. 14.6 Sections in the (001) plane of the one-phonon scattering surfaces for aluminium. The three surfaces correspond to the three acoustic branches of the dispersion relations and are plotted for phonon annihilation. The surface for the lowest-frequency branch is shown as a broken curve. (*After* Squires, 1978.)

However, for the one-phonon inelastic scattering of neutrons the locus of the end point of $\mathbf{k}_f$, known as the scattering surface, is much more complicated, and there are two such surfaces (one for phonon creation and one for phonon annihilation) for each branch of the dispersion relations. For a given $\mathbf{k}_i$, the shape of the scattering surfaces depends solely on the dispersion relation $\omega(\mathbf{q})$. For example, in the case of aluminium, with one atom in the unit cell, there are three branches of the dispersion relations, and six scattering surfaces, three of which are plotted in Fig. 14.6. This figure was calculated using the conservation relations, eqns (14.1) and (14.2), and knowing that the incident neutrons are of wavelength 6.7 Å and are parallel to the [100] direction of the crystal. As indicated in the figure, the scattering surfaces for neutrons can be extremely complicated and are often not simply connected. This is in contrast with the one-phonon scattering of X-rays, where the energy of the X-ray photon is many orders of magnitude larger than the phonon energy: eqn. (14.2) then becomes $k_i \approx k_f$ and the scattering surfaces for all branches are coincident with the Ewald sphere. Thermal diffuse scattering of X-rays takes place in those parts of reciprocal space that are intersected by all portions of this sphere.

For slower-than-sound neutrons, thermal diffuse scattering (TDS) is forbidden in the neighbourhood of the Bragg reflections, and this gives rise to gaps or 'windows' in the permitted energies of the scattered neutrons. At the two edges of these windows the TDS rises to a steep maximum: at one edge the maximum is due to phonon creation (Stokes process) and at the other edge to phonon annihilation (anti-Stokes process). These windows were observed in the experiment described by Carlile and Willis (1989) and did not involve the direct measurement of energy transfers. The theory of this curious effect, using eqns (14.1) and (14.2) to determine the topology of the scattering surfaces and the velocity of sound in the sample, is described by Schofield and Willis (1987) and by Popa and Willis (1994).

14.3 Instruments for inelastic scattering experiments

In the inelastic neutron scattering experiment, the quantity we measure is the double differential cross-section $\left[\mathrm{d}^2\sigma/(\mathrm{d}\Omega\mathrm{d}E_\mathrm{f})\right]_\mathrm{coh}$ in eqn. (14.4). The parameters of particular interest are the energy transfer $\hbar\omega$ and the momentum transfer $\hbar\mathbf{Q}$. Just as for elastic scattering experiments, these inelastic experiments are carried out either with a reactor source and fixed incident wavelength or with a pulsed source and a broadband of wavelengths. In the first case, the instrument normally employed is a triple-axis spectrometer and in the second case it is a time-of-flight spectrometer. The step-by-step mode of operation of a triple-axis spectrometer makes it the instrument of choice for obtaining information about excitations at a given point or along a given direction in $(\mathbf{Q}, \omega)$ space. The time-of-flight spectrometer (Section 14.3.2), with numerous detector channels, is more suited to a global overview of the excitations in a broad range of $(\mathbf{Q}, \omega)$ space.

14.3.1 The triple-axis spectrometer

Since its invention over 50 years ago, the triple-axis spectrometer (TAS) has been widely used to study thermal excitations (phonons) or magnetic excitations (magnons) in single crystals. The name TAS refers to the axes of the three components of the instrument: the monochromator, the analyser and the sample. The monochromator, which typically consists of a single crystal of a material such as copper, silicon, germanium, beryllium or pyrolytic graphite, selects a monoenergetic beam from the broad distribution of neutron energies emerging from the neutron source and, therefore, defines the wave vector $\mathbf{k}_\mathrm{i}$ of the neutrons striking the sample. The monochromator is surrounded by heavy shielding with a channel to allow the neutron beam to pass at the scattered angle $2\theta_\mathrm{M}$, so that $\mathbf{k}_\mathrm{i}$ may be varied by altering the Bragg angle θ_M. The analyser receives the neutrons scattered by the sample through an angle ϕ. These neutrons are then Bragg reflected by the analyser through an angle $2\theta_\mathrm{A}$, and those with a particular final wave vector $\mathbf{k}_\mathrm{f}$ [with $k_\mathrm{f} = \pi/(d_\mathrm{A}\sin\theta_\mathrm{A})$] are recorded in the detector. Figure 14.7 is a schematic diagram of the TAS showing the corresponding vectors in real and reciprocal space. An illustration of the IN14 spectrometer at the Institut Laue Langevin is given in Fig. 14.8.

A rather complicated scanning procedure with the TAS gives the spectrum of excitations at a given point $\mathbf{Q}$ in reciprocal space. At each point of this scan, the scattering triangle in Fig. 14.9 must close at the same point $\mathbf{Q}$ and the energy transfer is varied by altering the length of $\mathbf{k}_\mathrm{i}$ or $\mathbf{k}_\mathrm{f}$. The instrument operates in a step-by-step mode, in which the angles (ϕ, ψ) of the scattering triangle change during the scan. In fact, there are six experimental angles: the Bragg angle and scattering angle of the monochromator, the rotation angle ψ and scattering angle ϕ of the sample, and the Bragg angle and scattering angle of the

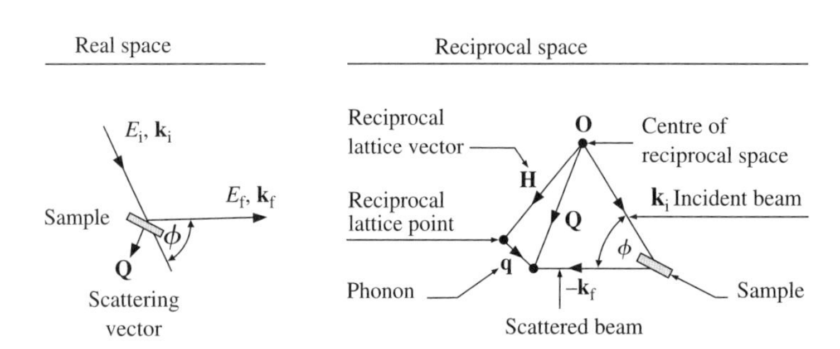

Fig. 14.7 (Upper panel) Schematic diagram of triple-axis spectrometer. (Lower panels) vectors in real and reciprocal space.

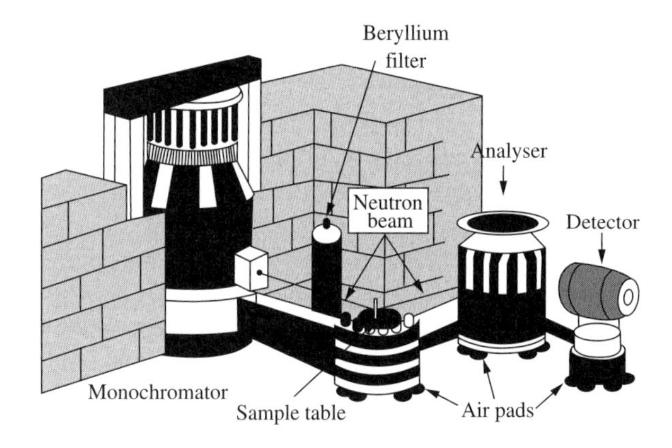

Fig. 14.8 The cold neutron three-axis spectrometer IN14 at the ILL.

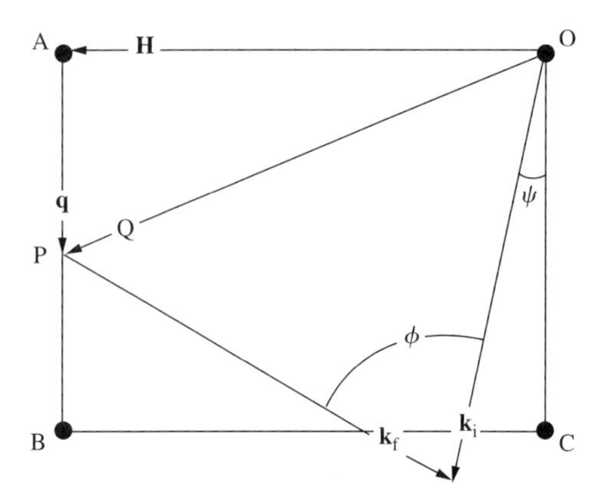

Fig. 14.9 A, B, C are points of the reciprocal lattice whose origin is O. $\mathbf{q}$ is the wave vector of the phonon which propagates along the symmetry direction AB. In scanning the energy transfer at each point P along AB, it is necessary to vary the lengths of $\mathbf{k}_i$ or $\mathbf{k}_f$.

analyser. The first two are linked in a 1:2 theta/two-theta ratio, as are the last two, so that there are four independent angles at the disposal of the experimentalist: θ_M, θ_A, ψ and ϕ. For scattering in a horizontal plane of reciprocal space, eqn. (14.1) has two variables, Q_x and Q_y, and eqn. (14.2) has one variable, ω. Hence any point in (Q_x, Q_y, ω) space can be selected by adjusting any three of the four angles: this point can be probed using any number of combinations of $\mathbf{k}_i$ and $\mathbf{k}_f$, although, in practice, the particular choice is dictated by intensity and resolution considerations. Many aspects of the power of TAS instruments, and the optimization of their experimental parameters, are described in the book by Shirane *et al.* (2002), although some of the advanced focusing devices in use today are not mentioned there.

The *constant-Q method* is largely used for both phonon and magnon measurements, but measurements at constant energy transfer $\hbar\omega$ are also used when the dispersion curve is very steep (as for magnons in the transition metals). Figure 14.10 illustrates the procedure for measuring optic and acoustic phonons along a symmetry direction. The energy scan is achieved by keeping k_i constant while varying the angles ϕ and ψ to keep $\mathbf{Q}$ constant; hence, the incident wave vector $\mathbf{k}_i$ moves around a circular arc centred on the origin at O. To keep $\mathbf{Q}$ constant in a fixed E_f scan the analyser and the sample must rotate together around the monochromator assembly. The entire procedure is controlled by computer, although in Brockhouse's pioneering work each measuring point in reciprocal space was laboriously plotted by hand following manual

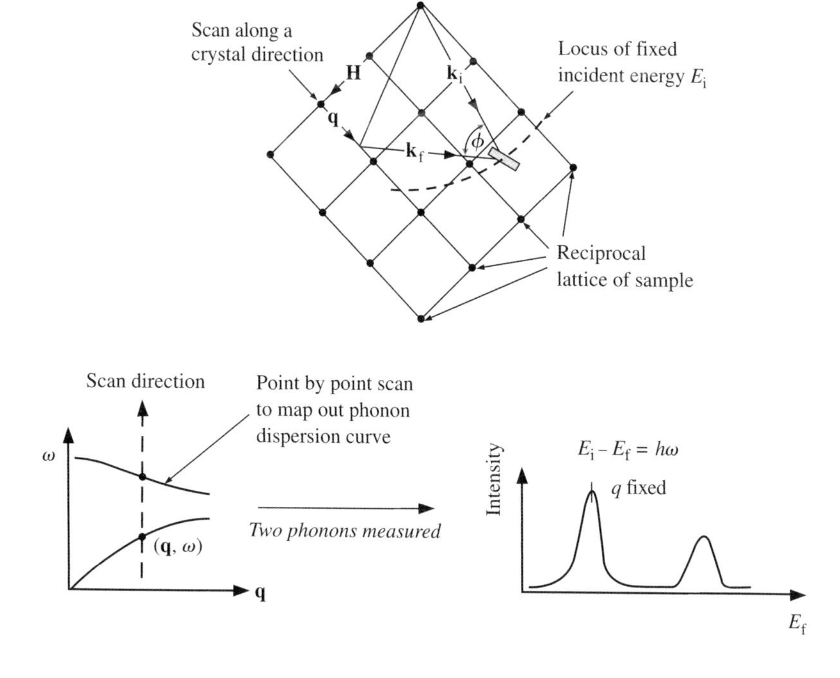

Fig. 14.10 TAS measurement at constant **Q** and at a fixed incident energy E_i.

adjustment of the three axes! Working with a fixed final wave vector $\mathbf{k}_f$ has the advantage that the experimental intensities are directly comparable with the quantity $|G_j(\mathbf{Q},\mathbf{q})|^2$ defined in eqn. (14.3).

In the absence of degeneracy, a neutron energy spectrum (or neutron group) for a given value of $\mathbf{Q}$ will have the same number of peaks as there are modes of vibration. Thus for n atoms per unit cell there will be $3n$ phonon peaks, and for complex crystals this gives rise to very complicated spectra. To interpret these spectra, a model is postulated for the force constants of the crystal, and this is used to calculate both the phonon frequencies and the polarization vectors for a wide range of $\mathbf{Q}$. The calculation requires the setting up of the dynamical matrix of the Born–von Kármán treatment. The eigenvalues of this matrix yield the frequencies of the normal modes of vibration and the eigenvectors give the polarization vectors of the atoms vibrating in these modes, for example, see the book by Dove (1993). The intensity scattered by each mode is then calculated using the structure-factor formula given in eqn. (14.3). If the lattice-dynamical model gives a reasonable fit, the observations can be used to refine the force constants.

Historically, the TAS was used to measure phonon or magnon dispersion curves, but this activity is pursued much less today than before. The TAS has been found to be a very versatile machine in both fixed-$\mathbf{Q}$ and fixed-energy modes of operation. It has an extremely good signal-to-noise ratio and can focus on either scattering vector $\mathbf{Q}$ or energy $\hbar\omega$ provided large single crystals are available. Many effects can be studied which give intensities only 10^{-6} of the nuclear Bragg peaks: such effects are polaron fluctuations (related to soft phonon modes or lattice instabilities), magnetic fluctuations (as found in high-T_c superconductors) or incipient charge density waves. Examples of such studies are given in Section 14.4. The TAS is well suited to investigate these problems, especially if it operates on a cold neutron source. The problems can be described, in general, as electronic instabilities associated with a host of new materials (superconductors, heavy fermions, manganites, ferroics, *etc.*). The traditional TAS is limited to point-by-point measurements in $(\mathbf{Q}, \omega)$ space, but improved instruments, such as RITA (developed at Risø in Denmark but now at the Paul Scherrer Institute in Switzerland) and the flat-cone instrument at the ILL, have been designed to mitigate this difficulty by surveying simultaneously wide regions of $(\mathbf{Q}, \omega)$. With TAS measurements restricted to high-symmetry directions, it is possible that something important might be missed.

In magnetic inelastic scattering, magnons can be measured and then modelled with a series of exchange constants. The main interest today is in weak electronic signals that appear in the inelastic response function and which can be related to instabilities associated with the formation of charge or spin-density waves or with phenomena involving an interplay of phonons and electronic effects. For decades, the supreme instrument for carrying out these experiments has been the TAS operating at a steady-state neutron source. With the advent of powerful spallation neutron sources and the dramatic growth in computer power, time-of-flight spectrometers offer an alternative approach in which data are gathered in a large manifold of $(\mathbf{Q}, \omega)$ space and analysed in parallel rather than on a point-to-point basis.

14.3.2 Time-of-flight spectrometers

Time-of flight spectrometers may be divided into two classes:

1. *Direct-geometry spectrometers* in which the incident energy $E_{\rm i}$ is selected by a crystal or chopper, and the final energy $E_{\rm f}$ is measured by time-of-flight.
2. *Indirect-geometry spectrometers* in which the incident energy $E_{\rm i}$ is measured by time-of-flight and the final energy $E_{\rm f}$ is set by a crystal or filter.

Both types of spectrometer are used on pulsed sources and continuous sources. With a steady-state reactor source further pulsing devices are required such as choppers.

Direct geometry

A schematic diagram of a direct-geometry spectrometer is shown in Fig. 14.11, in which a Fermi chopper selects a single neutron energy from a white incident beam. The energy $E_{\rm i}$ of the incoming neutron is related to its momentum $\hbar k_{\rm i}$ by

$$E_{\rm i} = \frac{\hbar^2 k_{\rm i}^2}{2m_{\rm n}} \tag{14.6}$$

The time taken for these neutrons to travel the distance L_1 from the chopper to the sample is

$$\tau_{\rm i} = \frac{L_1}{v_1} \tag{14.7}$$

where $v_{\rm i}$ is the neutron velocity. But the momentum $\hbar k_{\rm i}$ is $m_{\rm n} v_{\rm i}$ and so

$$\tau_{\rm i} = \frac{m_{\rm n} L_1}{\hbar k_{\rm i}} \tag{14.8}$$

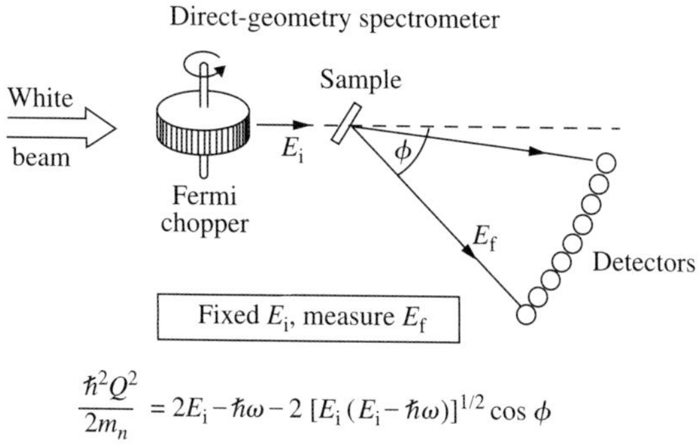

$$\frac{\hbar^2 Q^2}{2m_n} = 2E_{\rm i} - \hbar\omega - 2\,[E_{\rm i}\,(E_{\rm i} - \hbar\omega)]^{1/2} \cos\phi$$

Fig. 14.11 Schematic diagram of the direct-geometry spectrometer.

The total time-of-flight τ_{tot} from the chopper to the detector is measured directly in the experiment, and so the time τ_{f} taken to travel from the sample to the detector is given by the difference:

$$\tau_{\text{f}} = \tau_{\text{tot}} - \tau_{\text{i}} \tag{14.9}$$

Knowing the time τ_{f}, the wave number k_{f} of the scattered neutron is

$$k_{\text{f}} = \frac{m_{\text{n}} L_2}{\hbar \tau_{\text{f}}} \tag{14.10}$$

The final energy is

$$E_{\text{f}} = \frac{\hbar^2 k_{\text{f}}^2}{2m_{\text{n}}} \tag{14.11}$$

and the energy transfer $\hbar\omega$ is

$$\hbar\omega = E_{\text{i}} - E_{\text{f}} \tag{14.12}$$

with E_{i} given by (14.6) and E_{f} by (14.11).

To derive the Q value corresponding to this value of $\hbar\omega$, we use the expression

$$Q^2 = k_{\text{i}}^2 + k_{\text{f}}^2 - 2k_{\text{i}}k_{\text{f}} \cos\phi \tag{14.13}$$

which becomes, for fixed E_{i},

$$\frac{\hbar^2 Q^2}{2m_{\text{n}}} = 2E_{\text{i}} - \hbar\omega - 2\left[E_{\text{i}}\left(E_{\text{i}} - \hbar\omega\right)\right]^{1/2} \cos\phi \tag{14.14}$$

Direct-geometry spectrometers have proved to be especially effective in the general exploration of $(\mathbf{Q}, \omega)$ space. At the ISIS Facility alone, there are several such instruments (HET, MARI, MAPS, MERLIN, LET) and we shall refer briefly to just one of them. The multi-angle position sensitive (MAPS) spectrometer is a direct-geometry instrument with a huge array of 40,000 detector elements, which gives almost continuous coverage of the solid angle in the forward direction. The pixel size in reciprocal space is smaller than the resolution volume defined by other instrumental parameters. With MAPS, there is complete freedom to carry out scans along any direction in reciprocal space and to project data on to any plane in this space.

Figure 14.12 shows the layout of the instrument. An under-moderated pulse of neutrons is produced by a water moderator at ambient temperature, and a Fermi chopper ahead of the moderator then produces a monochromatic pulse of neutrons at the sample. The frequency and phasing of the Fermi chopper can be selected within the energy range from 15 to 2,000 meV. The sample environment equipment includes closed-cycle refrigerators, a 7.5 T superconducting cryomagnet, cryostats, furnaces and pressure cells. The user has complete freedom to view sections of reciprocal space as a function of any two of the four dimensions of $(\mathbf{Q}, \omega)$. In this way, the user can control the resolution of both

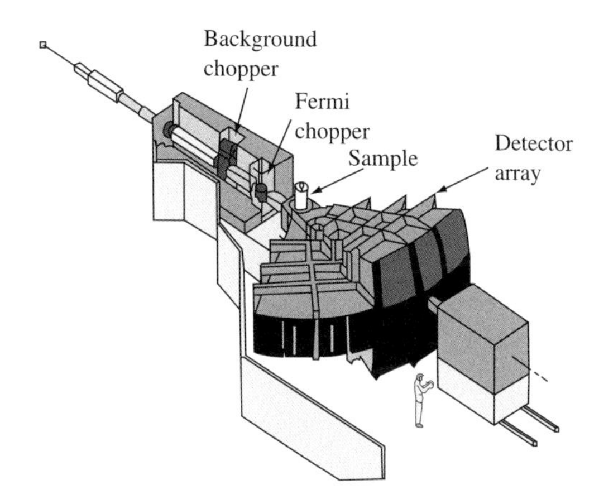

Fig. 14.12 Schematic layout of the MAPS spectrometer at ISIS.

the momentum and the energy transfer in accordance with the criteria appropriate to his/her experiment. MAPS can be used to study systems with energy scales from a few meV to 500 meV and with interactions in one, two or three dimensions. Examples of such systems include layered compounds with two-dimensional spin waves, high-temperature superconductors, colossal magnetoresistive manganites, *etc.* Some of these studies are described in Section 14.4.

Indirect geometry

Figure 14.13 is a schematic diagram of an indirect-geometry spectrometer. A white beam is scattered by the sample and a single-crystal analyser set at the Bragg angle θ_A selects a single energy E_f from the scattered beam. Choppers are used to remove unwanted high- and low-energy neutrons, leaving an energy band in the incident beam extending typically from 1 to 250 meV. Thus, it is possible to make use *simultaneously* of most of the thermal neutrons produced in a pulse rather than neutrons of just a single energy at a time.

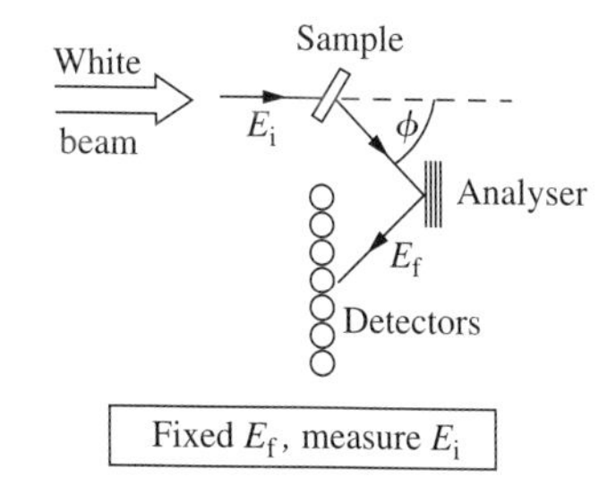

$$\frac{\hbar^2 Q^2}{2m_n} = 2E_f + \hbar\omega\ 2\,[E_f(E_f + \hbar\omega)]^{1/2} \cos\phi$$

Fig. 14.13 Schematic diagram of the indirect-geometry spectrometer.

The wave number k_f of the detected neutrons is

$$k_f = \frac{\pi}{d_A \sin\theta_A} \tag{14.15}$$

and the time τ_f taken by these neutrons to travel the combined distance L_2 from sample to analyser and analyser to detector is

$$\tau_f = \frac{m_n L_2}{\hbar k_f} \tag{14.16}$$

Knowing τ_f and the measured time-of-flight τ_{tot}, the incident time-of-flight τ_i is obtained from eqn. (14.7) and the incoming wave number of the detected neutrons from

$$k_i = \frac{m_n L_1}{\hbar \tau_i} \tag{14.17}$$

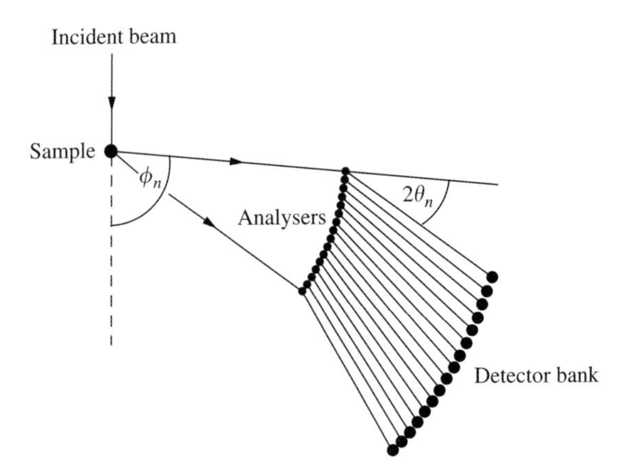

Fig. 14.14 Geometry of the PRISMA spectrometer.

where L_1 is the source-to-sample distance. The energy transfer $\hbar\omega$ is derived from the values of $k_{\rm i}$, $k_{\rm f}$ and the corresponding Q value, for fixed $E_{\rm f}$, is given by

$$\frac{\hbar^2 Q^2}{2m_{\rm n}} = 2E_{\rm f} - \hbar\omega - 2\left[E_{\rm f}\left(E_{\rm f} + \hbar\omega\right)\right]^{1/2}\cos\phi \tag{14.18}$$

The procedure for converting the time-of-flight spectrum to $S(\mathbf{Q}, \omega)$, using the relations above, is described in detail by Steigenberger *et al.* (1991) with reference to the indirect-geometry spectrometer PRISMA at ISIS. This instrument possesses 16 analyser–detector systems and each system detects neutrons scattered through a different angle ϕ_n where $n = 1 \rightarrow 16$ (see Fig. 14.14). Because of the open geometry in an indirect-geometry instrument, together with the fact that the sample is bathed in a white neutron beam, there is a high background of radiation superimposed on a weak signal from inelastic scattering. This feature is a very serious limitation of such instruments.

The flexibility of a direct-geometry spectrometer, with numerous position-sensitive detectors covering a large solid angle, guarantees that the instrument is the chosen work-horse at pulsed sources. Its advantages over the equivalent reactor-based TAS are the easier control of energy and momentum resolution and the comprehensive survey of reciprocal $(\mathbf{Q}, \omega)$ space, leading to the possibility of unexpected serendipitous discoveries.

14.4 Some examples of coherent inelastic scattering studies of single crystals

14.4.1 Phonon dispersion curves

Much of the work on phonon dispersion curves is summarized in the book of Bilz and Kress (1979). The book is a 'phonon atlas' containing phonon dispersion curves of more than 100 materials, including half

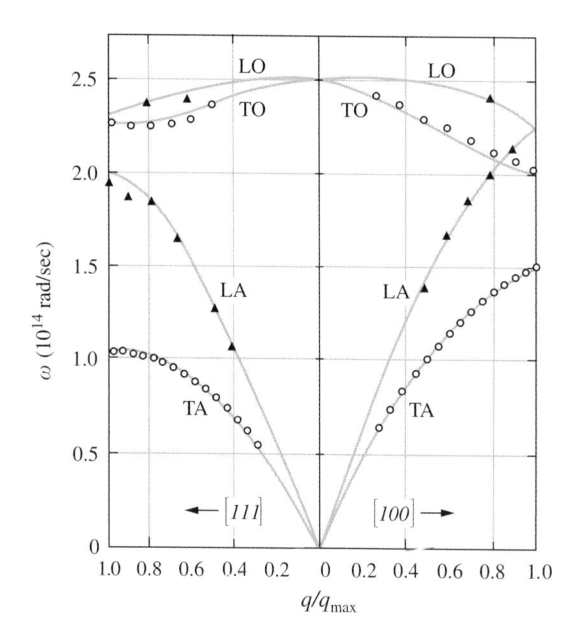

Fig. 14.15 Dispersion curves for diamond along the [100] and [111] directions. LO = longitudinal optic branch; TO = transverse optic; LA = longitudinal acoustic; LO = longitudinal optic. (*After* Warren *et al.*, 1967.)

the elements, many diatomic inorganic compounds and a few molecular crystals.

Figure 14.15 shows some early results of Warren *et al.* (1967), who measured the dispersion curves along the [111] and [100] directions for the acoustic and optic branches of diamond. There are two atoms in the cubic unit cell, and so there are three acoustic and three optic phonon branches; only four branches are seen in Fig. 14.15, as the transverse branches are doubly degenerate.

A more ambitious study was carried out by Pintschovius *et al.* (1989), who examined the lattice dynamics of a single crystal La_2NiO_4. There are seven atoms in the unit cell and, hence, 21 branches for each crystallographic direction. The measurements are shown in Fig. 14.16(a) and the calculated dispersion curves in Fig. 14.16(b). A complicated force field was required for these calculations: it reproduced the essential features of the observed curves, and yet significant discrepancies still remained. Modelling the dispersion relations of simple ionic solids such as the alkali halides is difficult enough, and in view of the long experimental times required for this work it is perhaps not surprising that the field is no longer highly fashionable.

14.4.2 Phase transitions

In the study of phase transitions, the TAS has played an important and unique role, which is now being augmented by the direct- and indirect-geometry instruments at spallation sources. A famous study of phase transitions is the discovery of the so-called *incommensurate* phase transitions, as exemplified by the study of K_2SeO_4 by Iizumi *et al.* (1977).

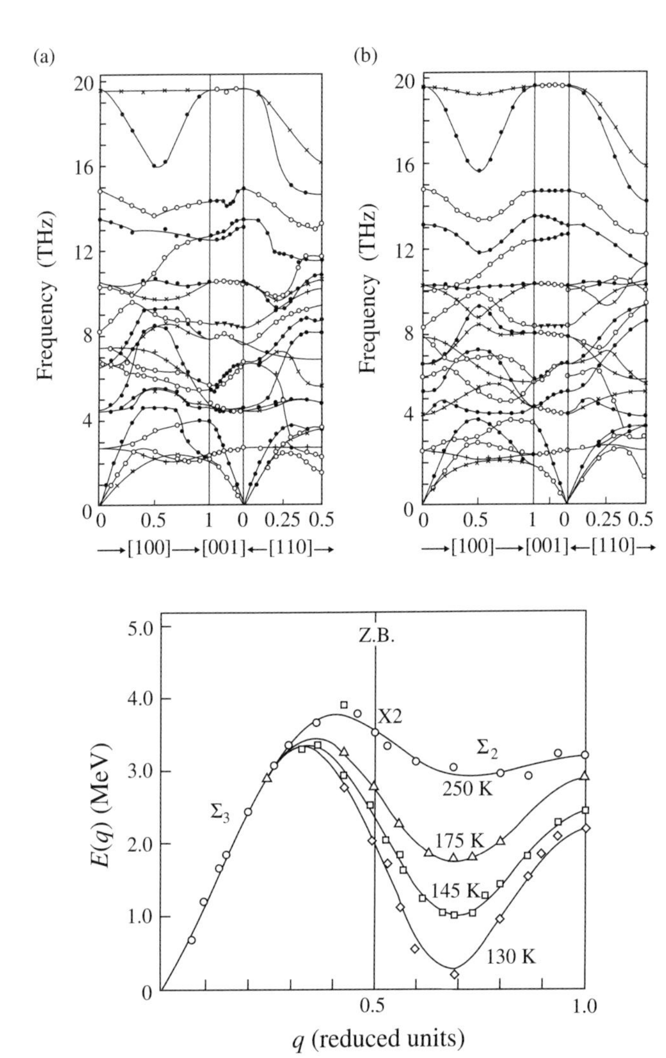

Fig. 14.16 (a) Experimental and (b) calculated dispersion curves for La_2NiO_4. (*After* Pintschovius *et al.*, 1989.)

Fig. 14.17 Dispersion relations of the Σ_2 soft mode and the Σ_3 acoustic mode in K_2SeO_4 plotted in an extended zone. Z.B. indicates the original zone boundary. The solid lines show the results of fitting force constant models to the data.

There is a phase transition in K_2SeO_4 at $T_c = 130$ K, and below this temperature the superlattice reflections are incommensurate with the lattice periodicity. The dispersion relations for the soft mode at temperatures above T_c are shown in Fig. 14.17. The phase transition does not result in a doubling of the unit cell, for which the phonons would have a minimum value at the zone boundary ($q = 0.5$ in Fig. 14.17), instead there is a minimum at a value of $q \sim 2/3$. This unusual situation, which occurs for many materials, is a result of competing interactions. Although the fourth nearest-neighbour force constants would favour an instability at $q = 1/2$, other competing interactions drive the instability to a higher wave vector. The solid lines show that the observed behaviour of the phonons with changing temperature can be reproduced adequately by choosing appropriate force constants. This type of phase transition is particularly prevalent in ferroelectric materials.

A comprehensive review of structural phase transitions, and the importance of inelastic neutron scattering in revealing their properties, has been compiled by Bruce and Cowley (1980).

14.4.3 Superconductivity

Neutrons do not couple directly to superconductivity, and yet, as we have shown in the case of flux lattices (Chapter 10), valuable information on their symmetry can be obtained by small-angle neutron scattering. Here, we give a different perspective using inelastic scattering from a single crystal. In the BCS theory of *s*-wave superconductivity, the lattice vibrations (phonons) mediate the interaction between electrons to form Cooper pairs. Thus, the measurements of phonons and the phonon density of states are important in understanding the mechanism of superconductivity.

The first example concerns the electron–phonon interaction in the conventional (*i.e. s*-type) superconductor Nb_3Sn. In such a system, below the critical temperature $T_c = 18$ K, phonons with energies *less* than the superconducting energy gap are energetically incapable of decaying by excitation of electron-hole pairs. This means that for such low-energy phonons their width below T_c should be narrower than the width above T_c, because the width of a phonon mode is inversely proportional to its lifetime and one of the possible decay channels governing the lifetime has been removed below T_c. The measurements represent a convolution of the real phonon width with the instrumental resolution, and so the latter must be as good as possible. Axe and Shirane (1973) used focusing techniques with thermal neutrons to achieve good resolution, and Fig. 14.18 shows the line widths of acoustic phonons above and below T_c. One phonon is shown in Fig. 14.18, as measured above and below T_c, and the effect on the width of the superconducting transition is relatively large.

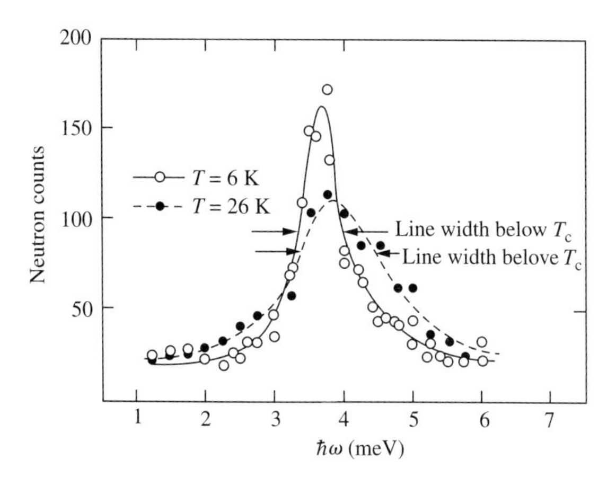

Fig. 14.18 The line width of low-energy acoustic phonons in Nb_3Sn propagating along [110]. The phonons broaden appreciably with increasing temperature.

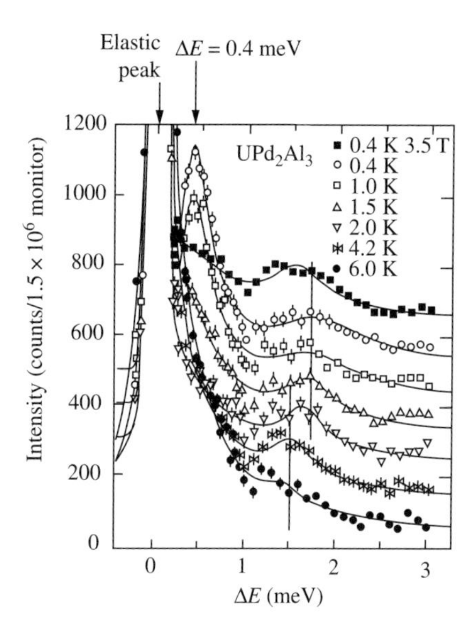

Fig. 14.19 The inelastic scattering in UPd_2Al_3 above and below the superconducting temperature of 1.8 K.

The second example, also obtained with a TAS instrument, is taken from a study by Metoki *et al.* (1998) of the material UPd_2Al_3, in which there is a strong coupling between magnetism and superconductivity. This compound is a heavy fermion superconductor ($T_c = 1.8$ K) with the Cooper spin pairing driven by magnetic fluctuations. The cold-source TAS worked with incident neutrons of about 5 meV energy. Figure 14.19 shows an extra *magnetic* peak at an energy transfer of 0.4 meV: the peak appears below 1.0 K at the wave vector $\mathbf{Q} = (0, 0, 1/2)$. This behaviour is believed to be due to excitations out of the superconducting state that break the spin-pairing mechanism. A magnetic field of 3.5 T (1 Tesla $= 10^4$ gauss), which is close to the upper critical field required to suppress the superconductivity, greatly reduces the intensity of the magnetic peak. Above T_c the low-energy scattering is observed to be quasielastic and strongly concentrated at the antiferromagnetic wave vector. The broad peak at about 1.4 meV is the heavily damped magnetic spin wave and is present both above and below T_c. It should be recalled that in Fig. 14.19, the elastic channel contains not only incoherent and background scattering, but also the large magnetic Bragg peak. These measurements were performed with a cold-source triple-axis machine at the Japanese JRR-3M reactor in Tokai. A more complete report on neutron studies of this material is given by Hiess *et al.* (2006).

In a third example, we consider experiments (using the MAPS direct-geometry instrument at ISIS) on the compound $YBa_2Cu_3O_{6.6}$, which is a superconductor at 63 K. Since the discovery by Müller and Bednorz (1986) of the first ceramic superconductor, *viz.* a La-based cuprate perovskite, an enormous amount of neutron work has been done to characterize such materials and to provide data that can be used in developing an appropriate theory, which still remains elusive. At finite energy transfers *magnetic* signals are seen in both the Ba doped La_2NiO_4 and

La_2CuO_4, but they appear at incommensurate positions in the lattice. In the case of the higher T_c superconductor $YBa_2Cu_3O_7$, the nature of the magnetic fluctuations was not apparent for several years, but the use of a multidetector type of instrument, initially HET (Mook *et al.*, 1998) and more recently MAPS (Hayden *et al.* 2004), showed that at certain energies the fluctuations were incommensurate and strongly resembled those occurring in the earlier La-based ceramic superconductors. In both types of compounds, the superconductivity is carried in the copper–oxygen planes, and the neutron experiments reveal two-dimensional magnetic fluctuations in these planes.

The mechanism responsible for the high superconducting transition temperature is still controversial and there are other neutron studies in which different interpretations are proposed, for example, Tranquada *et al.* (1995, 2004), Stock *et al.* (2005) and Reznik *et al.* (2006).

14.4.4 Magnon dispersion curves

The magnon dispersion curves for the transition elements are of particular interest as the 3*d* electrons are itinerant and the description of the spin waves gives a good test of theory. Figure 14.20 shows experimental results obtained with PRISMA on a single crystal of iron doped with silicon. The crystal was grown from the isotope ^{54}Fe which has a very small nuclear incoherent scattering cross-section. Figure 14.20(a) shows scans in reciprocal space along the [110] direction passing through the point $Q_0 = 200$. Both phonon and magnon dispersion relations are measured simultaneously. The spin-wave dispersion is nearly quadratic and can be observed up to energies of 50 meV. Transverse acoustic phonons are measured for $Q < Q_0$ and longitudinal acoustic phonons for $Q > Q_0$. All measurements were made in neutron energy loss, for which the Bose population factor is more favourable than for energy gain. The 16 parabolic lines in Fig. 14.20(a) are the paths through $(\mathbf{Q}, \omega)$ space covered by the individual detectors. The path for one of the detectors (no. 4) is highlighted in Fig. 14.20(a) and the corresponding intensity spectrum is shown in Fig. 14.20(b).

Because the measurement of the spin waves in iron is so difficult in terms of obtaining large energy transfers whilst still maintaining relatively small values of $|\mathbf{Q}|$ (thus ensuring that the form factor does not get too small), this experiment presents a special challenge at either a reactor or spallation source. An earlier study at the IPNS spallation source (Loong *et al.*, 1984) made measurements up to 160 meV energy transfer, but only in one direction. Yethiraj *et al.* (1991) used an inverted-geometry instrument at the Los Alamos spallation source to confirm the anisotropy observed earlier in the magnon intensities. The re-building of the TAS instrument at the ILL hot source has provided data to energy transfers of ~300 meV, showing that for this particular problem a hot-source triple axis instrument will give the best results (Paul *et al.*, 1988).

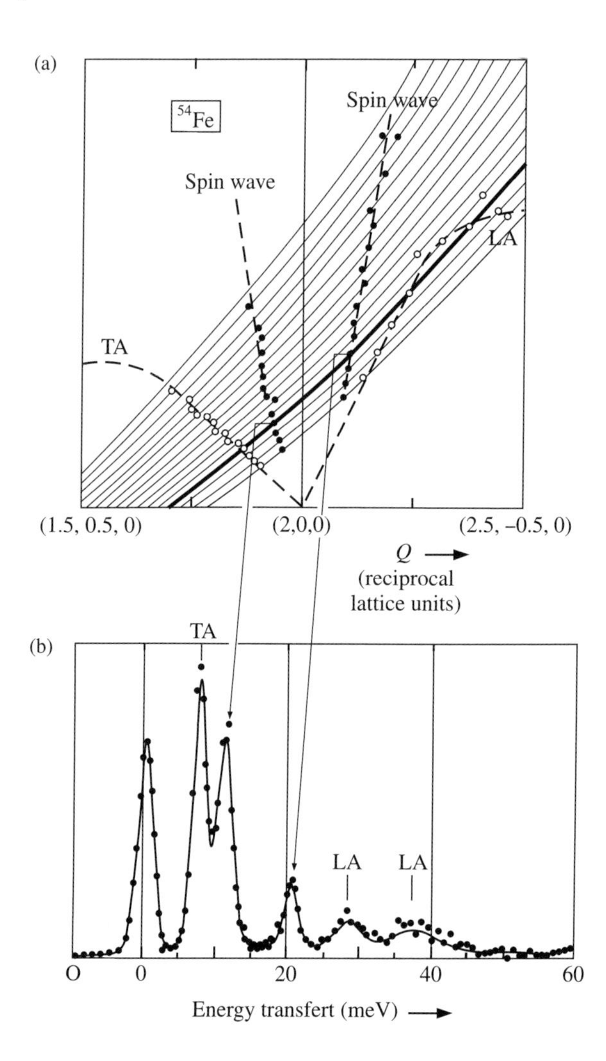

Fig. 14.20 (a and b) Dispersion relations for acoustic phonons and spin waves in iron, as measured by a multi-analyser time-of-flight spectrometer. (*After* Steigenberger *et al.*, 1991.)

Energy transfers as high as 300 meV have been measured with the direct-geometry instrument HET at ISIS. La_2CuO_4 is the parent compound of the high-T_c superconductors: it is a localized antiferromagnet, but when doped with 10% Sr it becomes a superconductor with $T_c \approx 35$ K. The two-dimensional spin waves in this layered compound have been examined at the temperatures of 10 and 295 K (Coldea *et al.*, 2001), and Fig. 14.21 gives the dispersion in the two-dimensional (h, k) zone. The dispersion relation shows that there are interactions beyond the Heisenberg nearest-neighbour term. Moreover, the nearest-neighbour exchange is antiferromagnetic while the next nearest-neighbour exchange is *ferromagnetic*, in contradiction to earlier theoretical predictions (Annett *et al.*, 1989).

Finally, we refer to the study by Boothroyd *et al.* (2003) of the spin dynamics of $La_{5/3}Sr_{1/3}NiO_4$. The inelastic magnetic and non-magnetic scattering of this system were separated by using a TAS together with

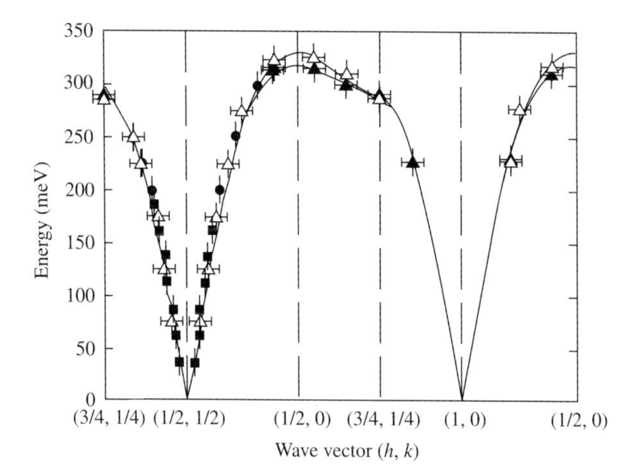

Fig. 14.21 Magnon dispersion relation of La_2CuO_4 along a high-symmetry direction in the two-dimensional Brillouin zone. Temperature $T = 10$ K (open symbols) and 295 K (closed symbols).

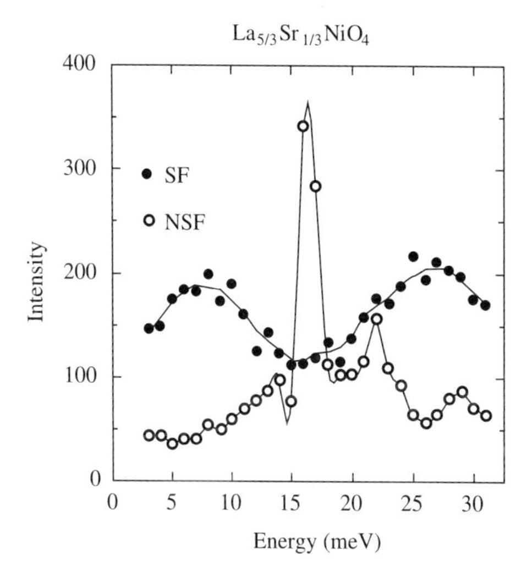

Fig. 14.22 Spin flip (SF) and non-spin flip (NSF) scattering in $La_{5/3}Sr_{1/3}NiO_4$ at 14 K. The measurements were made as a function of energy at constant **Q** $(=4/3, 4/3, 0)$. There are sharp phonon peaks at 16.5 meV and 22 meV in the NSF channel which are not present in the SF channel. (*After* Boothroyd *et al.*, 2003.)

polarization analysis. The scattered neutrons were spin flipped (SF) by magnetic excitations, but were non-spin flipped (NSF) by non-magnetic excitations. Figure 14.22 shows sharp phonon peaks in the NSF channel, which are not present in the SF channel.

14.5 Comparison of TAS and TOF methods of measuring phonon dispersion relations

Many authors have argued for or against using a time-of-flight spectrometer in preference to a TAS in studying excitations in single crystals.

For example, Dorner (1994) maintains that for studying the dispersion relations of crystals with many dispersion branches, TAS is the instrument of choice even when used with a neutron spallation source: this conclusion has been challenged by Hagen *et al.* (1996). Here, we shall simply refer to some of the problems associated with the two types of instrument.

The TAS employs a crystal to monochromate the incident beam, and this gives rise to order contamination and to a strong TDS background, particularly at high values of E_i. Order contamination results in the appearance of wavelengths $\lambda/2, \lambda/3\ldots$, in the incident beam, in addition to the primary wavelength λ, and this produces 'spurions' in the data unless higher orders of reflection are removed with a suitable filter, such as beryllium or pyrolytic graphite. A second problem is the occurrence of multiple scattering (Renninger reflections) which can produce a strong beam in an unwanted direction. Third, any Bragg reflection from the sample must be avoided. If this does occur, the intensity produced in the unwanted sequence—Bragg scattering (from the monochromator), Bragg scattering (sample), inelastic scattering (analyser)—may be comparable with the intensity from the desired sequence—Bragg scattering (monochromator), inelastic scattering (sample), Bragg scattering (analyser).

The main difficulty with time-of-flight instruments is that the scans are not necessarily along a required direction in reciprocal space, *e.g.* along a symmetry direction in the crystal. We have seen above that this drawback is overcome, to some extent, in the multianalyser machine PRISMA. The deconvolution of the data for the effect of the resolution function is more difficult with time-of-flight instruments than with the TAS on account of the greater flexibility of the TAS in traversing $(\mathbf{Q}, \omega)$ space.

As shown by the examples above, the trend today is not so much to measure the dispersion curves (either phonon or magnon) but to search for 'extra' effects associated with phase transitions, superconductivity, or other electronic instabilities. These effects often give rise to weak signals whose location in reciprocal space is not known in advance. All the instruments discussed in this chapter play a role in these studies and those at reactor and spallation sources are complementary rather than competitive. For example, the multidetector direct-geometry instruments at the spallation sources are particularly good at searching in reciprocal space for weak effects as they cover a wide range of $(\mathbf{Q}, \omega)$ space. On the other hand, the TAS instruments at reactors often have a better signal-to-noise ratio, as they focus on a particular $\mathbf{Q}$ value.

An assessment of the possibilities offered by operating chopper spectrometers with pulsed spallation sources is given in a review article by Aeppli *et al.* (1997). These authors believe that, in future studies of coherent inelastic neutron scattering, the main limitations will be the availability of suitable computer software and the cost of large detector banks to cover as large a solid angle as possible.

References

Aeppli, G., Hayden, S. and Perring, T. (1997) *Phys. World* **10** 33–37. "*Seeing the spins in solids.*"

Annett, J. F., Martin, R. M., McMahan, A. K. and Satpathy, S. (1989) *Phys. Rev. B* **40** 2620. "*Electronic Hamiltonian and antiferromagnetic interaction in* La_2CuO_4."

Axe, J. D. and Shirane, G. (1973) *Phys. Rev. Lett.* **30** 214–216. "*Influence of superconducting energy gap on phonon linewidths in* Nb_3Sn."

Bilz, H. and Kress, W. (1979) "*Phonon dispersion relations in insulators.*" Springer-Verlag, Berlin.

Blackman, M. (1937). *Proc. R. Soc. Lond. Ser. A* **159** 416–431. "*On the vibrational spectrum of a three-dimensional lattice.*"

Boothroyd, A. T., Prabhakaran, D., Freeman, P. G., Lister, S. J. S., Enderle, M., Hiess, A. and Kulda, J. (2003) *Phys. Rev. B* **67** 100407(R). "*Spin dynamics in sripe-ordered* $La_{5/3}Sr_{1/3}NiO_4$."

Born, M. and Huang, K. (1954) "*The dynamical theory of crystal lattices.*" Clarendon Press, Oxford.

Born, M. and von Kármán, T. (1912). *Phys. Z.* **13** 279–309. "*Uber Schwingungen in Raumgittern.*"

Born, M. and von Kármán, T. (1913). *Phys. Z.* **14** 15–19. "*Zur Theorie der spezifischen Wärme.*"

Brockhouse, B. N. and Stewart, A. T. (1958) *Rev. Mod. Phys.* **30** 236–249. "*Normal modes of aluminum by neutron spectrometry.*"

Bruce, A. D. and Cowley, R. A. (1980) *Adv. Phys.* **29** 219–321. "*Structural phase transitions.*"

Carlile, C. J. and Willis, B. T. M. (1989) *Acta Cryst. A* **45** 708–715. "*The pulsed neutron diffraction method of studying acoustic phonons in barium fluoride and calcium fluoride.*"

Cochran, W. (1973). "*The dynamics of atoms in crystals.*" Edward Arnold, London.

Coldea, R., Hayden, S. M., Aeppli, G., Perring, T. G., Frost, C. D., Mason, T. E., Cheong, S. W. and Fisk, Z. (2001) *Phys. Rev. Lett.* **86** 5377–5380. "*Spin waves and electronic interactions in* La_2CuO_4."

Debye, P. (1912). *Ann. Phys. (Leipzig)* **39** 789–839. "*Zur Theorie der spezifischen Wärme.*"

Dorner, B. (1994) *J. Neutron Res.* **2** 115–127. "*A comparison of time-of-flight and three axis spectrometer techniques for the study of excitations in single crystals.*"

Dorner, B., Burkel, E., Illini, Th. and Peisl, J. (1987) *Z. Phys. B Condens. Matter* **69** 179–183. "*First measurement of a phonon dispersion curve by inelastic X-ray scattering.*"

Dove, M. T. (1993) "*Introduction to lattice dynamics.*" Cambridge Univ. Press, Cambridge.

Einstein, A. (1907). *Ann. Phys. (Leipzig)* **22** 180–190. "*Die Planckschc Theorie der Strahlung und die Theorie der spezifisichen Wärme.*"

Hagen, M., Steigenberger, U., Petrillo, C. and Sacchetti, F. (1996) *J. Neutron Res.* **3** 69–92. "*Future opportunities and present possibilities for coherent inelastic single crystal measurements on pulsed neutron sources.*"

Hayden, S. M., Mook, H. A., Penchang Dai, Perring, T. G. and Dogan, F. (2004) *Nature* **429** 531. "*The structure of the high-energy spin excitations in a high-transition-temperature superconductor.*"

Hiess, A., Bernhoeft, N., Metoki, N., Lander, G. H., Roessli, B., Sato, N., Aso, N., Haga, Y., Koike, Y., Komatsubara, T. and Onuki, Y. (2006) *J. Phys. Condens. Matter* **18** R437. "*Magnetization dynamics in the normal and superconducting phases of UPd_2Al_3: survey in reciprocal space using neutron inelastic scattering.*"

Iizumi, M., Axe, J. D., Shirane, G. and Shimaoka, K. (1977) *Phys. Rev. B* **15** 4392–4411. "*Structural phase transformation in K_2SeO_4.*"

Laval, J. (1939). *Bull. Soc. Fr. Miner.* **62** 137–253. "*Etude experimentale de la diffusion des rayons X par les cristaux.*"

Loong, C.-K., Carpenter, J. M., Lynn, J. W., Robinson, R. A. and Mook, H. A. (1984) *J. Appl. Phys.* **55** 1895–1897. "*Neutron scattering study of the magnetic excitations in ferromagnetic iron at high energy transfers.*"

Lovesey, S. W. (1984) "*Theory of neutron scattering from condensed matter. Volume 1: Nuclear scattering.*" Clarendon Press, Oxford.

Metoki, N., Haga, Y., Koike, Y. and Onuki, Y. (1998) *Phys. Rev. Lett.* **80** 5417–5420. "*Superconducting energy gap observed in the magnetic excitation spectra of a heavy fermion superconductor UPd_2Al_3.*"

Mook, H. A., Pengcheng Dai, Hayden, S. M., Aeppli, G., Perring, T. G. and Dogan, F. (1998) *Nature* **395** 580. "*Spin fluctuations in $YBa_2Cu_3O_{6.6}$.*"

Müller, K. A. and Bednorz, J. G. (1987) *Science* **237** 1133–1139. "*The discovery of a class of high-temperature superconductors.*"

Paul, D. M., Mitchell, P. W., Mook, H. A. and Steigenberger, U. (1988) *Phys. Rev B* **38** 580. "*Observation of itinerant-electron effects on the magnetic excitations of iron.*"

Pintschovius, L., Bassat, J. M., Gervais, F., Chevrier, G., Reichardt, W. and Gomf, F. (1989) *Phys. Rev. B* **40** 2229–2238. "*Lattice dynamics of La_2NiO_4.*"

Popa, N. C. and Willis, B. T. M. (1994). *Acta Cryst. A* **50** 57–63. "*Thermal diffuse scattering in time-of-flight neutron diffractometry.*"

Reznik, D., Pintschovius, L., Ito, M., Iikubo, S., Sato, M., Fujita, M., Yamada, K., Gu, G. D. and Tranquada, J. M. (2006) *Nature* **440** 1170. "*Electron-phonon coupling reflecting dynamic charge inhomogeneity in copper oxide superconductors.*"

Schofield, P. and Willis, B. T. M. (1987) *Acta Cryst. A* **43** 803–809. "*Thermal diffuse scattering in time-of-flight neutron diffraction.*"

Shirane, G., Shapiro, S. M. and Tranquada, J. M. (2002) "*Neutron scattering with a triple-axis spectrometer: basic techniques.*" Cambridge Univ. Press, Cambridge.

Squires, G. L. (1978) "*Introduction to the theory of thermal neutron scattering.*" Cambridge Univ. Press, Cambridge.

Steigenberger, U., Hagen, M., Caciuffo, R., Petrillo, C., Cilloco, C. F. and Sacchetti, F. (1991) *Nucl. Instrum. Methods Phys. Res. B* **53** 87–96. "*The development of the PRISMA spectrometer at ISIS.*"

Stock, C., Buyers, W. J. L., Cowley, R. A., Clegg, P. S., Coldea, R., Frost, C. D., Liang, R., Peets, D., Bonn, D., Hardy, W. N. and Birgenau, R. J. (2005) *Phys. Rev. B* **71** 024522. "*From incommensurate to dispersive spin fluctuations: the high-energy inelastic spectrum in superconducting* $YBa_2Cu_3O_{6.5}$."

Tranquada, J. M., Sternlieb, B. J., Axe, J. D., Nakamura, Y. and Uchida, S. (1995) *Nature* **375** 561. "*Evidence for stripe correlations of spins and holes in copper oxide superconductors.*"

Tranquada, J. M., Woo, H., Perring, T. G., Goka, H., Gu, G. D., Xu, G., Fujita, M. and Yamada, K. (2004) *Nature* **429** 534. "*Quantum magnetic excitations from stripes in copper oxide superconductors.*"

Warren, J. L., Yarnel, J. L., Dolling, G. and Cowley, R. A. (1967) *Phys. Rev.* **158** 805–808. "*Lattice dynamics of diamond.*"

Willis, B. T. M. and Pryor, A. W. (1975) "*Thermal vibrations in crystallography.*" Cambridge Univ. Press, Cambridge.

Yethiraj, M., Robinson, R. A., Sivia, D. S., Lynn J. W. and Mook, H. A. (1991) *Phys. Rev B* **43** 2565–2574. "*Neutron-scattering study of the magnon energies and intensities in iron.*"

Inelastic neutron scattering spectroscopy

15

15.1 Introduction

In the previous chapter, we have seen how coherent inelastic neutron scattering can be used for the study of collective excitations, such as phonons and magnons, in single crystals. This spectroscopic technique was pioneered in the 1950s by the Canadian physicist, Bertram Brockhouse, and much of the earlier work was confined to the field of physics. However, the interest today in inelastic neutron scattering (INS) has shifted to include the fields of chemistry, materials science, geology and biology, where work is carried out in areas such as polymers, hydrogen bonding and catalysis. In a paper presented at an IAEA meeting in Trombay (IAEA Bombay 1959), Brockhouse described a time-of-flight rotating crystal spectrometer, which would nowadays be used for molecular spectroscopy. *Incoherent* INS spectroscopy probes the local environment of the scattering atom and is complementary to the optical techniques of infrared and Raman spectroscopy. The frequencies examined cover a large range stretching from 10^9 Hz (allowing the study of slow diffusive motions in solids) to 10^{14} Hz (for the study of molecular vibrations). Diffusional studies are described in Chapter 16. In the present chapter, we are concerned with vibrational neutron spectroscopy, which is reviewed comprehensively in the book by Mitchell *et al.* (2005). Early work was reviewed by John White (1973), a pioneer in this field. We shall exclude references to non-collective excitations, such as local magnetic excitations and crystal field transitions, which are of particular interest to physicists.

The majority of measurements in molecular spectroscopy are undertaken using infrared spectroscopy in absorption to identify directly the energies involved in vibrational transitions. Raman spectroscopy is also used: this is a technique in which the vibrational spectrum is examined by measuring the energy difference between the incident radiation and the radiation scattered inelastically. INS resembles Raman scattering with photons replaced by neutrons but with no neutron selection rules.

In Fig. 15.1, we show a hypothetical INS spectrum of a molecular crystal, which illustrates in schematic form the different kinds of possible excitations across a wide frequency range. If the scattering atom

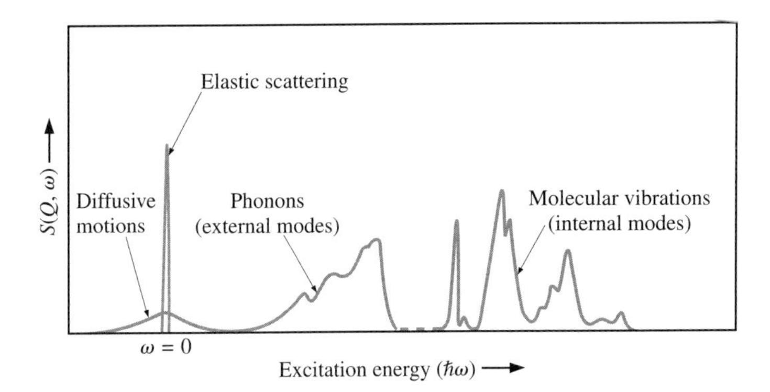

Fig. 15.1 Inelastic neutron scattering spectrum from a hypothetical molecular crystal. (*After* Eckert, 1992.)

is localized in space, then the scattering is purely elastic, *i.e.* $\omega = 0$, giving a line in the spectrum which reflects the resolution function of the instrument. If the atom jumps from site to site, within a time frame to which the instrument is sensitive, the elastic line is broadened by an amount that reflects the jump time of the atom. This is known as quasi-elastic scattering and an analysis of the line width as a function of scattering angle, or momentum transfer Q, gives information about the jump length and jump time in translational diffusion processes. The ratio of the purely elastic contribution to the sum of the elastic and quasi-elastic components is called the elastic incoherent structure factor (EISF). For rotational diffusion, the EISF has a maximum value at $Q = 0$ and falls to a minimum at a value of Q determined by the radius of gyration of the molecule. Scattering at very small energy transfers, typically ± 1 meV, tells us about the diffusion of atoms and molecules on time scales from 10^{-12} sec down to 10^{-7} sec. In the energy range from near-elastic up to 1 eV ($\sim$8,000 cm^{-1}), incoherent INS spectroscopy can be used to investigate internal molecular vibrations and the nature of the forces between the atoms of molecular systems. Molecular spectroscopy is one of the most valuable tools used in chemistry and will be our main concern in this chapter. In the next chapter, we shall cover quasi-elastic studies, together with excitations that occur at much lower frequencies, including tunnelling spectroscopy. The same instruments are employed for the studies in both chapters.

There are very few INS spectrometers in the world that can even approach the resolution of ordinary 'bench-top' infrared and Raman spectrometers. What then are the advantages of using INS spectrometers? An obvious advantage is that neutrons readily penetrate most materials, but this is not the principal advantage. Unlike infrared and Raman spectroscopy, INS has no selection rules and all modes are allowed, so that INS can be used to measure the full frequency spectrum of vibrational modes, some of which may be forbidden in optical spectra. Indeed, in the early period of INS its main purpose was to fill in the 'gaps' left by optical work. Information on atomic displacements is extracted directly from measured neutron intensities, whereas this is much more difficult to obtain from optical spectra because the electron distribution changes during the period of the vibration. The optical intensity arises from changes in the electronic properties (the dipole moment in infrared spectroscopy and the polarizability in Raman spectroscopy) during the period of the vibration. The neutron spectrum relates directly to the vibrational frequency and amplitude of the nucleus of the atom. The theoretical framework for vibrational analysis, known as Normal Coordinate Analysis, leads to the derivation of the eigenvalues (frequencies) and eigenvectors (atomic displacements) of these vibrations. This method assumes an empirical force field (see Section 14.1) and the preferred current practice is to derive eigenvalues and eigenvectors instead from *ab initio* calculations. [A comparison of the two methods is given by Parker *et al.* (2001).] Today with the significant improvements in the energy resolution of neutron spectrometers, most notably at pulsed sources, a complete set of eigenvalues and eigenvectors can be derived from the neutron data alone. The inclusion of both frequency and intensity measurements in the analysis offers a more sensitive test of the intermolecular forces than the analysis of frequency data alone. Finally, INS is especially suited to the study of systems containing hydrogen: the incoherent scattering from protons is more than an order of magnitude stronger than from any other nucleus. Neutron scattering is, therefore, particularly effective in studying large amplitudes of vibration of hydrogen atoms (*e.g.* in hydrogen-bonded systems or metal hydrides) or for identifying hydrogen motions in complex systems, such as biological systems.

We can illustrate the restrictions imposed by the selection rules for optical data by considering fullerene, C_{60}. The molecule has $(3N-6)$ modes of vibration (excluding the six rigid translational and rotational modes), where N is the number of atoms, making 174 modes *in toto*. On account of the molecular symmetry there are degeneracies of these modes, leading to 46 distinct modes. The Raman technique can excite only ten modes and the infrared method only four. Thus, more than 70% of the vibrations of the fullerene molecule are inaccessible to optical spectroscopy. Neutrons probe the dynamics of atomic nuclei directly and there are no restrictions on the observation of the vibrations. Neutron spectroscopy is sensitive to all 46 resonances in the spectrum and so is the spectroscopic choice for investigating highly symmetric molecules such as fullerene C_{60}. However, the spectrum is likely to include strong overlap between individual spectral lines, simply because of the intrinsic line width and the richness of the spectrum. In modern spectrometers the line width is limited more by effects such as poor crystallinity, dispersion of the modes and inhomogeneous broadening than by the resolution of the instrument.

An important difference between INS and Raman scattering is that INS spectroscopy involves a transfer of momentum $\hbar Q$ as well as a transfer of energy in the scattering process, whereas the scattering of light takes place at the centre of the Brillouin zone ($\mathbf{Q}=0$). Neutron experiments are usually performed with polycrystalline samples and this means

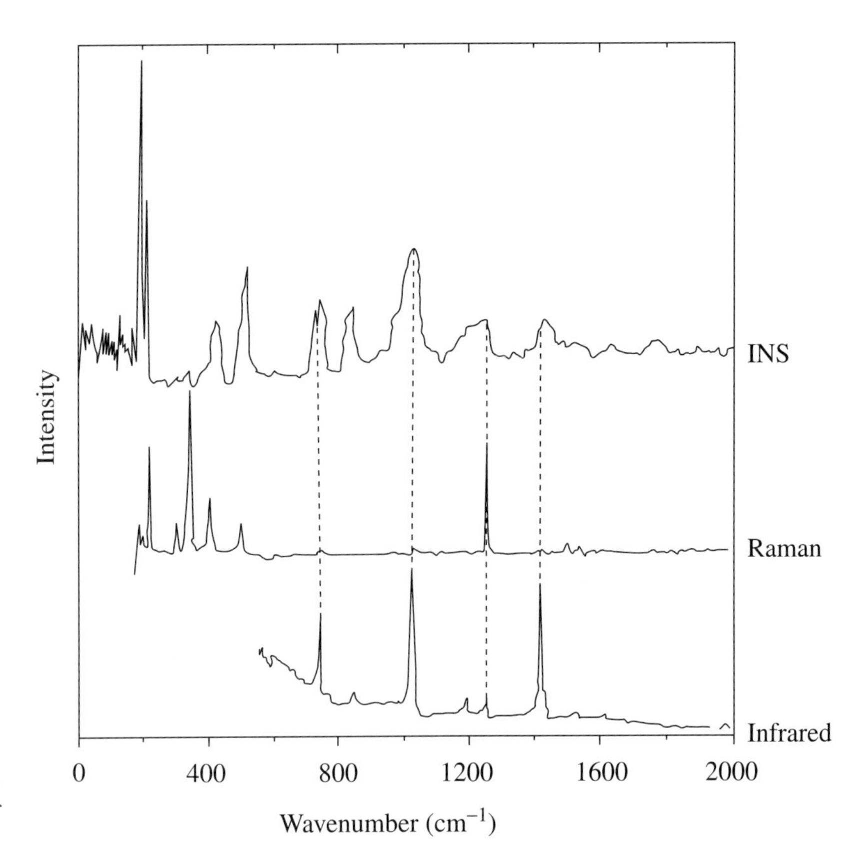

Fig. 15.2 The INS, Raman and infrared spectra of K[Pt(C_2H_4)Cl_3]. *After* Mitchell *et al.* (2005).

that the scattered intensity is a powder average over **Q**. Furthermore, INS instruments tend to have a fixed relationship between momentum and energy transfer, and so different frequencies are observed at different values of the momentum transfer and depend on the (**Q**, ω) trajectory of the instrument. Some of the discrepancies observed in the frequencies given by INS and optical spectroscopy are due to this **Q**-dependence of INS spectra. In addition, unless the molecule is bound, as in a crystal, the INS spectrum will experience excessive broadening on account of the recoil of the molecule in the scattering process. This means that the study of liquids by INS is particularly difficult and the majority of INS work is on polycrystalline samples.

The complementary nature of INS spectroscopy and the optical spectroscopies are illustrated in Fig. 15.2. This shows the INS spectrum of Zeise's salt, K[Pt(C_2H_4)Cl_3], together with the corresponding Raman and infrared spectra. Some peaks in the INS spectrum are not seen in the Raman and infrared spectra because of the optical selection rules, whereas the Pt–Cl stretch at 336 cm^{-1} gives an intense Raman band but only a weak INS feature. Vibrations involving hydrogen atoms have high INS intensity, such as the torsion of the C_2H_4 ligand at 190 cm^{-1}. The low-energy region (<190 cm^{-1}) is more accessible to INS spectroscopy than to the optical techniques.

Table 15.1 Cross-sections of selected elements.

Natural element	σ_{coh} (barns)	σ_{incoh} (barns)
Hydrogen	1.76	80.27
Deuterium	5.60	2.04
Carbon	5.55	0.00
Nitrogen	11.01	0.50
Oxygen	4.23	0.00
Aluminium	1.50	0.01
Chlorine	11.53	5.30
Silicon	2.16	0.01
Iron	11.22	0.40

15.2 Inelastic molecular neutron spectroscopy

In the neutron experiment, we are interested in the scattering function $S(\mathbf{Q}, \omega)$, which consists of a mixture of coherent and incoherent components:

$$S(\mathbf{Q}, \omega) = S_{\text{coh}}(\mathbf{Q}, \omega) + S_{\text{incoh}}(\mathbf{Q}, \omega) \tag{15.1}$$

The relative importance of these two components of $S(\mathbf{Q}, \omega)$ depends on the relative magnitudes of the cross-sections, σ_{coh} and σ_{incoh}, of the scattering nuclei. We recall from Section 2.4 that $\sigma_{\text{coh}} = 4\pi\langle b\rangle^2$, where $\langle b\rangle$ is the scattering length of the element, averaged over all its isotopes and spin states, and that σ_{incoh} is determined by the mean square deviation of the scattering length about its mean, or $\sigma_{\text{incoh}} = 4\pi\left(\langle b^2\rangle - \langle b\rangle^2\right)$. Table 15.1 lists the coherent and incoherent cross-sections for a few selected atoms. For carbon, oxygen, nitrogen, aluminium, silicon and iron, the scattering is almost entirely coherent, and for deuterium and chlorine there is a mixture of both coherent and incoherent scattering. The scattering cross-section for hydrogen is almost entirely incoherent and is extremely large compared with the values of σ_{incoh} of the other elements in the table. σ_{incoh} is much larger for hydrogen than for deuterium, and so isotopic substitution of D for H can be used to determine the scattering from a particular part of a molecule. For example, in an experiment to study the vibrational modes of methanol, CH_3OH, the deuterated compound CH_3OD can be examined to determine the hindered rotational motion of the CH_3 group and the CD_3OH compound to find the OH bending motions. It is not surprising that most of the molecular spectroscopy carried out with neutrons is incoherent inelastic scattering in which the scattering is dominated by the hydrogen atoms in the sample.

The double differential scattering cross-section can be separated into the components of coherent and incoherent scattering:

$$\frac{\mathrm{d}^2\sigma}{\mathrm{d}\Omega\mathrm{d}E_{\mathrm{f}}} = \frac{k_{\mathrm{f}}}{k_{\mathrm{i}}}\left[S_{\text{coh}}(\mathbf{Q}, \omega) + S_{\text{incoh}}(\mathbf{Q}, \omega)\right] \tag{15.2}$$

where the scattering functions, $S_{\text{coh}}(\mathbf{Q}, \omega)$ and $S_{\text{incoh}}(\mathbf{Q}, \omega)$, refer to the properties of the sample alone, and the term $k_{\text{f}}/k_{\text{i}}$ refers to the properties of the neutron probe. $S_{\text{coh}}(\mathbf{Q}, \omega)$ is associated with collective excitations such as lattice vibrations or phonons, and a detailed map of collective excitations in $(\mathbf{Q}, \omega)$ space provides the maximum amount of information on the intermolecular interactions. To obtain such a map requires an inordinate amount of time except for the simplest systems, and such studies are of more interest to physicists than to chemists. The incoherent part of the cross-section in eqn. (15.2) is associated with the motions of single particles, where there is no correlation between the motions of different atoms or molecules, and it is these motions that are of particular significance in molecular spectroscopy.

For a molecule bound in a crystal, the frequencies of the internal vibrational modes have no significant dispersion and the molecular system can be treated as a collection of harmonic oscillators. The double differential cross-section for incoherent one-phonon scattering is then given by

$$\begin{aligned} \frac{\mathrm{d}^2\sigma_{\text{incoh}}}{\mathrm{d}\Omega \mathrm{d}E_{\text{f}}} &= \frac{k_{\text{f}}}{k_{\text{i}}} \sum_{\mathbf{q},j} \delta\left(\hbar\omega \mp \hbar\omega_j\left(\mathbf{q}\right)\right) \frac{\hbar\left(\bar{n} + \frac{1}{2} \pm \frac{1}{2}\right)}{2\omega_j\left(\mathbf{q}\right)} \\ &\quad \times \sum_r \frac{\left(\sigma_{\text{incoh}}\right)_r}{4\pi m_r} \left|\mathbf{Q} \cdot \mathbf{U}_r^j\left(\mathbf{q}\right)^2\right| \exp(-Q^2\left\langle U^2\right\rangle) \end{aligned} \tag{15.3}$$

where $\omega_j(\mathbf{q})$ is the frequency of the vibrational mode j with wave vector $\mathbf{q}$, $\mathbf{U}_r^j(\mathbf{q})$ is the vector displacement of atom r in this mode, and $\exp(-Q^2\langle U^2\rangle)$ is the Debye–Waller factor associated with atom r where $\langle U^2\rangle$ is the mean-square amplitude of vibration of the atom. The upper sign in $\pm$ refers to phonon emission and the lower sign to phonon absorption. The phonon modes have Bose–Einstein population factors of $\bar{n} + 1$ for energy loss and $\bar{n}$ for energy gain, where

$$\bar{n} = \left[\exp\left(\frac{\hbar\omega_j\left(\mathbf{q}\right)}{k_{\text{B}}T}\right) - 1\right]^{-1} \tag{15.4}$$

At low temperatures $\bar{n} \rightarrow 0$ and there is no scattering with neutron energy gain: experiments are then restricted to neutron energy loss. Equation (15.3) shows that the intensity scattered by the mode j is proportional to the square of the amplitude of vibration $\mathbf{U}_r^j(\mathbf{q})$ of each atom r in that mode. The polarization term $\left|\mathbf{Q} \cdot \mathbf{U}_r^j\right|$ in eqn. (15.3) is effectively a selection rule which suppresses the scattering from those modes where the atomic displacement is normal to the scattering vector $\mathbf{Q}$.

For a powdered sample with one atom in the primitive cell, we can replace $(\mathbf{Q}.\mathbf{U})^2$ by $\frac{1}{3}Q^2\left\langle U^2\right\rangle$, and the incoherent scattering cross-section then reduces to

$$\frac{\mathrm{d}^2\sigma_{\text{incoh}}}{\mathrm{d}\Omega\mathrm{d}E_{\mathrm{f}}} = \frac{1}{4\pi}\frac{k_{\mathrm{f}}}{k_{\mathrm{i}}}\sigma_{\text{incoh}}\frac{Q^2\left\langle U^2\right\rangle}{2m}(\bar{n}+1/2\pm 1/2)\exp\left(-Q^2\left\langle U^2\right\rangle\right)N\frac{Z(\omega)}{\omega} \tag{15.5}$$

$3N$ is the number of phonon states and $Z(\omega)$ is the normalized density of states with $\int_0^\infty Z(\omega)\,\mathrm{d}\omega = 1$. An experimental measurement on a powdered sample gives, therefore, the energy range of the vibrational spectrum. For a hydrogenous sample, the scattering from hydrogen dominates all other scattering processes, and the intensity of scattering is proportional to the density of states of each mode with frequency ω multiplied by the mean-square amplitude of the hydrogen atom in that mode.

15.2.1 Simplified theory

For the study of hydrogenous materials at very low temperatures (neutron energy loss only), it is sufficient to write eqn. (15.5) in the form

$$\frac{\mathrm{d}^2\sigma_{\text{incoh}}}{\mathrm{d}\Omega\mathrm{d}E_{\mathrm{f}}} = \frac{k_{\mathrm{f}}}{k_{\mathrm{i}}}S_{\text{incoh}}(\mathbf{Q},\omega) \tag{15.6}$$

where

$$S_{\text{incoh}}(\mathbf{Q},\omega) = \sigma_{\text{incoh}}Q^2\left\langle U^2\right\rangle\exp\left(-Q^2\left\langle U^2\right\rangle\right)\delta(\omega-\omega_0) \tag{15.7}$$

Equation (15.7) is the scattering function for the fundamental frequency ω_0 of a harmonically bound oscillator, where σ_{incoh} is the incoherent scattering cross-section of hydrogen. At low Q, the intensity is controlled by the pre-exponential factor $Q^2\left\langle U^2\right\rangle$ and is proportional to Q^2, but at high Q the Debye–Waller factor takes over and the intensity falls off after passing through a maximum at $Q^2\left\langle U^2\right\rangle = 1$. Figure 15.3 illustrates the variation of $S_{\text{incoh}}(Q,\omega)$ with momentum transfer $\hbar Q$ and energy

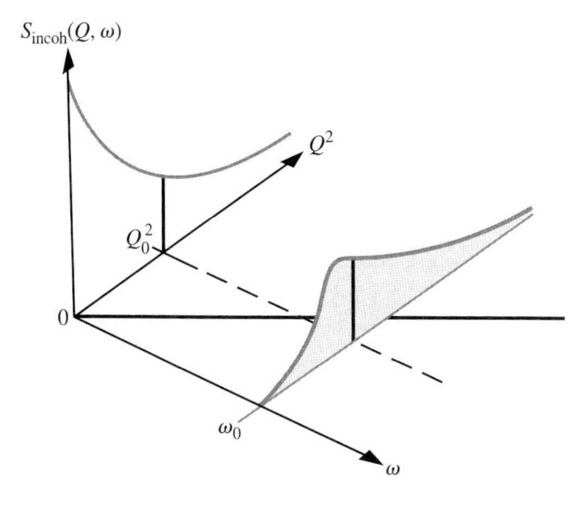

Fig. 15.3 The incoherent scattering function for a harmonic oscillator. (*After* Tomkinson, 1992.)

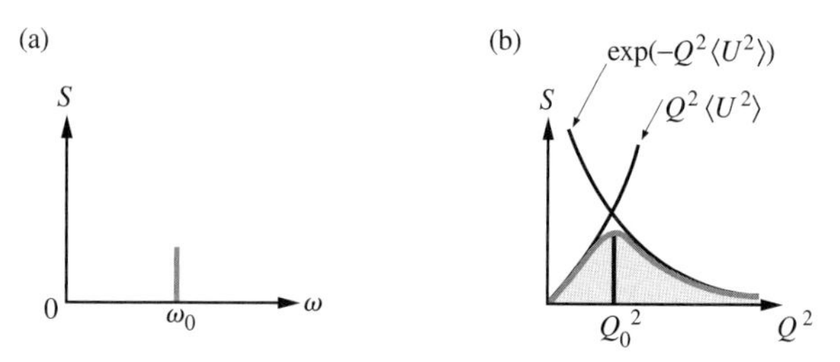

Fig. 15.4 Two ideal situations: (a) variation of S with frequency at constant Q^2 and (b) variation of S with Q^2 at constant frequency. (*After* Tomkinson, 1992.)

transfer $\hbar\omega$. If we perform an experiment in which the intensity is measured as a function of frequency while Q^2 is kept constant at the value $(Q_0)^2 = 1/(\langle U^2\rangle)$, we record two sharp peaks: the first one at $\omega = 0$ is the elastic peak and the second at $\omega = \omega_0$ is the fundamental peak at a frequency ω_0 [Fig. 15.4(a)]. In the second experiment, the frequency is kept constant at $\omega = \omega_0$ and the intensity is measured as a function of momentum transfer: in this case, the intensity increases from zero at $Q^2 = 0$ and rises to a peak at $Q = Q_0$ before falling again [Fig. 15.4(b)].

Overtones occur when the oscillator is excited from the ground state to energy levels beyond the first excited state. The scattering function for the nth order overtone is given by the expression

$$S_n(Q, n\omega) = \frac{1}{n!}\left(Q^2\langle U^2\rangle\right)^n \exp\left(-Q^2\langle U^2\rangle\right)\delta(\omega - \omega_0) \qquad (15.8)$$

Substitution of $n = 0$ and $n = 1$ into this equation yields the intensities of the elastic and fundamental lines. The mean-square displacement $\langle U^2\rangle$ is inversely proportional to the oscillator mass, and so overtone sequences are particularly prominent for hydrogen and fall away dramatically for heavier atoms.

So far, we have treated the case of an isolated molecule that is not allowed to recoil. In a solid, the molecule undergoes whole-body displacements described by the external modes of vibration. These vibrations are governed by intermolecular forces, which are much weaker than the forces within the molecule. Thus, the external modes have lower frequencies than the internal modes, and the mixing of the two types of modes gives rise to a scattering function that can be calculated approximately by considering the molecule to be an Einstein oscillator vibrating about its centre of mass with frequency $\omega = \omega_{\text{ext}}$. For this simple model, the scattering function in eqn. (15.5) is rewritten in the form

$$S_{\text{incoh}}(\mathbf{Q}, \omega) = Q^2\langle U^2\rangle \exp\left(-Q^2\langle U^2\rangle + \langle U^2\rangle_{\text{ext}}\right)\delta(\omega - \omega_0) \quad (15.9)$$

The effect of introducing whole-body displacements $\langle U^2\rangle_{\text{ext}}$ is to modify the scattering function in the manner illustrated in Fig. 15.5. The 'lost' intensity caused by the term $\langle U^2\rangle_{\text{ext}}$ in eqn. (15.9) reappears as a wing to the oscillator band at $\omega = \omega_0 = \omega_{\text{ext}}$. Thus in molecular crystals, we can consider the internal and external vibrations separately, discussing the external modes in terms of phonons, but allowing for the combination of external and internal modes. The phonon wings can be troublesome,

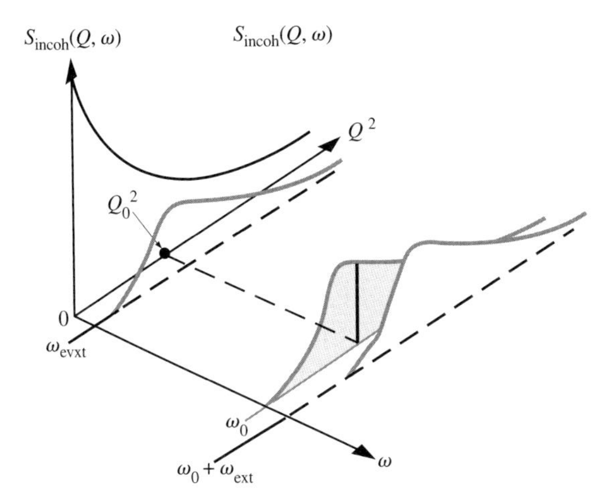

Fig. 15.5 Change in the scattering function S produced by embedding the isolated molecule of frequency ω_0 in an Einstein solid of frequency ω_{ext}.

masking weak features of the spectrum or making it difficult to assign the vibrations of complex spectra. These difficulties become more acute as the temperature is raised and as Q^2 increases, and so it is best to operate the spectrometers at low temperatures and low momentum transfers. In neutron diffraction studies, the measured Bragg intensities yield the so-called *atomic displacement parameters.* If these parameters are isotropic, then the diffraction experiment gives the quantity $\langle U^2 \rangle_{Bragg}$, which is the sum of the displacements from the internal and external modes of vibration:

$$\langle \mathrm{U}^2 \rangle_{\mathrm{Bragg}} = \langle \mathrm{U}^2 \rangle_{\mathrm{int}} + \langle \mathrm{U}^2 \rangle_{\mathrm{ext}} \qquad (15.10)$$

In diffraction work it is not possible, as it is in spectroscopic studies, to derive the internal and external components separately, but eqn. (15.10) provides a useful link between the two techniques.

An example of phonon wings is given in Fig. 15.6, showing the INS spectrum of sodium bifluoride, $NaHF_2$. There is a strong internal mode at 1,200 cm^{-1} and this contributes to the phonon wings causing a dispersion of the antisymmetric stretch at 1,400 cm^{-1}. The 'lost' intensity of the internal mode is transferred to the phonon wings and the combined total intensity is conserved. The calculated frequencies in Fig. 15.6 are for the isolated molecule.

The vibrational spectra of molecules require the calculation of the second derivatives of the energy with respect to the coordinates of the atoms about their equilibrium positions. There are many computer codes for calculating the $3N \times 3N$ dynamical matrix from these second derivatives, where N is the number of atoms. As an example, Fig. 15.7 shows the case of the benzene molecule. The lower trace is the observed INS spectrum and the upper curve the calculated trace using a specifically shaped phonon wing.

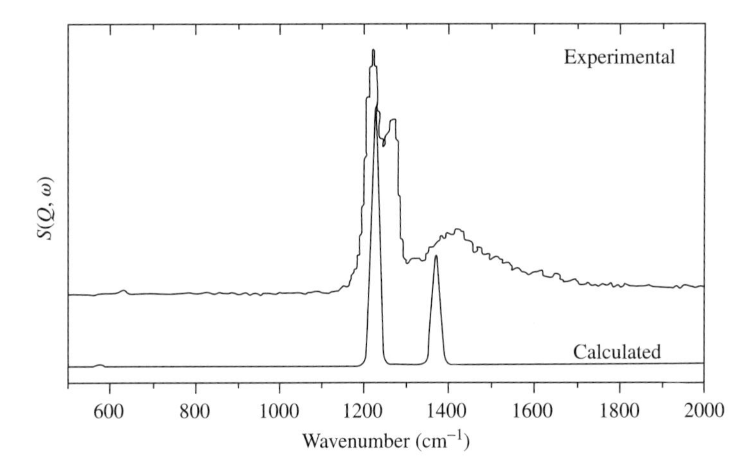

Fig. 15.6 INS spectrum of $NaHF_2$: observed (upper curve) and simple *ab initio* calculation for the isolated molecule (below). (*After* Mitchell *et al.* 2005).

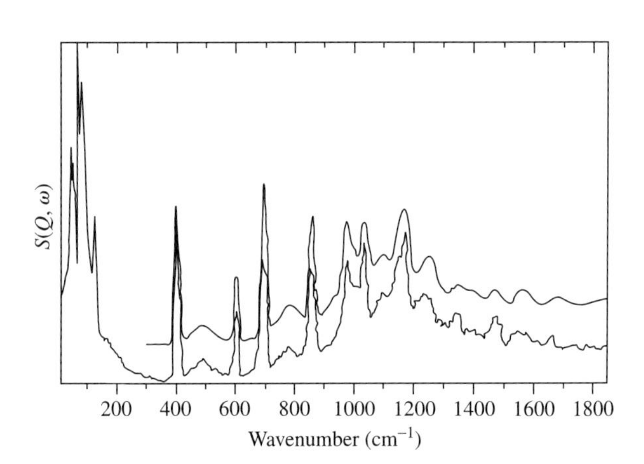

Fig. 15.7 INS spectrum of benzene: observed below; calculated above. (*After* Mitchell *et al.* 2005.)

15.3 Instrumentation for neutron spectroscopy

INS instruments used for molecular spectroscopy are designed to give a high count rate and good energy resolution at the expense of having little control over the momentum transfer. This last characteristic is justified on the grounds that most internal molecular vibrations show very small dispersion with Q. It is desirable that the complete frequency range of vibrations (20–4,000 cm^{-1}) is covered with the same instrument. This range includes the high-frequency region of bond stretches and bond-angle deformations and the low-frequency region of molecular deformations. There are different designs of INS spectrometers for reactor and pulsed neutron sources.

15.3.1 Direct-geometry spectroscopy

Figure 15.8 is a schematic view of the IN6 cold neutron time-focusing spectrometer at the Institut Laue Langevin. An intense beam of neutrons is extracted from the reactor source and passes down a long neutron guide to the monochromator assembly, consisting of three pyrolytic graphite monochromators. The monochromators use the full height (20 cm) of the guide and focus the beam at the sample position. The second-order reflection is removed from the monochromatic beam by passing it through a beryllium filter cooled to liquid-nitrogen temperature. The beam is pulsed by a chopper, and at the sample some neutrons gain or lose energy as they are scattered in many directions to a detector bank consisting of 337 detectors. The number of neutrons in each pulse arriving at the detectors in a given time is recorded, and from this time-of-flight spectrum the dynamical characteristics of the sample are determined. An overlap chopper in the incident beam serves to eliminate frame overlap, whereby faster neutrons from one pulse overtake the slower neutrons from a previous pulse.

To improve count rates on this spectrometer, 'time focusing' is used to enforce neutrons from the three monochromators, which have exchanged the same amount of energy with the sample, to arrive at the detector bank simultaneously regardless of their initial and final energies. Time focusing (Fig. 15.9) is achieved by scanning the monochromator crystals in turn during every pulse with a Fermi chopper. The angles and

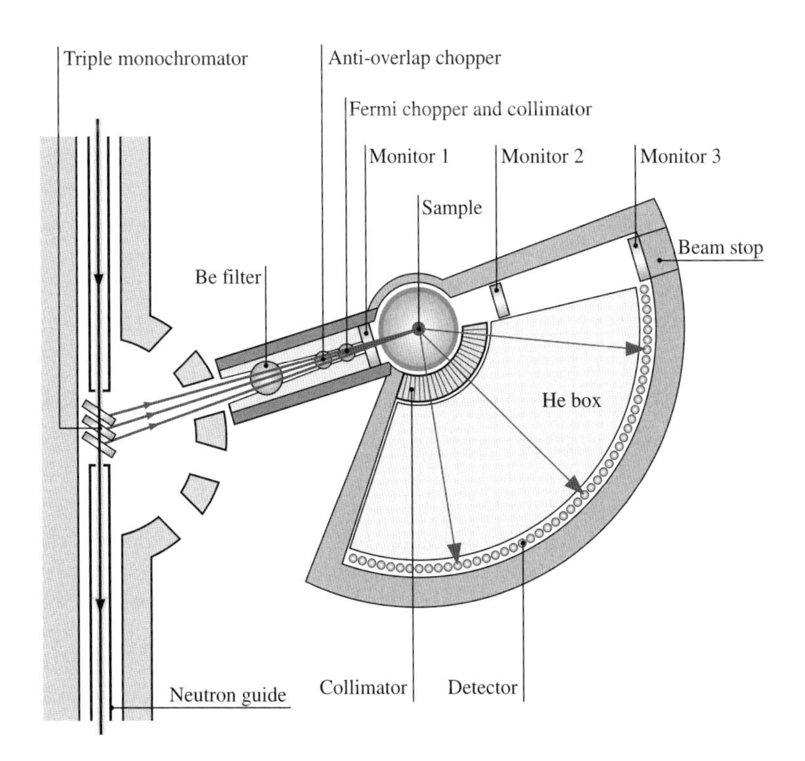

Fig. 15.8 IN6 spectrometer with vertically focusing monochromators.

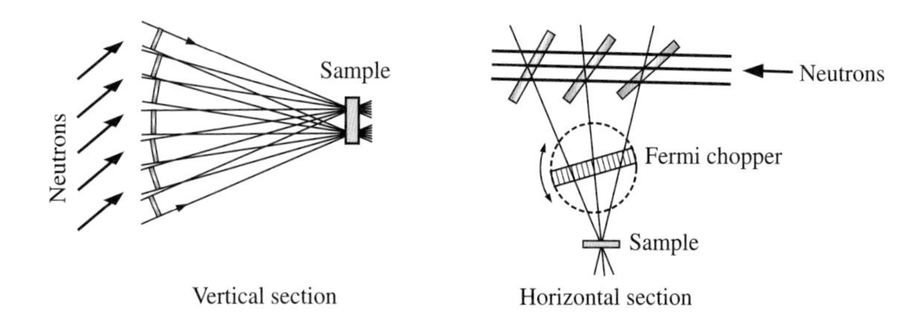

Fig. 15.9 Principle of time focusing on IN6.

positions of the crystals are arranged so that the slow neutrons take a shorter path and the fast neutrons a longer path.

15.3.2 Indirect-geometry spectroscopy

Instruments using indirect geometry, that is a white incident beam and a fixed final energy, cover a wide frequency range. The TOSCA instrument at the ISIS pulsed neutron source is a high-resolution time-of-flight spectrometer designed to operate in the region 32–4,000 cm^{-1} (4–500 meV). The scattered neutrons are accepted at either forward-scattering (45°) or back-scattering (135°) angles and then enter the analysers consisting of arrays of pyrolytic graphite crystals. These crystals are arranged such that neutrons with an energy of 3.5 meV are reflected into the detectors and the remainder are absorbed in the shielding. TOSCA has very good resolution at low energies and gives a clear definition of the phonon wings: these wings are combinations between intramolecular and intermolecular modes, which are of greater importance at high values of Q and must be correctly interpreted.

A major problem with crystal analyser spectrometers is that they select not only the desired wavelength λ at energy E_i, but also higher orders $\lambda/2, \ldots, \lambda/n$ at energies $4E_\text{i}, \ldots, n^2E_\text{i}$. In many INS spectrometers, these higher-order Bragg reflections are removed with a beryllium filter cooled to liquid-nitrogen temperature. The filter is transparent to neutrons of energy less than 40 cm^{-1} and higher-energy neutrons are scattered by Bragg reflections from the beryllium and are not transmitted (see Fig. 5.9).

15.4 Spectroscopic studies using inelastic neutron scattering

We have seen that hydrogen atoms scatter more strongly than other atoms by at least an order of magnitude. The INS spectrum is dominated by hydrogen atoms, not only because of their large scattering cross-section, but also because of their large amplitudes of motion. Consequently, INS spectroscopy is particularly valuable for the study of

(a) Methyl group torsions and hydrogen-bonded systems where hydrogen atoms undergo large amplitude motions.

(b) Metal–hydrogen vibrations in metallorganic and coordination complexes.

(c) Hydrogen-atom motions in organic and biological systems.

(d) Hydrogen-atom motions in optically opaque systems, such as metal hydrides and the surfaces of metallic catalysts.

Some of these works have been stimulated by the search for hydrogen storage materials; the pores in nanomaterials, for example, are possible sites for storing hydrogen. INS studies of dihydrogen rotational spectroscopy are reviewed by Mitchell *et al.* (2005).

Non-hydrogenous materials have also been studied in spite of their relative insensitivity to neutron scattering and the need to use large samples. In particular, extensive work has been undertaken on carbon in its allotropic forms of diamond, graphite and fullerene, and also on fullerene derivatives.

There have been relatively few experiments on biological materials. This is partly because large samples ($\sim$100 mg) and long counting times ($\sim$1 day) are required, due to the relatively low flux of the neutron beam. Smith (1991) has reviewed some results of experiments relating to protein dynamics where comparisons are made between simulations and the observed incoherent INS. The simulations make use of empirical potential energy functions, which include terms corresponding to deformations of bond lengths and bond angles, the hindered rotation of dihedral angles, and van der Waals interactions and electrostatic interactions between non-bonded atoms. Using a potential-energy function, the equations of motion are solved to calculate atomic trajectories in the protein. In the molecular dynamics method, the trajectories are calculated in femtosecond (10^{-15} sec) time steps and are built up over intervals of $\sim$100 fsec. $g(\mathbf{r},t)$ is obtained from this dynamical model and, after Fourier inversion, compared with the experimental $S(\mathbf{Q}, \omega)$.

The INS spectra of increasingly complex systems can be analysed by *ab initio* quantum chemistry. The calculations are based on a knowledge of the structure and an assumed force field for the system. The article by Hudson (2001) describes INS as a tool in molecular vibration spectroscopy and as a test of *ab initio* methods. Appendix 4 of the book by Mitchell *et al.* (2005) contains a list of about 500 systems, which have been analysed by high-resolution vibrational INS spectroscopy. Here we shall refer to just two systems that illustrate the power of INS spectroscopy.

In the first (Parker *et al.*, 1997), Mg_2FeH_6 was examined by infrared and Raman spectroscopy in addition to INS. At room temperature Mg_2FeH_6 crystallizes in the space group Fm3m and all the ions lie on special crystallographic sites. This high degree of symmetry leads to the absence of some of the infrared and optical vibrations. By combining the three types of vibrational spectroscopy, it was possible to observe and assign all of the internal modes of vibration and most of the external

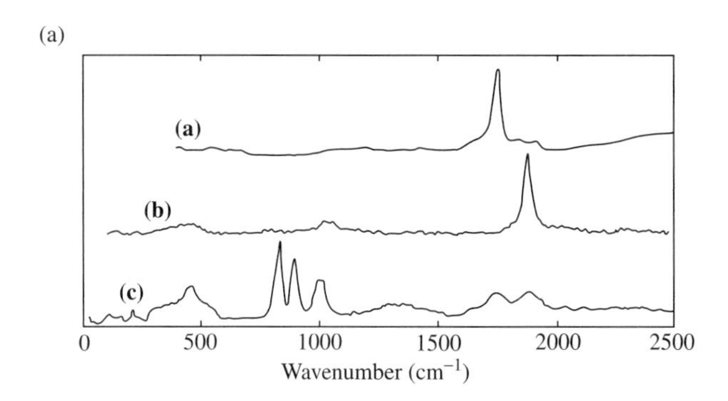

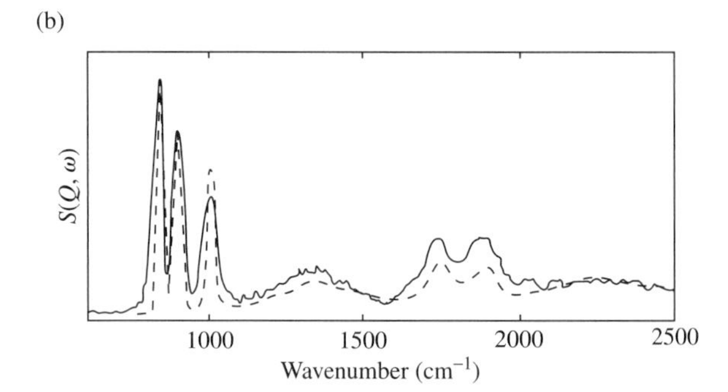

Fig. 15.10 (a) Vibrational spectra of Mg_2FeH_6: **(a)** infrared, **(b)** Raman and **(c)** INS. (*After* Parker *et al.* 1997.) (b) Mg_2FeH_6: comparison of the observed INS spectrum (solid line) with calculated fit (dashed line). (*After* Parker *et al.* 1997.)

modes. Figure 15.10(a) shows the three spectra and Fig. 15.10(b) compares the observed INS spectrum with that calculated from an empirical force field.

In a subsequent study (Parker *et al.*, 2003) of Rb_2PtH_6 and Rb_2PtD_6, a complete set of assignments was given for both the protonated and deuterated compounds, and this led to an explanation of the unusual infrared spectrum of the rubidium platinum hydride. A significant aspect of the work was the use of information at different values of Q: this information is not available with infrared and Raman spectroscopies. The analysis was carried out with Density Functional Theory (DFT), and this DFT code can be employed for analysing the INS data from *any* molecular system.

INS spectroscopy is limited by the available flux of the neutron source and by the worldwide scarcity of instruments. In the near future, these drawbacks will be ameliorated by the commissioning of new facilities: the second target station at ISIS (Rutherford Laboratory, UK), the Spallation Neutron Source SNS (Oakridge, USA) and the spallation source J-PARC (Tokai, Japan). INS spectroscopy has two special advantages compared with optical spectroscopies, *viz.* its ability to locate features that are forbidden by optical selection rules, and the ease of calculating the spectrum with modern computer codes: these unique features will ensure that INS spectroscopy has a bright future.

References

Eckert, J. (1992) *Spectrochim. Acta* **48A** 271–283. "*Theoretical introduction to neutron scattering spectroscopy.*"

Hudson, B. S. (2001) *J. Phys. Chem. A* **105** 3949–3960. "*Inelastic neutron scattering: a tool in molecular vibrational spectroscopy and a test of ab initio methods.*"

Mitchell, P. C. H., Parker, S. F., Ramirez-Cuesta, A. J. and Tomkinson, J. (2005) "*Vibrational spectroscopy with neutrons.*" World Scientific, Singapore.

Parker, S. F., Williams, K. P. J., Bortz, M. and Yvon, K. (1997) *Inorganic Chem.* **36** 5218–5221. "*Inelastic neutron scattering, infrared and Raman spectroscopic studies of* Mg_2FeH_6 *and* Mg_2FeD_6."

Parker, S. F., Wilson, C. C., Tomkinson, J., Keen, D. A., Shankland, K., Ramirez-Cuesta, A. J., Mitchell, P. C. H., Florence, A. J. and Shankland, N. (2001) *J. Phys. Chem. A* **105** 3064–3070. "*Structure and dynamics of maleic anhydride.*"

Parker, S. F., Bennington, S. M., Ramirez-Cuesta, A. J., Auffermann, G., Bronger, W., Herman, H., Williams, K. P. J. and Smith, T. (2003) *J. Am. Chem. Soc.* **125** 11656–11661. "*Inelastic neutron scattering and Raman spectroscopies and periodic DFT studies of* Rb_2PtH_6 *and* Rb_2PtD_6."

Smith, J. C. (1991) *Quart. Rev. Biophys.* **24** 227–291. "*Protein dynamics: comparison of simulations with inelastic neutron scattering experiments.*"

Tomkinson, J. (1992) "*Inelastic incoherent neutron scattering spectroscopy of hydrogen vibrations in metals and molecules.*" in *Neutron scattering at a pulsed source*, edited by Newport, R. J., Rainford, B. D. and Cywinski, R., Chapter 18. Adam Hilger, Bristol.

White, J. W. (1973) "*Neutron inelastic scattering and molecular spectroscopy.*" in *Chemical applications of neutron scattering*, edited by Willis, B. T. M. Oxford Univ. Press, Oxford.

Quasi-elastic scattering and high-resolution spectroscopy

16

16.1 Introduction

Quasi-elastic neutron scattering (QENS) refers to those inelastic processes that are *almost* elastic. The term is usually considered to mean a broadening of the elastic line in the energy spectrum rather than the appearance of discrete peaks representing inelastic events. In the context of this chapter, it is also appropriate to include low-energy inelastic processes of energies up to 1 meV: this is because the energy and momentum transfers in both cases are similar and, therefore, the two kinds of measurement are usually carried out with the same type of spectrometer.

Just as for other kinds of neutron scattering, QENS contains both coherent and incoherent scattering components. The coherent component yields information about interference phenomena between atoms, such as lattice distortions or short-range order. Incoherent scattering relates to scattering by individual atoms: if the atoms or molecules undergo stochastic motions (translational or rotational diffusion) during the scattering event, this single-particle scattering is accompanied by the transfer of energy to or from the neutrons. The motions are not quantized, and so there is a continuous distribution of energy giving a broadening of the sharp line arising from elastically scattered neutrons. The energy gain or loss can amount to an appreciable fraction of the kinetic energy of the neutron, and the plot of neutron intensity *versus* energy transfer $\hbar\omega$ gives a spectrum centred on the elastic position ($\hbar\omega = 0$) but broadened in energy compared with the instrumental width. Thus QENS can be used to study diffusion in solids, where an individual particle performs a random walk over the crystal lattice. In this case, the incoherent component of scattering gives the jump rates, jump lengths and jump directions of the diffusing particle.

In this chapter, we restrict ourselves to single-particle diffusion in solids with only one mobile component. The book by Beé (1988) provides an introduction to the principles of QENS and its application to solid-state diffusion. A more comprehensive survey of QENS applied to

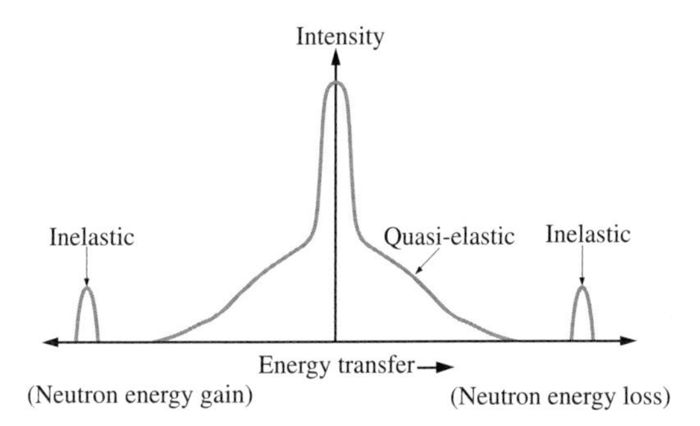

Fig. 16.1 Elastic, quasi-elastic and inelastic scattering regions.

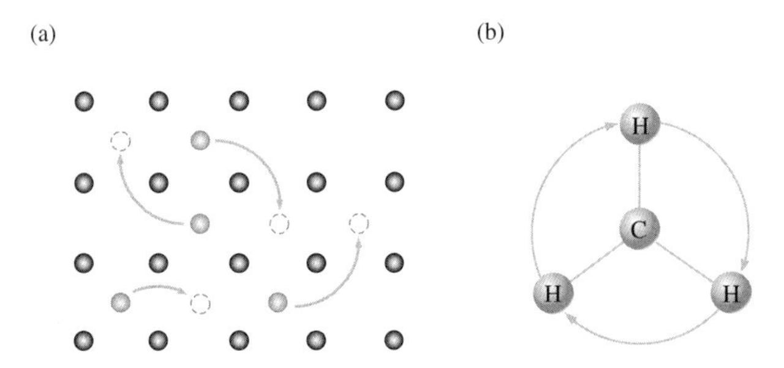

Fig. 16.2 Two types of diffusional motion: (a) translational motion and (b) rotational motion of CH_3 group.

solid-state diffusion in metals, alloys, intercalation compounds and solid ionic conductors is given in the book by Hempelmann (2000).

The scattering close to the elastic line can thus be divided into different regions as indicated in Fig. 16.1. The *elastic* line itself is due to scattering from atoms which are localized in space, and the *inelastic* lines result from scattering from atoms which vibrate in a periodic manner and with a fixed frequency. The *quasi-elastic* line arises when neutrons scatter from atoms that are moving in a random (stochastic) manner.

Diffusional motions can further be subdivided into two categories: *translational diffusion* and *rotational diffusion* (see Fig. 16.2). In translational diffusion, the centre of mass moves within the crystalline lattice of the material under investigation. The diffusing entity is not restricted to any fixed volume in the lattice and is free to explore the whole volume of the sample. Rotational diffusion does not involve any mass transport. Instead a molecule or part of a molecule rotates and changes its orientation with its centre of mass remaining fixed in space but with the molecule as a whole moving randomly in time. Spatially, these reorientations (*e.g.* $2\pi/3$ rotations of an NH_4 group) can be in register with the symmetry of the lattice or in register with the symmetry of the molecule itself.

16.2 Energy resolution and time scales

Translational and rotational diffusion often occur simultaneously and on different time scales. The energy resolution of a spectrometer ΔE

Table 16.1 Time scales for quasi-elastic spectroscopy.

Time scale (sec)	Resolution ΔE (μeV)	Spectroscopic technique
10^{-11}	10–100	Direct-geometry time-of-flight
10^{-9}	0.3–20	Backscattering crystal analyser
10^{-7}	0.005–1	Spin-echo

is linked to the time scale τ of the diffusive motion by the Heisenberg uncertainty principle

$$\Delta E \cdot \tau \geq h/2\pi \tag{16.1}$$

We see that in order to observe rapid motions, relatively poor energy resolution is sufficient but for slow motions good resolution is needed. Nevertheless, the energy resolution required for quasi-elastic spectroscopy is always high, and is typically better than 1% in $\Delta E/E$ and 100 μeV in ΔE. Different spectroscopic techniques achieve different resolutions (and hence observe diffusive motion on different time scales).

Table 16.1 links the time scale of the motion, the energy resolution required to observe the motion, and the instrumental technique necessary to achieve this resolution. The time scales that can be probed in QENS correspond closely to the frequency of diffusive motions of atoms and molecules in solids and liquids. The resolutions needed are high by the normal standards of neutron scattering and instruments using cold neutrons ($E < 5$ meV) are necessary to obtain these resolutions. A further consequence of achieving high resolution is the relatively modest data rate of such measurements, even though we are observing elastic scattering. Probing longer time scales with higher resolution is paid for dearly in terms of loss of intensity.

The momentum transferred during the scattering process is also an important variable as it gives information on jump distances and on the geometry of these jumps. Just as for the non-commuting variables of energy transfer and time [eqn. (16.1)], so are the momentum transfer and distance variables linked by the Heisenberg uncertainty principle. For neutrons with wavelength λ, which are elastically scattered through an angle 2θ, the momentum transferred is

$$Q = (4\pi/\lambda)\sin\theta \tag{16.2}$$

and for cold neutrons ($E < 5$ meV or $\lambda > 4$ Å) a Q range from 0.05 to 2.5 Å^{-1} can be explored. This range corresponds to distances from 1 to 100 Å, as illustrated in the Table 16.2.

The distances probed by neutron quasi-elastic spectroscopy correspond to typical distances encountered in studying diffusive jumps at the atomic level in solids and liquids.

Table 16.2 Distances probed in quasi-elastic experiments.

Momentum transfer (Å^{-1})	Distance (Å)	Regime
0.05	100	Continuous or macroscopic diffusion
5	1	Atomic or microscopic diffusion

16.3 Types of spectrometer

From the conservation of momentum we have

$$\mathbf{Q} = \mathbf{k}_\mathrm{i} - \mathbf{k}_\mathrm{f} \tag{16.3}$$

and from the conservation of energy we have

$$\begin{aligned} \hbar\omega &= E_\mathrm{i} - E_\mathrm{f} \\ &= \frac{\hbar^2}{2m_n}\left(k_\mathrm{i}^2 - k_\mathrm{f}^2\right) \end{aligned} \tag{16.4}$$

By convention, the energy transfer is defined as positive when the neutron loses energy and the sample gains energy. Our objective, therefore, is to determine the intensity of scattering processes with an energy transfer $\hbar\omega$ as a function of momentum transfer $\mathbf{Q}$.

There are three ways to determine the energy of a neutron: by measuring its velocity, its wavelength or its spin precession rate in a magnetic field. To measure any of these three parameters a different type of spectrometer is required, each with a specific energy resolution.

16.3.1 Direct geometry time-of-flight spectrometer

In this spectrometer, the incident neutron energy is selected or monochromated using a rotating mechanical chopper, which also serves to pulse the neutron beam in time. The pulse of neutrons strikes the sample and the scattered neutrons are timed over a known distance to the detectors that are set on the arc of a circle centred on the sample at a range of scattering angles. In this manner, the distribution of scattered neutron velocities and the energy spectrum are determined.

As the incident energy is known (usually by a calibration of the elastic line by time-of-flight), the measured energy spectrum can be mapped on to an energy transfer scale. A direct-geometry spectrometer on a pulsed neutron source only requires the chopper to monochromate the beam by opening at the correct time (or phase) with respect to the origin of the neutron pulse issued by the target.

The range of resolutions attainable with these spectrometers is limited by the accuracy with which the neutron velocity can be measured. This is determined by the maximum spinning speeds of the mechanical choppers and the uncertainties in the various flight paths; in practice, the best

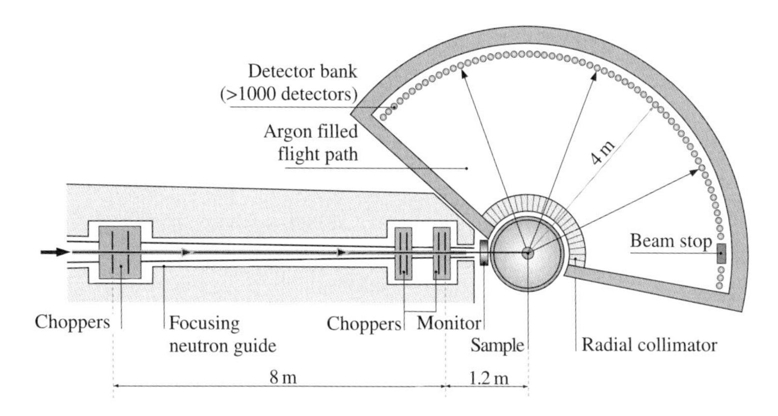

Fig. 16.3 The cold neutron multichopper spectrometer IN5 at the ILL, Grenoble.

spectrometers, when selecting very low neutron energies, can routinely achieve resolutions from 10 μeV to 100 μeV. Such a spectrometer is the multichopper spectrometer IN5 at Institut Laue Langevin, Grenoble, which is illustrated in Fig. 16.3. The set of six phased disc choppers spin at up to 20,000 rpm and they both monochromate and pulse the incident beam. In a disc chopper (see Section 5.4.1), the rim of the chopper is coated with a gadolinium absorber except for a transparent neutron window, which is equal to the size of the incident beam. The phasing of the first and last choppers in IN5, which are separated by a distance of 8 m, determines the energy selected. The purpose of the second chopper is to eliminate the long wavelength harmonics, which would otherwise pass through the final chopper after a further revolution. The third and fourth choppers spin at an integral fraction of the speed of the other choppers and eliminate a certain number of pulses, say 1 in 2, or 3 in 4. If the full frequency of pulses were allowed to fall on to the sample, then it would be possible for fast neutrons to be upscattered inelastically in a subsequent frame and to overtake slow neutrons downscattered in the first frame ('frame overlap'). When the neutron pulse strikes the sample, a time-of-flight scan, lasting say 10 msec and divided into many hundreds of narrow time bins each of say 10 μsec, is triggered and the time of arrival of a neutron at one of the detectors situated on an arc 4 m from the sample is recorded in a specific time bin in the memory of the time-of-flight analyser. The final spectrum is built up from about a million of such time scans.

The resolution function of IN5, which is almost exactly triangular in shape, is determined by the shape in time of the chopper aperture cutting an incident beam of the same width. This shape is a unique property of IN5 and is of immense help to experimenters in assessing by eye the line broadening due to diffusional processes. The majority of other spectrometers have Gaussian resolution functions that are less convenient, whilst some have the even worse Lorentzian resolution function with its long tails. Spectrometers on pulsed sources tend to have asymmetric resolution functions due to folding in of the decaying pulse produced by the moderator. Modern computer codes for fitting quasi-elastic peaks can cope with these awkward line shapes whereas the eye cannot.

16.3.2 Inverted-geometry spectrometers

The inverted-geometry spectrometer measures the incident neutron energy by time-of-flight over a known distance and selects from the scattered spectrum a single neutron energy using an energy analyser. In practice, this means using a white incident neutron beam and analysing the wavelength of the scattered beam by Bragg diffraction from a single crystal.

The resolution is determined by differentiating Bragg's law ($\lambda = 2d \sin\theta$) to give

$$\frac{\Delta E}{E} = 2\frac{\Delta\lambda}{\lambda} = 2\left[(\cot\theta \cdot \Delta\theta)^2 + \left(\frac{\Delta d}{d}\right)^2\right]^{1/2} \tag{16.5}$$

The dominant term in most cases is $\cot\theta \cdot \Delta\theta$, since for most single crystals the uncertainty $\Delta d/d$ in the lattice parameter is about 10^{-4} and can be ignored. As the Bragg angle approaches 90°, $\cot\theta$ tends to zero and the energy resolution of the analyser becomes extremely good, and is 0.5 μeV for perfect silicon and 10 μeV for graphite. Thus, the crystal-analyser inverted-geometry spectrometer is often referred to as a back-scattering spectrometer.

The best known of these spectrometers at the ILL is IN10, which uses an oscillating single crystal of silicon to produce, by the Doppler effect, a white incident neutron beam. IN10 possesses a large array of silicon crystals in exact backscattering to analyse the scattered beam from the sample. IN10 provides resolutions of 0.3–1.0 μeV over an energy transfer range of ±15 μeV. A second spectrometer, IN13, at the ILL uses a different technique to scan the incident neutron energy with a CaF_2 crystal monochromator in a variable-temperature furnace and to analyse the scattered neutron beam with an array of CaF_2 crystals again in exact backscattering. IN13 provides resolutions of 10–25 μeV over an energy transfer range of +320 to −300 μeV.

IRIS is the pulsed source analogue of the backscattering spectrometer and is illustrated in Fig. 16.4. A broad band of neutron wavelengths is selected from the neutron pulse emitted by the liquid-hydrogen moderator and chopped by a rotating disc chopper at a distance of 6.4 m. The required energy resolution in the incident beam is achieved by allowing this band of neutrons to disperse in time over a long distance as it drifts along a curved neutron guide to the sample situated at 36 m from the moderator. Neutrons whose energies change to precisely the analyser energy after scattering from the sample are analysed by Bragg scattering from a pyrolytic graphite crystal array in near backscattering geometry ($2\theta = 175°$). A continuous range of scattering angles from 15° to 165° is covered simultaneously by the analyser–detector

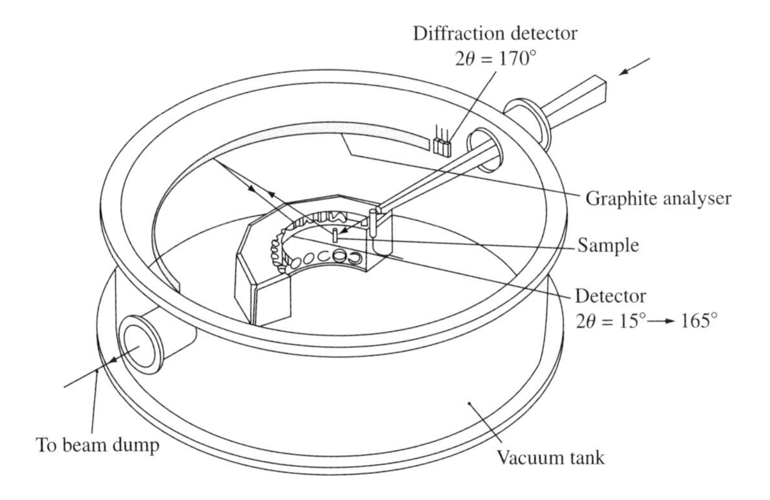

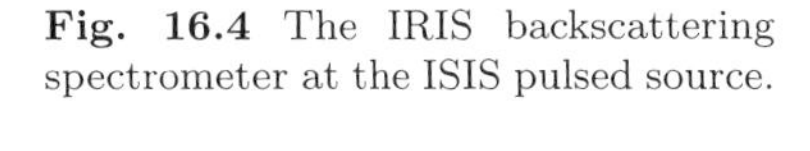

Fig. 16.4 The IRIS backscattering spectrometer at the ISIS pulsed source.

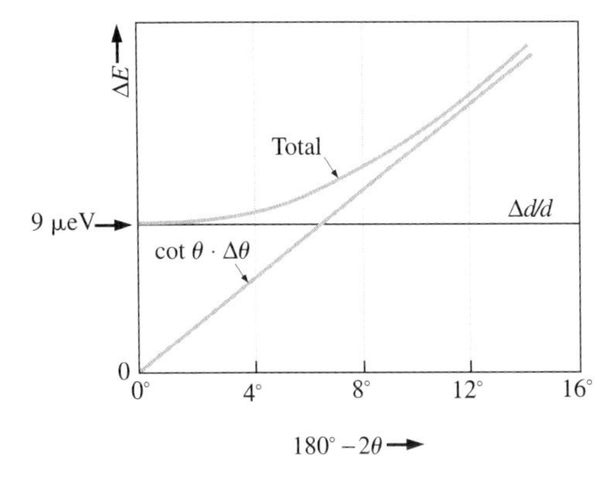

Fig. 16.5 Energy resolution ΔE of 002 reflection of pyrolytic graphite in back scattering, showing the two contributions to ΔE.

array. Using the 002 reflection from the graphite analyser, a resolution of 15 μeV is provided over an energy transfer range from +4 meV to −1 meV.

For pyrolytic graphite $\Delta d/d$ is $\sim 2.5 \times 10^{-3}$, an order of magnitude larger than for other materials, and thus for the 002 reflection at 1.82 meV in backscattering an analyser energy resolution of 9 μeV is obtained. The $\cot\theta \cdot \Delta\theta$ term in the resolution eqn. (16.1) only exceeds the $\Delta d/d$ term at Bragg angles $\leq 85°$ as shown in Fig. 16.5. Thus, it is not necessary to go to exact backscattering to achieve high resolution and this allows the analyser geometry to be simplified. The IRIS analyser is set at a Bragg angle of 87.5° on a Rowland circle to obtain focusing conditions. This eliminates the need for a beam modulation chopper as on IN10 to discriminate against neutrons scattered directly into the detector. The matched incident and scattered energy resolutions give a final resolution of 15 μeV for the 002 reflection. By rephasing the disc chopper, a wavelength window can be selected which illuminates the 004 reflection to give a doubling of the Q-range at a resolution of 50 μeV. The performance of IRIS is summarized in Table 16.3. [Carlile and Adams 1992, Telling *et al.* 2002]

Table 16.3 Properties of IRIS spectrometer.

Reflection	Energy of analyser (meV)	ΔE (μeV)	Q ($\mathring{A}^{-1}$)	Relative intensity
002	1.82	15	0.25–1.85	1.0
004	7.2	50	0.5–3.7	0.6
006	16.3	100	0.75–5.5	0.08

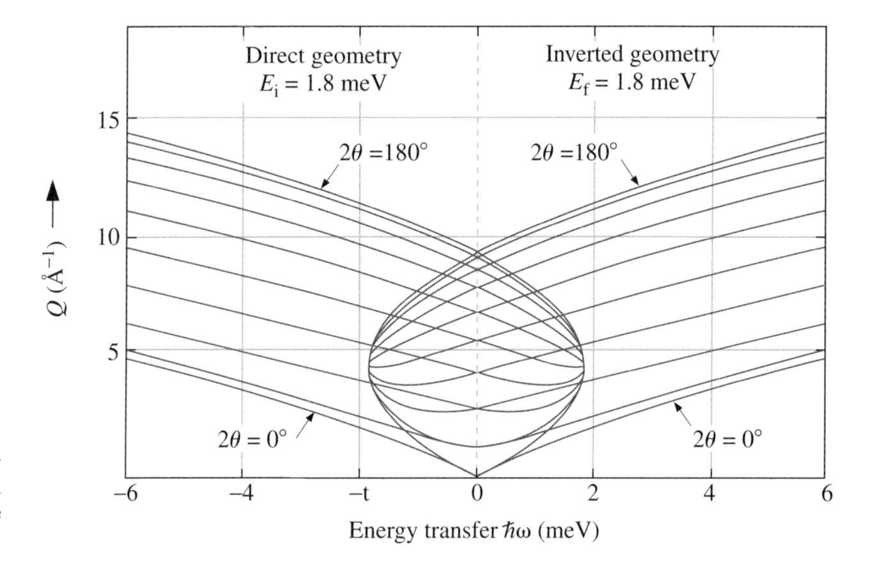

Fig. 16.6 Accessible regions of ($\mathbf{Q}$, ω) space for direct- and indirect-geometry spectrometers. The curves correspond to fixed values of the scattering angle 2θ in the range $0° < 2\theta < 180°$.

Observable range of Q

The equations which govern the conservation of energy and momentum of the neutron in the scattering process also determine the range in ($\mathbf{Q}$, ω) space available for measurements at particular values of neutron energy and scattering angle. For fixed incident energy (as in direct-geometry spectrometers) and fixed scattered energy (as in inverted-geometry spectrometers), the curves have identical shapes but are reversed with respect to each other as shown in Fig. 16.6. It is clear that a direct-geometry spectrometer cannot operate in neutron energy-loss at energy transfers greater than the value of the incident neutron energy. (The final neutron energy can never be less than zero!) Similarly, an inverted-geometry spectrometer cannot operate in neutron energy-gain at energy transfers exceeding the value of the neutron energy measured by the analyser. At both these limits, all scattering angles converge to the same value of Q. On balance, this favours the indirect-geometry spectrometers since, for the small values of the incident energy and the analyser energy needed to obtain high resolutions, the range in neutron energy-loss of the direct-geometry spectrometer is quite small. It is necessary also to observe in neutron energy-loss when samples are cooled to very low temperature and only the ground states are populated.

16.3.3 Neutron spin-echo spectrometer

The two types of spectrometer just described utilize, respectively, the particle nature of the neutron, that is its velocity, in the case of the direct-geometry spectrometer, and the wave nature of the neutron, that is its wavelength, in the case of the backscattering spectrometer, in determining the exchange of energy between the neutron and the sample. The neutron spin-echo spectrometer uses a third property of the neutron, *viz.* its magnetic moment, in order to detect this energy change.

The basic principle of the spectrometer was first formulated by Mezei (1972, 1979) and can be summarized as follows. A neutron has a spin of 1/2, and so in a magnetic field it has two states, nominally up-spin +, and down-spin −. Up-spin neutrons can be separated from down-spin neutrons by a polarizing technique (see Section 9.3): the neutron beam then consists of only one spin state and is said to be polarized. A polarized neutron when traversing a known distance in a magnetic field undergoes a number of precessions about the magnetic field direction and this number is determined by the velocity of the neutron, the magnitude of the magnetic field and its direction with respect to the neutron spin (Fig. 16.7). Typically, for cold neutrons and magnetic fields of 0.3 T (a few hundred gauss), the number of precessions over a distance of 1 metre is about 10^5. Thus, the velocity of a neutron can be measured with an accuracy of better than 1 in 10^5 by observing the rate of precession before and after scattering, and the energy exchanged in the scattering process can be determined to a corresponding accuracy.

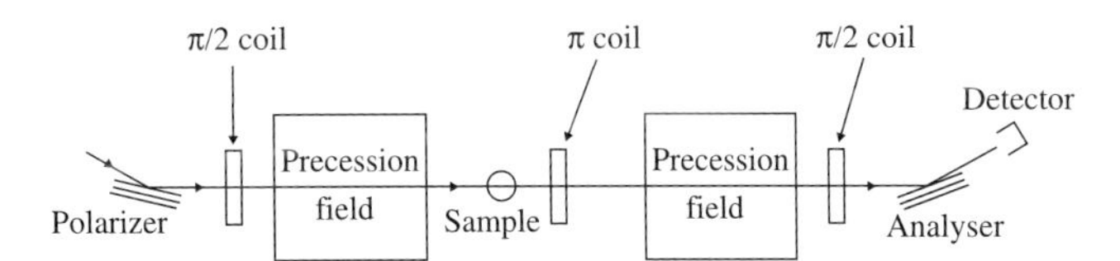

Fig. 16.7 Principle of the spin-echo spectrometer. The $\pi/2$ coil flips the neutrons by 90°, and the π coil reverses the neutron spins by flipping them through 180°.

To apply the technique in practice is extremely complex, and it has taken 10 years of development before the application of the principle resulted in an operational spectrometer IN11 at the ILL in Grenoble (Fig. 16.8). In this spectrometer, the absolute number of precessions is not measured: instead the behaviour of the neutron after scattering is compared with its behaviour prior to scattering whilst drifting over the same distances and under the same magnetic field. The only difference between the incident flight path and the scattered flight path is that the direction of neutron polarization is reversed at the scattering sample. Thus, the direction of precession in the scattered beam is in the opposite sense to that in the incident beam. If after a time t, the field is suddenly reversed (or the spins are flipped by π with respect to the field), then the Larmor precession is reversed and the initial polarization is recovered after a time $2t$. This time is proportional to the line

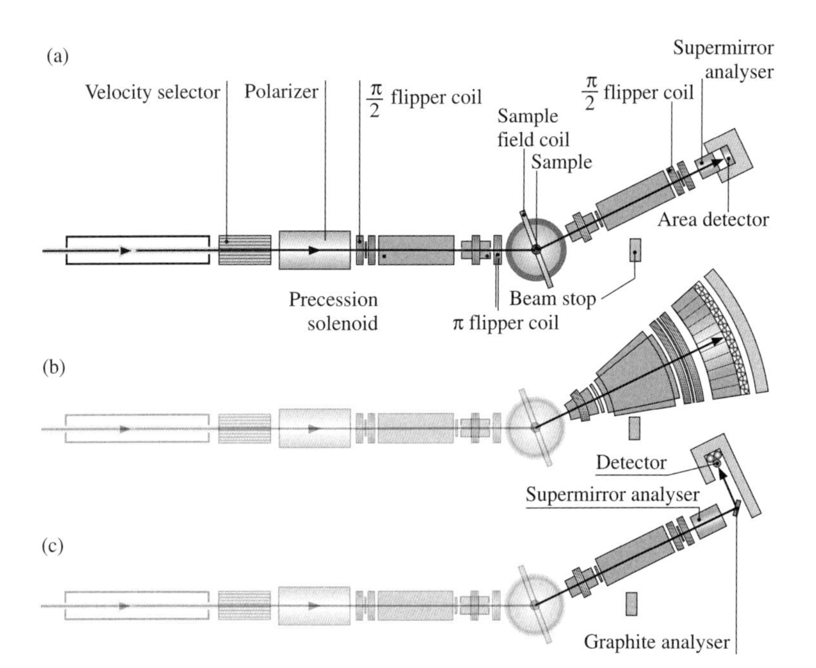

Fig. 16.8 The spin-echo spectrometer IN11 at the ILL, Grenoble. (a) Single detector option. (b) Multidetector option. (c) Triple-axis option.

integral Bl, where B is the field and l is the path length. In the IN11 spectrometer, the line integral (0.3 T·m) is the same before and after the sample; for totally elastic coherent scattering, the neutron will arrive at the detector in precisely the same state in which it entered the spectrometer and will therefore result in an 'echo'. This echo takes place for all velocities in the primary beam, and the NSE spectrometer can operate with a monochromatic beam possessing a wavelength spread $\Delta\lambda/\lambda$ as high as 15%. Effectively, the beam monochromation is decoupled from the instrumental energy resolution.

For quasi-elastic or inelastic scattering, the spins of the incident and the scattered neutrons will Larmor precess at different rates and 'fan out': the spins will not mirror each other exactly and the echo signal will be diminished. The level of the echo as a function of magnetic field contains information on the scattering process that can be measured directly. The spectrometer is capable of achieving an energy resolution equivalent to 5 neV (or 0.005 μeV) and is able to probe diffusive behaviour over a time scale of around 10^{-7} sec. Thus, the spin-echo spectrometer has an energy resolution that cannot be matched by any other instrument.

16.4 Examples of studies by high-resolution spectroscopy

16.4.1 Diffusion

In general, for an atom which diffuses within a fixed volume, the incoherent scattering law $S(Q, \omega)$ in the elastic region is separable into two components comprising a purely elastic component and a quasi-elastic component centred on $\omega = 0$. The law can be written as

$$S(Q,\omega) = \underset{\text{(elastic)}}{A_0(Q)\,\delta(\omega)} \;+\; \underset{\text{(quasi-elastic)}}{A_1(Q)\,L(\omega)} \tag{16.6}$$

where $\delta(\omega)$ is the delta function and $L(\omega)$ is a Lorentzian function. The first term contains information on the geometry of the diffusing entity and the Lorentzian term information on the time scale of its diffusion. The fall in the elastic intensity as a function of Q can be represented by a structure factor called the elastic incoherent structure factor (EISF), which is defined as

$$\text{EISF} = \frac{A_0(Q)}{A_0(Q) + A_1(Q)} = \frac{\text{elastic intensity}}{\text{total intensity}} \tag{16.7}$$

For purely translational diffusion, the EISF is everywhere zero except at $Q = 0$. For the rotational diffusion of a particle, the EISF is unity at $Q = 0$ and falls to a minimum at a Q value which is inversely related to the radius of gyration of the rotating particle and is thereafter oscillatory in character. The precise shape of the EISF is indicative of the geometry of the diffusional process and is particularly sensitive to the model in the Q region beyond the first minimum.

An example of this diffusional behaviour is given by the molecular crystal adamantane $C_{10}H_{16}$, which has the molecular structure shown in Fig. 16.9. Examination of the shape of the molecule shows that it is a tetrahedron sitting within the constraints of a cubic crystalline lattice. The tetrahedron can assume only two different but equivalent configurations within this lattice. Combinations of rotations of the molecule can occur about only three basic axes, the 2-fold, 3-fold and 4-fold axes. A C_2 diad-rotation by 180° and a C_3 triad-rotation by 120° lead to an identical position, whereas a C_2 rotation by 180° or C_4 tetrad-rotation by 90° move the molecule to a different position. It is possible to differentiate between these two alternatives by examining the EISF determined from the quasi-elastic spectra shown in Fig. 16.10. By fitting various models

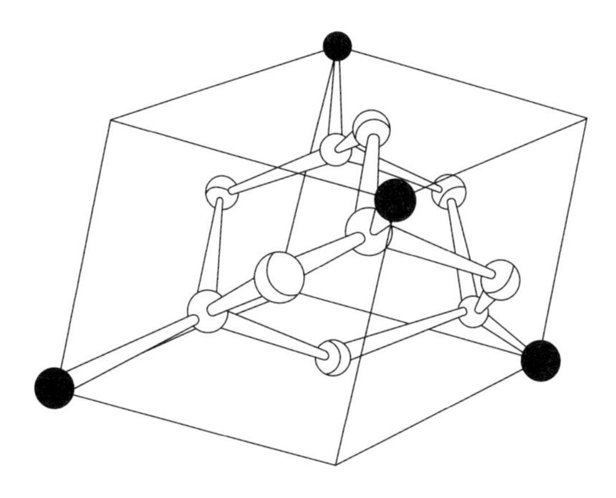

Fig. 16.9 Molecular structure of adamantane $C_{10}H_{16}$ showing carbon atoms only.

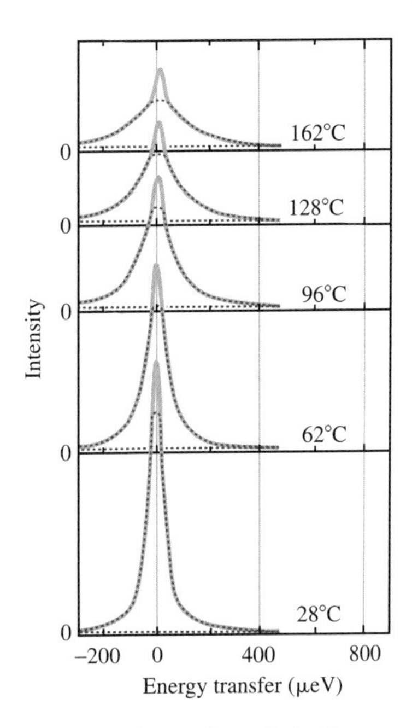

Fig. 16.10 A set of quasi-elastic spectra from adamantane at different temperatures, showing the separation of elastic and quasi-elastic lines. The spectra indicate rotational diffusion within a limited volume. [Amoureux *et al.* 1980]

to the EISF, it is clear that the C_3 model or the isotropic rotational model does not fit the observations and that either C_4 or C_2 processes dominate. Note that only data at Q-values greater than the first minimum in the structure factor are significant in determining the precise rotational behaviour of the molecule.

Further quasi-elastic measurements on single crystal samples of adamantane allowed the model to be refined revealing that only C_4 rotations occur. An analysis of the temperature dependence of the scattering led to a measurement of the activation energy for the process. For adamantane, the diffusional process follows an Arrhenius behaviour with a rotational jump rate $1/\tau$ given by

$$1/\tau = 5.2 \times 10^{12} \exp(-1350/T) \text{ rotations/sec} \qquad (16.8)$$

The measured activation energy corresponding to the temperature 1,350 K is 116 meV.

16.4.2 Tunnelling spectroscopy

Molecular groups, which at room temperature can rotate by virtue of their thermal motion, are frozen into well-defined positions at low temperature. The corresponding potential well is usually parabolic or harmonic in shape. The height of the potential barrier V varies considerably from one material to another but for high-barrier molecular groups, where V is much greater than the rotational constant B, a set of well-defined hindered rotational (or librational) energy levels occur. [Note that $B = \hbar^2/2I$, where I is the moment of inertia of the molecule or molecular group.] As the ratio V/B falls, there is an increasing probability of the proton on one site overlapping with its neighbour. In this case, the librational levels of the two protons are degenerate and the energy level splits to remove this degeneracy. The splitting is referred to as tunnel splitting, the implication being that the proton on one site is able to tunnel through the potential barrier to the neighbouring site.

For large values of V/B where the overlap is small the tunnel splitting is also small and is in the range 0.1–5 μeV. Tunnelling in these high-barrier materials, such as ammonium perchlorate NH_4ClO_4 (see Fig. 16.11), can only be observed on spectrometers with exceptionally good energy resolution such as IN10. As V/B decreases the overlap increases and so does the tunnel splitting, and splittings from 5 μeV to 1 meV have been observed in a wide range of materials. A spectrum of the tunnel splitting in 4-methyl pyridine as measured on IRIS is shown in Fig. 16.12 with tunnelling levels at 520 μeV. At this value of the height of the potential barrier almost free rotations are occurring. The peak for energy gain of the molecule is more intense than for energy loss because of the effect of the Bose–Einstein population factor (see Section 14.2) at low temperature.

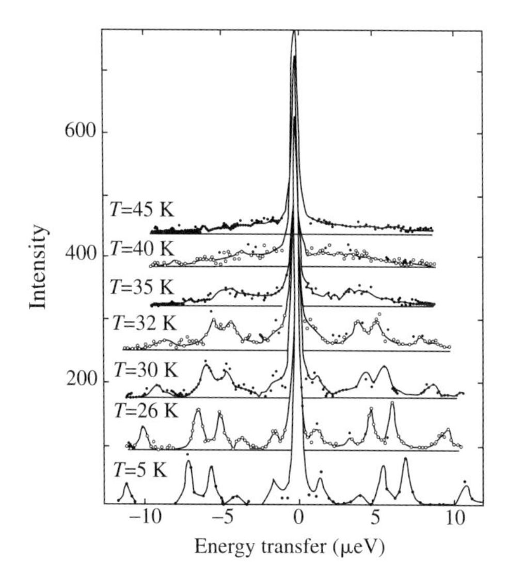

Fig. 16.11 Tunnel splitting in NH_4 ClO_4 at different temperatures. (*After* Prager and Press, 1981.)

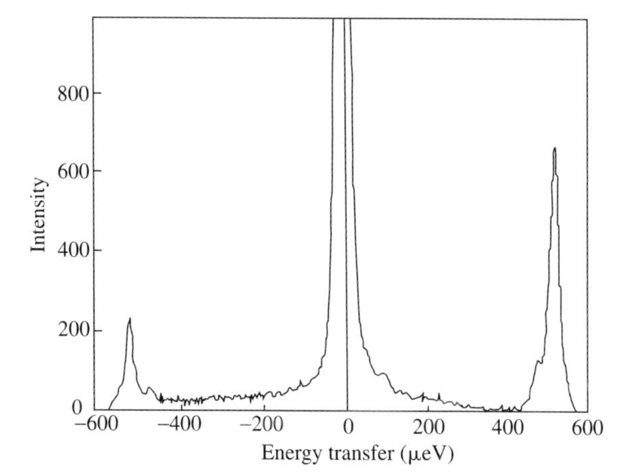

Fig. 16.12 Tunnel splitting in 4-methyl pyridine at 4 K. (*After* Carlile, unpublished data.)

The phenomenon of rotational tunnelling was first rendered accessible to neutron spectroscopy with the development of high-resolution neutron spectrometers in the 1970s. Since then the field developed rapidly, proving to be a powerful method of studying molecular tunnelling. It is documented in the proceedings of biennial workshops and in a number of reviews. The most comprehensive review to date is that published by Prager and Heidemann (1997); these authors list numerous experimental studies with neutrons, and place them in the context of other techniques, such as Raman, infrared and NMR spectroscopies, which reveal different aspects of tunnelling.

16.4.3 Fast dynamical studies

The molecular dynamics of proteins and its relation to biological function have been extensively studied by neutron scattering. The most recent review (Gabel *et al.*, 2002) includes results obtained by QENS on oriented purple membrane.

QENS has also been used in a geological context to investigate the fast dynamics of H_2O in silicate glasses (Indris *et al.*, 2005). A knowledge of these dynamical processes is important in the geosciences for understanding, for example, the mechanisms of volcanic eruptions. For the water-rich glasses examined between 2 K and 420 K, it was shown that the dynamical processes set in at lower temperatures in calcium-bearing glass than in the sodium-bearing glass.

Finally, an example of the application of neutron spin-echo spectroscopy in studying the dynamics of polyethylene is described by Schleger *et al.* (1998). The motion of polymer chains is restricted by the conformations of the surrounding chains. In the reptation model of de Gennes (1971) it is postulated that the diffusion of a polymer chain takes place through a network of impenetrable and immovable obstacles which form a tube around the chain. During the lifetime of this tube, any lateral diffusion of the chain is quenched, and diffusion can only occur by little 'wiggles' of surplus length in the chain. By making NSE measurements in the Fourier time range of $0.3 < t < 175$ nsec, it was shown that reptation is the dominant relaxation mechanism in entangled linear polymers, and that other competing relaxation models are ruled out.

References

Amoureux, J. P., Bée, M. and Virlet, J. (1980) "*Anisotropic molecular reorientations of adamantane in its plastic solid phase*" Mol. Phys. **41** 313–324.

Bée, M. (1988) "*Quasielastic neutron scattering and solid state diffusion: Principles and applications in solid-state chemistry, biology and materials science.*" Adam Hilger, Bristol.

Carlile, C. J. and Adams, M. A. (1992) "*The design of the IRIS in elastic neutron spectrometer and improvements to its analysers*" Physica B **182** 431–440.

Gabel, F., Bicout, D., Lehnert, U., Tehei, M., Weik, M. and Zaccai, G. (2002) *Quart. Rev. Biophys.* **35** 327–367. "*Protein dynamics studied by neutron scattering.*"

de Gennes, P. G. (1971) *J. Chem. Phys.* **55** 572–579. "*Reptation of a polymer chain in the presence of fixed obstacles.*"

Hempelmann, R. (2000) "*Quasielastic neutron scattering and solid state diffusion.*" Oxford Univ. Press, Oxford.

Indris, S., Heitjans, P., Behrens, H., Zorn, R. and Frick, B. (2005) *Phys. Rev. B* **71** 064205–064209. "*Fast dynamics of H_2O in hydrous aluminosilicate glasses studied with quasielastic neutron scattering.*"

Mezei, F. (1972) *Z. Physik.* **255** 146–160. "*Neutron spin echo: a new concept in polarised thermal-neutron techniques.*"

Mezei, F. (1979) *Nucl. Instrum. Methods* **164** 153–156. "*The application of neutron spin echo on pulsed neutron sources.*"

Prager, M. and Heidemann, A. (1997) *Chem. Rev.* **97** 2933–2966. "*Rotational tunnelling and neutron spectroscopy: a compilation.*"

Prager, M. and Press, W. (1981) *J. Chem. Phys.* **75** 494. "*Rotational tunnelling in NH_4ClO_4.*"

Schleger, P., Farago, B., Lartigue, C., Kollmar, A. and Richter, D. (1998) *Phys. Rev. Lett.* **81** 124–127. "*Clear evidence of reptation in polyethylene from neutron spin-echo spectroscopy.*"

Telling, M. T. F., Campbell, S. I., Abbley, D. D., Cragg, D. A., Balchin, J. J. P. and Carlile, C. J. (2002) *Appl. Phys* **A74**(Suppl.) S61–S63.

Appendix A

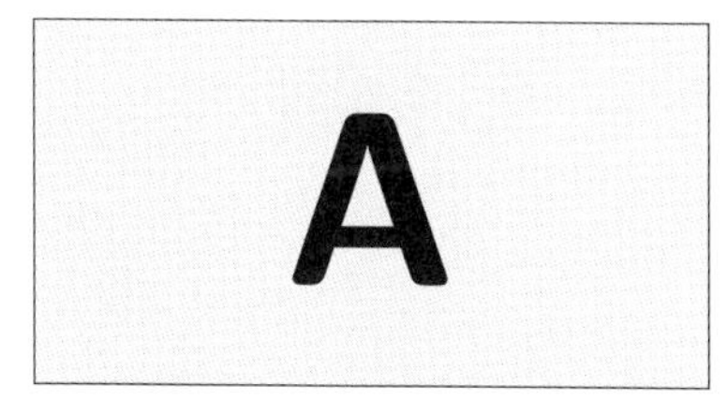

A.1 Glossary of special terms

Absorption cross-section

The effective area presented by the nucleus to a neutron when the neutron is removed by an absorption process such as an (n,γ) reaction.

Acoustic branch

The vibrational spectrum of a crystal is described in terms of the dispersion relations, that is, the frequency *versus* the wave vector $\mathbf{q}$ of the thermal modes of vibration. For N atoms in the unit cell, there are $3N$ branches of these dispersion relations. Three of these branches tend towards zero frequency when $\mathbf{q} \to 0$ and are called acoustic branches.

Acoustic mode

A mode of vibration belonging to an acoustic branch.

Atomic mass unit (a.m.u.)

A relative mass unit, based on the mass of the most abundant isotope of carbon, C^{12}, being defined as exactly 12 mass units.

Barn

Unit for measuring cross-sections: $1\text{b} = 10^{-28}\ \text{m}^2$.

Born approximation

The approximation used in first-order perturbation theory. It assumes that the incident neutron (or particle) wave function is only slightly perturbed by the scattering potential and that the scattering is s-wave scattering.

Born–von Kármán model

Lattice dynamical model in which the restoring force on an atom is governed by its displacement relative to neighbouring atoms.

Coherent scattering

Scattering in which there is a correlation between the space and/or time relationships of different atoms.

Coriolis effect

The Coriolis force arises from the rotation of the earth and causes a small aberration of the image from a liquid mirror.

Cross-section

The effective area presented by the sample to a neutron in a scattering or absorption process.

Debye model

A model for the normal modes of vibration of a crystal, in which all modes are treated as acoustic waves with a maximum frequency ω_{D} determined by the number of possible modes.

Debye temperature

Temperature θ_{D} corresponding to the cut-off frequency ω_{D}; that is, $\hbar\omega_{\mathrm{D}} = k_{\mathrm{B}}\theta_{\mathrm{D}}$.

Delayed neutrons

The small number of neutrons in a fission process which are not released immediately but up to several seconds later, thus facilitating the control of a nuclear reactor.

Down scattering

Scattering in which the scattered neutron loses energy by transferring it to the sample.

Einstein model

A simple model for the normal modes of vibration of a crystal, in which all modes are assumed to have the same frequency.

Elastic scattering

Scattering with no change in energy of the scattered neutron.

Epithermal neutrons

Neutrons slowed down by a moderator but not in equilibrium with the moderator temperatures and having energies still well above thermal energies.

Fast reactor

A reactor in which, on account of the high content of fissile material and lack of a moderator, the nuclear chain reaction is sustained by fast neutrons.

Fertile isotope

An isotope which does not undergo fission with thermal neutrons but with fast neutrons [E>1MeV] and can produce a fissile isotope by neutron capture.

Fermi

Unit for measuring scattering amplitudes: 1 fm $= 10^{-15}$ m.

Fermi chopper

A cylindrical chopper consisting of a slot cut across its diameter and with its axis of rotation perpendicular to the incident beam.

Fermi pseudopotential

An artificial potential of very short range which gives, using first-order perturbation theory, the required result of isotropic (s-wave) neutron scattering by a single nucleus.

Frame overlap

The condition in a scattering experiment on a time of flight instrument, whereby the slower neutrons in a pulse are overtaken by the faster neutrons of a later pulse.

Incoherent scattering

Scattering in which there is no correlation between the space and time relationships of different atoms. Incoherent scattering gives information only about individual atoms.

Inelastic scattering

Scattering in which there is an exchange of energy between the scattered neutron and the sample.

Lorentz factor

The geometrical term in the expression for the intensity of a Bragg reflection, which accounts for the rate of traversing the Ewald sphere.

Magnon dispersion relation

The relation between the frequency and wave vector of magnons.

Magnon

The quantum of energy in a spin wave.

Micelle

An aggregate of surfactant molecules in a liquid. The aggregate consists of a hydrophobic region in the centre of the micelle which encloses a hydrophilic region in contact with the solvent.

Moderator

A substance in the core of a fission reactor or close to the target of a spallation source which reduces the energy of neutrons to that of the ambient temperature of the moderator.

Neutron bottle

A container which traps ultracold neutrons by total internal reflection.

Neutron cross-section

The quantity measured in a scattering experiment. Different types of cross-section are defined in Chapter 2.

Neutron flux

The number of neutrons passing through unit area per second.

Neutron guide

A device for transporting thermal neutrons away from the neutron source without losing intensity by the inverse square law.

Normal mode of vibration

A wave of thermal vibration in a crystal. It is characterized by its frequency, wave vector and polarization properties.

Optic mode of vibration

A normal mode with non-zero frequency when the wave number is zero.

Phonon

A quantized mode of vibration in a crystal. It is also used to denote the quantum of energy in a normal mode.

Phonon dispersion relation

The relation between the frequency and wave vector of phonons.

Placzek corrections

Corrections to the measured intensity due to inelasticity effects and a departure from the static approximation.

Polarization direction

The alignment direction of the spin of a neutron with respect to an applied magnetic field. A fully polarized beam consists of neutrons which are all 'up' or 'down', whereas an unpolarized beam has an equal number of 'up' and 'down' neutrons.

Rotational diffusion

Diffusional motion where the centre of mass remains stationary but the molecular group rotates.

Quark

In the quark model of the neutron, the neutron is made up of three quarks held together by the exchange of vector bosons or 'gluons'. The theory describing the interaction of quarks and gluons is known as Quantum Chromodynamics or QCD.

Quasi-elastic scattering

Inelastic processes which are close to the elastic line of the spectrum. The term is normally taken to refer to a broadening of the elastic line rather than to discrete inelastic events.

Scattering length

Amplitude of scattering of a nucleus.

Spallation

A nuclear reaction occurring when high-energy particles, such as protons, bombard a heavy atom target. The nuclei of the target atoms are chipped, or spalled, yielding neutrons and a variety of other particles.

Specular reflection

The mirror-like reflection of neutrons (or light) from a perfectly smooth surface, in which the angle of incidence is equal to the angle of reflection. If the surface is rough the reflection is both specular and diffuse, that is, spread over several scattering angles.

Spin wave

The elementary excitations of an ordered spin system in a magnetic crystal.

Static approximation

The assumption that, in the scattering of a neutron wave by two particles, the wave travels so fast that the two particles have not moved between the scattering events.

Supermirror

A multilayer system used for monochromating or polarizing a neutron beam. Many alternate layers of a magnetic material (*e.g.* iron) and a non-magnetic material (*e.g.* silver) are evaporated to produce an artificial 'crystal' with a gradually increasing layer spacing.

Surfactant

A substance, such as a detergent, which is a wetting agent. When added to a liquid, it reduces the surface tension and thereby increases the spreading.

Synchrotron

A device for accelerating charged particles (protons or electrons). It employs radiofrequency fields, spaced between a circular array of magnets, to accelerate the particles repeatedly in a fixed orbit. The r.f. accelerating field and the magnetic field are kept in synchrony with the motion of the particles.

Tesla

The SI unit of magnetic field (magnetic flux density). $1\ \mathrm{T} = 10^4$ gauss (CGS unit).

Thermal neutrons

Neutrons which are in thermal equilibrium with their surroundings. If the surroundings are at room temperature, the speed of the neutrons is about 2 km/sec.

Thermal reactor

A reactor in which the neutron chain reaction is maintained by thermal (slow) neutrons.

Translational diffusion

Diffusional motion where the centre of mass of the atom or molecule moves.

Ultracold neutrons

Neutrons of very long wavelength (~1000Å) and extremely low energy ($<10^{-6}$ eV or 1 μ eV).

Up scattering

Scattering in which the scattered neutron gains energy by transfer from the dynamic modes in the sample.

Virial coefficients

These appear as coefficients in the expansion of the pressure of a many-particle system in powers of the density. The second virial coefficient depends only on the pair-wise interaction between the particles. The word 'virial' comes from the Latin word *vis* for force.

Wave vector

A vector whose direction is the direction of propagation of the wave and whose magnitude is $2\pi \div$ its wavelength ($2\pi/\lambda$). The magnitude is the *wave number*.

Appendix B

B

B.1 Neutron scattering lengths and cross-sections of the elements

The following table lists the values of scattering lengths and cross-sections for the natural elements. It is derived from the more comprehensive table, which has been compiled by Sears (1992) and which contains data for individual isotopes as well as for the elements. Reference should be made also to Rauch and Waschkowski (2000). For most nuclides, the scattering lengths and cross-sections are independent of the neutron wavelength. The absorption cross-sections σ_{abs} are inversely proportional to velocity v (the so-called $1/v$ law), and by convention are tabulated for $v = 2,200$ m/sec, which corresponds to a neutron wavelength of $\lambda = 1.798$ Å.

Symbol	Quantity	Unit
b_{coh}	Bound coherent scattering length	Fermi ($=10^{-15}$ m)
σ_{coh}	Bound coherent scattering cross-section	Barn ($=10^{-28}$ m^2)
σ_{incoh}	Bound incoherent scattering cross-section	Barn
σ_{tot}	Total bound scattering cross-section	Barn
σ_{abs}	Absorption cross-section for 2,200 m/sec neutrons	Barn

Element	Atomic number	b_{coh}	σ_{coh}	σ_{incoh}	σ_{tot}	σ_{abs}
H	1	−3.741	1.7568	80.27	82.03	0.3326
He	2	3.26	1.34	0.00	1.34	0.00747
Li	3	−1.90	0.454	0.92	1.37	70.5
Be	4	7.79	7.63	0.0018	7.63	0.0076
B	5	5.30	3.54	1.70	5.24	767.0
C	6	6.646	5.550	0.001	5.551	0.00350
N	7	9.36	11.01	0.50	11.51	1.90
O	8	5.803	4.232	0.00	4.232	0.00019
F	9	5.654	4.017	0.0008	4.018	0.0096

(continued)

Continued

Element	Atomic number	b_{coh}	σ_{coh}	σ_{incoh}	σ_{tot}	σ_{abs}
Ne	10	4.566	2.620	0.008	2.628	0.039
Na	11	3.63	1.66	1.62	3.28	0.530
Mg	12	5.375	3.631	0.08	3.71	0.063
Al	13	3.449	1.495	0.0082	1.503	0.231
Si	14	4.149	2.1633	0.004	2.167	0.171
P	15	5.13	3.307	0.005	3.312	0.172
S	16	2.847	1.0186	0.007	1.026	0.53
Cl	17	9.5770	11.526	5.3	16.8	33.5
Ar	18	1.909	0.458	0.225	0.683	0.675
K	19	3.67	1.69	0.27	1.96	2.1
Ca	20	4.70	2.78	0.05	2.83	0.43
Sc	21	12.29	19.0	4.5	23.5	27.5
Ti	22	−3.438	1.485	2.87	4.35	6.09
V	23	−0.3824	0.0183	5.08	5.10	5.08
Cr	24	3.635	1.660	1.83	3.49	3.05
Mn	25	−3.73	1.75	0.40	2.15	13.3
Fe	26	9.45	11.22	0.40	11.62	2.56
Co	27	2.49	0.779	4.8	5.6	37.18
Ni	28	10.3	13.3	5.2	18.5	4.49
Cu	29	7.718	7.485	0.55	8.03	3.78
Zn	30	5.680	4.054	0.077	4.131	1.11
Ga	31	7.288	6.675	0.16	6.83	2.75
Ge	32	8.185	8.42	0.18	8.60	2.20
As	33	6.58	5.44	0.060	5.50	4.5
Se	34	7.970	7.98	0.32	8.30	11.7
Br	35	6.795	5.80	0.10	5.90	6.9
Kr	36	7.81	7.67	0.01	7.68	25.0
Rb	37	7.09	6.32	0.5	6.8	0.38
Sr	38	7.02	6.19	0.06	6.25	1.28
Y	39	7.75	7.55	0.15	7.70	1.28
Zr	40	7.16	6.44	0.02	6.46	0.185
Nb	41	7.054	6.253	0.0024	6.255	1.15
Mo	42	6.715	5.67	0.04	5.71	2.48
Tc	43	6.8	5.8	0.5	6.3	20.0
Ru	44	7.03	6.21	0.4	6.6	2.56
Rh	45	5.88	4.34	0.3	4.6	144.8
Pd	46	5.91	4.39	0.093	4.48	6.9
Ag	47	5.922	4.407	0.58	4.99	63.3
Cd	48	4.87	3.04	3.46	6.50	2520.0
In	49	4.065	2.08	0.54	2.62	193.8
Sn	50	6.225	4.870	0.022	4.892	0.626
Sb	51	5.57	3.90	0.00	3.90	4.91
Te	52	5.80	4.23	0.09	4.32	4.7
I	53	5.28	3.50	0.31	3.81	6.15
Xe	54	4.92	3.04			23.9
Cs	55	5.42	3.69	0.21	3.90	29.0
Ba	56	5.07	3.23	0.15	3.38	1.1
La	57	8.24	8.53	1.13	9.66	8.97
Ce	58	4.84	2.94	0.00	2.94	0.63
Pr	59	4.58	2.64	0.015	2.66	11.5
Nd	60	7.69	7.43	9.2	16.6	50.5

(continued)

Element	Atomic number	b_{coh}	σ_{coh}	σ_{incoh}	σ_{tot}	σ_{abs}
Pm	61	12.6	20.0	1.3	21.3	168.4
Sm	62	0.80	0.422	39.0	39.4	5922.0
Eu	63	7.22	6.75	2.5	9.2	4530.0
Gd	64	6.5	29.3	151.0	180.0	49700.0
Tb	65	7.38	6.84	0.004	6.84	23.4
Dy	66	16.9	35.9	54.4	90.3	994.0
Ho	67	8.01	8.06	0.36	8.42	64.7
Er	68	7.79	7.63	1.1	8.7	159.0
Tm	69	7.07	6.28	0.10	6.38	100.0
Yb	70	12.43	19.42	4.0	23.4	34.8
Lu	71	7.21	6.53	0.7	7.2	74.0
Hf	72	7.77	7.6	2.6	10.2	104.1
Ta	73	6.91	6.00	0.01	6.01	20.6
W	74	4.86	2.97	1.63	4.60	18.3
Re	75	9.2	10.6	0.9	11.5	89.7
Os	76	10.7	14.4	0.3	14.7	16.0
Ir	77	10.6	14.1	0.0	14.0	425.0
Pt	78	9.60	11.58	0.13	11.71	10.3
Au	79	7.63	7.32	0.43	7.75	98.65
Hg	80	12.69	20.24	6.6	26.8	372.3
Tl	81	8.776	9.678	0.21	9.89	3.43
Pb	82	9.405	11.115	0.0030	11.118	0.171
Bi	83	8.532	9.148	0.0084	9.156	0.0338
Po	84					
At	85					
Rn	86					
Fr	87					
Ra	88	10.0	13.0	0	13.0	12.8
Ac	89					
Th	90	10.31	13.36	0	13.36	7.37
Pa	91	9.1	10.4	0.1	10.5	200.6
U	92	8.417	8.903	0.005	8.908	7.57
Np	93	10.55	14.0	0.5	14.5	175.9
Pu	94					
Am	95	8.3	8.7	0.3	9.0	75.3
Cm	96					

References

Rauch, H. and Waschkowski, W. (2000) "*Landolt-Börnstein.*" Volume **1/16A**, Chapter 6. Springer-Verlag, Berlin.

Sears, V. F. (1992) *Neutron News* **3** 26–37. "*Neutron scattering lengths and cross-sections.*"

Index